ASTRONOMY 98/99
First Edition

Editor
David Dathe
Alverno College

David Dathe received his undergraduate degree in geology from Winona State University, his M.S. in geology from Northern Illinois University, and he is completing a Ph.D. in geology, also from Northern Illinois University. Currently, he is assistant professor of physical science at Alverno College in Milwaukee, Wisconsin, where he teaches integrated science, earth science, and geology, and he also coordinates Problem Solving. He is author of *Fundamentals of Historical Geology* and coauthor with Dr. Carla Montgomery of *Earth: Then and Now.*

A Library of Information from the Public Press
Dushkin/McGraw-Hill
Sluice Dock, Guilford, Connecticut 06437

Visit us on the Internet—http://www.dushkin.com/

The Annual Editions Series

ANNUAL EDITIONS, including GLOBAL STUDIES, consist of over 70 volumes designed to provide the reader with convenient, low-cost access to a wide range of current, carefully selected articles from some of the most important magazines, newspapers, and journals published today. ANNUAL EDITIONS are updated on an annual basis through a continuous monitoring of over 300 periodical sources. All ANNUAL EDITIONS have a number of features that are designed to make them particularly useful, including topic guides, annotated tables of contents, unit overviews, and indexes. For the teacher using ANNUAL EDITIONS in the classroom, an Instructor's Resource Guide with test questions is available for each volume. GLOBAL STUDIES titles provide comprehensive background information and selected world press articles on the regions and countries of the world.

VOLUMES AVAILABLE

ANNUAL EDITIONS
Abnormal Psychology
Accounting
Adolescent Psychology
Aging
American Foreign Policy
American Government
American History, Pre-Civil War
American History, Post-Civil War
American Public Policy
Anthropology
Archaeology
Astronomy
Biopsychology
Business Ethics
Child Growth and Development
Comparative Politics
Computers in Education
Computers in Society
Criminal Justice
Criminology
Developing World
Deviant Behavior
Drugs, Society, and Behavior
Dying, Death, and Bereavement
Early Childhood Education

Economics
Educating Exceptional Children
Education
Educational Psychology
Environment
Geography
Geology
Global Issues
Health
Human Development
Human Resources
Human Sexuality
International Business
Macroeconomics
Management
Marketing
Marriage and Family
Mass Media
Microeconomics
Multicultural Education
Nutrition
Personal Growth and Behavior
Physical Anthropology
Psychology
Public Administration
Race and Ethnic Relations

Social Problems
Social Psychology
Sociology
State and Local Government
Teaching English as a Second
 Language
Urban Society
Violence and Terrorism
Western Civilization,
 Pre-Reformation
Western Civilization,
 Post-Reformation
Women's Health
World History, Pre-Modern
World History, Modern
World Politics
GLOBAL STUDIES
Africa
China
India and South Asia
Japan and the Pacific Rim
Latin America
Middle East
Russia, the Eurasian Republics,
 and Central/Eastern Europe
Western Europe

Cataloging in Publication Data
Main entry under title: Annual Editions: Astronomy. 1998/99.
 1. Astronomy—Periodicals. I. Dathe, David, *comp.* II. Title: Astronomy.
ISBN 0–697–39298–8   520'.05

© 1998 by Dushkin/McGraw-Hill, Guilford, CT 06437, A Division of The McGraw-Hill Companies.

Copyright law prohibits the reproduction, storage, or transmission in any form by any means of any portion of this publication without the express written permission of Dushkin/McGraw-Hill, and of the copyright holder (if different) of the part of the publication to be reproduced. The Guidelines for Classroom Copying endorsed by Congress explicitly state that unauthorized copying may not be used to create, to replace, or to substitute for anthologies, compilations, or collective works.

Annual Editions® is a Registered Trademark of Dushkin/McGraw-Hill,
A Division of The McGraw-Hill Companies.

First Edition

Cover image ©1996 PhotoDisc, Inc.

Printed on Recycled Paper

Printed in the United States of America

Editors/Advisory Board

Members of the Advisory Board are instrumental in the final selection of articles for each edition of ANNUAL EDITIONS. Their review of articles for content, level, currentness, and appropriateness provides critical direction to the editor and staff. We think that you will find their careful consideration well reflected in this volume.

EDITOR

David Dathe
Alverno College

ADVISORY BOARD

Alfred N. Alaniz
San Antonio College

Arthur Alt
University of Great Falls

Harry J. Augensen
Widener University

Gordon Baird
University of Mississippi

E. Neale Blackwood
University of Charleston

Michael L. Broyles
Collin County Community College
Spring Creek

Anthony J. Buffa
California Polytechnic University

John W. Burns
Mt. San Antonio Community College

Donald F. Collins
Warren Wilson College

Neil Comins
University of Maine
Orono

Claire C. Correale
Burlington County College

Steve Danford
University of North Carolina
Greensboro

Ben de Mayo
State University of West Georgia

James De Vries
Lancaster Bible College

Bruce D. Dod
Mercer University

Jess Dowdy
Northeast Texas Community College

John J. Dykla
Loyola University

Dennis Englin
The Master's College

Richard M. Fuller
Gustavus Adolphus College

Helene S. Gabelnick
Harold Washington College

Michael K. Gainer
St. Vincent College

Francis G. Graham
Kent State University

William A. Hiscock
Montana State University

Karen B. Kwitter
Williams College

Staff

Ian A. Nielsen, Publisher

EDITORIAL STAFF

Roberta Monaco, Developmental Editor
Dorothy Fink, Associate Developmental Editor
Addie Raucci, Administrative Editor
Cheryl Greenleaf, Permissions Editor
Deanna Herrschaft, Permissions Assistant
Diane Barker, Proofreader
Lisa Holmes-Doebrick, Program Coordinator

PRODUCTION STAFF

Brenda S. Filley, Production Manager
Charles Vitelli, Designer
Shawn Callahan, Graphics
Lara M. Johnson, Graphics
Laura Levine, Graphics
Mike Campbell, Graphics
Joseph Offredi, Graphics
Juliana Arbo, Typesetting Supervisor
Jane Jaegersen, Typesetter
Marie Lazauskas, Word Processor
Kathleen D'Amico, Word Processor
Larry Killian, Copier Coordinator

To the Reader

In publishing ANNUAL EDITIONS we recognize the enormous role played by the magazines, newspapers, and journals of the *public press* in providing current, first-rate educational information in a broad spectrum of interest areas. Many of these articles are appropriate for students, researchers, and professionals seeking accurate, current material to help bridge the gap between principles and theories and the real world. These articles, however, become more useful for study when those of lasting value are carefully *collected, organized, indexed,* and *reproduced* in a *low-cost format,* which provides easy and permanent access when the material is needed. That is the role played by ANNUAL EDITIONS. Under the direction of each volume's *academic editor,* who is an expert in the subject area, and with the guidance of an *Advisory Board,* each year we seek to provide in each ANNUAL EDITION a current, well-balanced, carefully selected collection of the best of the public press for your study and enjoyment. We think that you will find this volume useful, and we hope that you will take a moment to let us know what you think.

The purpose of *Annual Editions: Astronomy 98/99* is to provide both a comprehensive and current overview of astronomy by including articles that discuss the latest discoveries and illustrate current ideas and models.

Several criteria were used in selecting the articles to include in this volume:

Audience. The primary criterion was the audience, and the assumption is that the reader is not an astronomer. The articles were selected to answer the question: What would an interested reader want to know about astronomy?

Readability. The most popular articles were chosen whenever possible. *Time, Newsweek,* and the *New York Times* are intended for general audiences. *Discover, Natural History, Science,* and *Scientific American* target general science audiences. *Astronomy* and *Sky & Telescope* cater to amateur (and professional) astronomers. Occasionally, however, more sophisticated articles from *Nature* were included, since reading articles intended for scientists and professional astronomers stretches mental muscles and provides insight into the world of professional astronomy.

Illustrative of astronomy as a science. Astronomy is a science, and, consequently, it reflects scientific methods, approaches, attitudes, and values. Many articles were selected to illustrate the scientific nature of astronomy. For example, in unit 5, the Big Bang theory subsection shows how a theory is established, provides evidence used to support theories, determines what problems exist with the theory, and suggests alternative interpretations. Articles in the subsection on the age paradox (calculated ages of stars indicate that they are older than the age of the universe calculated from cosmological parameters) show how a science deals with anomalies or inconsistencies. Two methods of calculating the age of the universe, each apparently reasonable, give two different ages that cannot both be true, and the essays show how astronomers are attempting to reconcile their findings. Also, the subsection on extraterrestrial life shows how science really operates compared to what is often shown on television and in the movies. Evidence for extraterrestrial life may already have been found in a meteorite from Mars. Of all the places in our solar system, Europa, one of Jupiter's moons, seems the most likely candidate to search for life. And Seth Shostak's article "When E.T. Calls Us" discusses astronomers' efforts to search for intelligent life outside our solar system.

This first edition of *Annual Editions: Astronomy 98/99* is divided into five units, which can conveniently be grouped into three areas. The first covers what we know by examining how data is acquired. Unit 1, Data-Gathering Techniques, looks at the wide variety of methods that astronomers use to gather data—visible light, X-rays, gamma rays, infrared light, and radio waves. The second area discusses what we know. Starting closest to home, Unit 2, The Earth and Moon, describes some aspects of our planet and the Moon. Unit 3, The Solar System, concerns the Sun, the planets, comets, meteorites, and asteroids. Unit 4, The Universe, involves the incredible variety of objects in our universe—stars, galaxies, black holes, and quasars. The last area then discusses our interpretations of the data. Unit 5, Ideas, Hypotheses, and Theories, concerns the nature of time and space, extraterrestrial life, the Big Bang theory, and the age paradox.

As this is the first edition of *Annual Editions: Astronomy 98/99,* any suggestions that you, the reader, can make to improve the next volume would be greatly appreciated. Please take a few minutes to provide your input regarding article selections and to make recommendations of other articles that you feel would be appropriate in the next edition. You may use the postage-paid article rating form at the end of this volume.

David Dathe

David Dathe
Editor

Contents

To the Reader ... iv
Topic Guide ... 2
Selected World Wide Web Sites ... 4

Overview ... 6

UNIT 1

Data-Gathering Techniques

The 12 articles in this section discuss how the space telescope, spectrum analysis, X-ray, gamma ray, infrared, and radio astronomy are used in the gathering of celestial data.

A. HUBBLE SPACE TELESCOPE

1. **In Golden Age of Discovery, Faraway Worlds Beckon,** John Noble Wilford, *New York Times*, February 9, 1997. ... 8
   More powerful telescopes on the ground and in space, especially the Hubble Space Telescope, and more sensitive electronic detection instruments are *sharpening the view of the cosmos*.

2. **Learning from Hubble's Deep Field,** Joshua Roth, *Sky & Telescope*, May 1997. ... 14
   Observations made by the *Hubble Space Telescope* have led to new ideas of *galaxy evolution*. The cataloging of over 3,000 extended objects within the Hubble Deep Field by brightness, color, and shape has helped astronomers understand how galaxies came into being. Further research is being conducted on how the intrinsic properties of galaxies have changed with time.

3. **Beyond the Hubble Space Telescope,** Harley A. Thronson Jr., Alan Dressler, and Douglas Richstone, *Nature*, June 6, 1996. ... 16
   The *Hubble Space Telescope* is scheduled to complete its primary scientific mission in 2005. What should be the next step? The authors present three initiatives that are recommended to *NASA* to continue scientific exploration of space.

B. SPECTRUM ANALYSIS

4. **The Spectral Types of Stars,** Alan M. MacRobert, *Sky & Telescope*, October 1996. ... 18
   What we know about stars is based on, in large part, information carried by their light. The modern spectral classification system is of vital knowledge to every astronomer. Important concepts to understand are the dissection of starlight, the establishment of various *spectral classes*, the luminosity classes, and how spectral type is used to produce the *Hertzsprung-Russell diagram*.

C. X-RAY ASTRONOMY

5. **Shrill Notes from the Stars,** Guy Miller, *Nature*, June 6, 1996. ... 21
   The Rossi X-ray Timing Explorer (RXTE) satellite allows astronomers to study rapid *X-ray* variability. Initial findings show that studying X-ray variability should help to indicate conditions near *neutron stars*.

6. **Recent Advances of X-Ray Astronomy,** Yasuo Tanaka, *Science*, January 7, 1994. ... 23
   Because Earth's atmosphere absorbs cosmic X-rays, astronomers were effectively blind in the X-ray astronomy wavelength band until satellite observations became possible. Important X-ray emitters in the galaxy include about 200 X-ray binaries. These include *neutron stars*, which appear as X-ray *pulsars*, and *black holes*. *Supernovae* are also strong X-ray emitters. Interstellar matter and the origin of the intense cosmic X-ray background have also been studied.

The concepts in bold italics are developed in the article. For further expansion please refer to the Topic Guide and the Index.

D. GAMMA-RAY ASTRONOMY

7. **In Line for a New Mission,** Peter J. T. Leonard, *Nature,* January 9, 1997. — 26

    Gamma rays can provide a great wealth of information on a variety of phenomena. The most direct way to measure the abundance of different elements and isotopes is through their nuclear emission lines. Study of gamma-ray emissions is helping to determine the composition of *solar flares* and *supernovae*.

8. **Gamma-Ray Bursts,** Gerald J. Fishman and Dieter H. Hartmann, *Scientific American,* July 1997. — 28

    Gamma-ray bursts represent the most powerful explosions in the universe. In this article, the authors examine the history of *gamma-ray observations* and what implications they may have on the evolution of the universe.

E. INFRARED ASTRONOMY

9. **An Infrared View of Our Universe,** Ian Gatley, *Astronomy,* April 1994. — 34

    Infrared light waves have longer wavelengths than visible light, which enables them to pass through gas and dust in space that block visible light. Detailed images of *nebulae* and *galaxies* made in the infrared thus provide information not otherwise available.

10. **Cool Gaze at Heartless Galaxies,** Gerry Gilmore, *Nature,* November 21, 1996. — 38

    The Infrared Space Observatory (ISO) has helped astronomers answer questions about both *star formation* and the nature of luminous infrared *galaxies*. In particular, observations have shown that ultraluminous infrared galaxies are powered solely by rapid star formation.

F. RADIO ASTRONOMY

11. **Radio Astronomy in the 21st Century,** Kenneth I. Kellermann, *Sky & Telescope,* February 1997. — 40

    Radio waves pass through Earth's atmosphere relatively unhindered and thus provide information on a variety of astronomical phenomena such as cosmic masers, beams of relativistic gas, *pulsars,* and radio bursts from the *Sun* and planets.

12. **A Radio Map of the Milky Way,** Paul W. Schuler III, *Sky & Telescope,* March 1994. — 47

    Paul Schuler, an amateur astronomer, describes how he built a satellite dish to observe the *radio emissions from neutral atomic hydrogen*. Schuler presents a radio map of the Milky Way, and he explains how the radio signal is processed. His observations are compared to those of a professional astronomer.

UNIT 2

The Earth and the Moon

The dynamics of Earth and our Moon are considered in the six selections in this section.

Overview — 50

A. THE EARTH

13. **Measuring the Size of the Earth,** Fernando Espinoza, *The Science Teacher,* April 1996. — 52

    One of the most important historical events in astronomy was the calculation of the *circumference of Earth*. The best-known method of measuring the size of Earth is attributed to Eratosthenes of Alexandria (250 B.C.).

14. **How the Earth Got Its Atmosphere,** Tobias Owen, *Ad Astra,* November/December 1995. — 53

    Earth retains its *atmosphere* because its mass is sufficient to produce a gravitational field that prevents most gases from escaping. Our abundant molecular oxygen is due to green plants. It is believed that other elements in our atmosphere may have been derived from *comets*.

15. **What Makes a Planet a Friend for Life?** Julie Paque, *Astronomy,* June 1995.   57
What enables a planet to sustain *life*? The answer depends, in part, on how you define life. Plausible conditions under which amino acids could have arisen on early Earth are discussed by Julie Paque. The possibility of organic molecules coming from space, such as by way of *comets,* is also considered.

16. **A Day in the Life of a Year,** Ian Stewart, *New Scientist,* January 6, 1996.   60
The relationship between Earth and the Sun is complex. Consequently, our methods of timekeeping are slightly imprecise, and the *calendar,* in particular, suffers from this slight imprecision.

B. THE MOON

17. **Moon Watching: An Experiment in Scientific Observation,** William P. Lovegrove, *The Physics Teacher,* February 1994.   63
William Lovegrove shows how students can incorporate the *scientific method* into astronomy observations. Generally, an astronomer, or any scientist, knows exactly what type of data to collect when the answer to a particular question is sought. However, this is not the case when confronting a new and unknown phenomenon.

18. **Charting the Moon by Eye,** Edmund A. Fortier, *Astronomy,* June 1997.   65
It is possible for the Moon to be accurately viewed and charted with the naked eye. Edmund Fortier points out, however, that *stalking the Moon's surface features* requires timing, precision, and skill.

UNIT 3

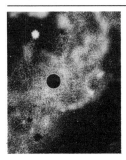

The Solar System

The 15 articles in this section focus on the makeup and interaction of our solar system: our Sun, the planets, comets, meteorites, and asteroids.

Overview   68

A. THE SUN

19. **Unsolved Mysteries of the Sun—Part 1,** Kenneth R. Lang, *Sky & Telescope,* August 1996.   70
In the first of a two-part series, Kenneth Lang discusses the Sun's energy-generating core, which presents the *solar neutrino* problem, as well as the unknown processes that heat the multimillion-degree *corona* and produce the *solar wind* in the Sun's outer atmosphere.

20. **Unsolved Mysteries of the Sun—Part 2,** Kenneth R. Lang, *Sky & Telescope,* September 1996.   74
Kenneth Lang, in the second of a two-part series, examines *solar-oscillation* data. Sound waves that reveal motions of the Sun's material are continuously studied. Currently, the Solar and Heliospheric Observatory (SOHO) spacecraft provides an ongoing "uninterrupted view of the Sun, 24 hours a day, 365 days a year."

B. THE PLANETS

21. **Atmospheric Dynamics on the Outer Planets,** Peter J. Gierasch, *Science,* July 19, 1996.   78
Atmospheric dynamics on Jupiter and the other outer planets are unlike anything on Earth. Vigorous and complicated atmospheric flows are common on *Jupiter, Uranus, Saturn,* and *Neptune.* Two mechanisms that contribute to such velocities are the latitudinal gradient of insolation and heating from below.

22. **Planet in a Bottle,** Jeff Kanipe, *New Scientist,* January 4, 1997.   80
Astronomers have been trying to construct physical models to explain the atmospheric dynamics of *Jupiter* in an attempt to explain both the atmosphere bands and the famous red spot. Astronomers also have been simulating winds on *Mars* in a wind tunnel.

23. **Magnetic Fields on Distant Moons Hint at Hidden Life,** William J. Broad, *New York Times,* May 20, 1997. 84
   Astronomers have discovered that *Io, Europa* and *Ganymede,* three of the larger moons of *Jupiter,* pulse with signs of magnetic fields. This suggests that beneath the surface ice the interiors are geologically, and perhaps biologically, active.

24. **Life in a Deep Freeze?** Jeffrey Kluger, *Time,* April 21, 1997. 87
   The recent discovery of *possible seas on Jupiter's moon, Europa,* suggests that life-sustaining factors could indeed be present.

25. **Water World,** Kathy A. Svitil, *Discover,* May 1997. 90
   Europa, one of the four largest moons of Jupiter, has an approximately 6-mile-thick ice sheet covering its surface. Calculations by planetary scientists suggest that tidal heating might allow some of the ice deep beneath the surface to actually exist as water. And where there's water, there's the possibility of life.

26. **The Stars of Mars,** Sharon Begley, *Newsweek,* July 21, 1997. 94
   The success of the *Mars exploration project* is discussed as scientists begin to digest the large amount of data that has been received from the Sojourner rover about the rock formations of the Ares Vallis site.

C. COMETS

27. **The Kuiper Belt,** Jane X. Luu and David C. Jewitt, *Scientific American,* May 1996. 97
   The *Kuiper belt* is a region beyond the orbit of *Neptune* that contains a varied number of small bodies and marks the beginning of the outer extremities of our solar system. Some *comets* may have originated here, as well as the planet *Pluto.*

28. **Comets That Changed the World,** Bradley E. Schaefer, *Sky & Telescope,* May 1997. 103
   Of all the various astronomical phenomena, probably none have influenced human history more than *comets.* Regarded in Roman times as a signal of the impending death of the emperor, comets have traditionally been labeled as predictors of catastrophe and destruction.

29. **The Comet's Gift: Hints of How Earth Came to Life,** William J. Broad, *New York Times,* April 1, 1997. 107
   After having the opportunity to observe the comet Hale-Bopp, scientists now believe that *comets may indeed have aided in establishing life on Earth.*

D. METEORITES

30. **Bits of Mars and Pieces of the Moon,** *Geotimes,* June 1996. 111
   Recent studies have confirmed that some *meteorites* that are found on Earth come from the *Moon* and from other planets, such as *Mars.* For billions of years inner solar system planets and their satellites have been "sharing bits and pieces" that are shorn off when meteorites or comets collide with them.

31. **Life from Ancient Mars?** J. Kelly Beatty, *Sky & Telescope,* October 1996. 112
   Certainly no other science story of 1996 generated more interest than the announcement of the discovery of what appeared to be primitive life preserved in a *meteorite* from *Mars.*

E. ASTEROIDS

32. **The Day the Dinosaurs Died,** Ron Cowen, *Astronomy,* April 1996. 114
   Did an *asteroid* strike Earth about 65 million years ago? The evidence certainly tends to support the idea. Did the asteroid's impact and the resultant climate and environmental changes cause the extinction of the dinosaurs? The evidence is more problematic in this area.

33. **Escaping the Ultimate Diaster—A Cosmic Collision,** John S. Lewis, *The Futurist,* January/February 1997. 119
   Popular books and movies have demonstrated many variations of the "killer asteroid" theme, that is, an impact with Earth by a *comet* or *asteroid.* Realistically, however, what are the chances of an impact?

The concepts in bold italics are developed in the article. For further expansion please refer to the Topic Guide and the Index.

UNIT 4

The Universe

In this section, 15 selections look at the universe: stars, stellar evolution, types of stars, black holes, the Milky Way Galaxy, dark matter, galaxy structure and evolution, and quasars.

Overview   122

A. STARS

34. **The Stellar Magnitude System,** Alan M. MacRobert, *Sky & Telescope,* January 1996.   124
    The *stellar magnitude system* is a mix of modern ideas and needs and historical frameworks. Two types of magnitude are needed: *visual magnitude,* the brightness of the star as we see it, and *absolute magnitude,* the actual brightness of the star.

35. **The Strange New Planetary Zoo,** Robert Naeye, *Astronomy,* April 1997.   127
    The discovery of *planets* orbiting solar-type *stars* has led to new ideas of how planets form. Our understanding of planet formation is based on only one solar system—our own.

B. STELLAR EVOLUTION

36. **Life and Times of a Star,** Ken Croswell, *New Scientist,* November 26, 1994.   132
    The *Hertzsprung-Russell diagram* is used by astronomers to study the tremendous diversity that exists among stars. The distribution of stars on such a plot reveals three types of stars: *main-sequence stars, supergiants,* and *white dwarfs.*

37. **Extreme Stars: At the Edge of Stellar Behavior,** James B. Kaler, *Astronomy,* January 1997.   136
    Stars may be characterized in three ways: size, temperature, and luminosity. These variations depend on a combination of the star's mass and age that determines the ultimate end for the star.

C. TYPES OF STARS

38. **Ka-Boom! How Stars Explode,** Robert Naeye, *Astronomy,* July 1997.   140
    Supernovae are the most spectacular end of stars. Now, scientists are gaining a better understanding of this dramatic process due to the ability of supercomputers to model the process.

D. BLACK HOLES

39. **New Findings Suggest Massive Black Holes Lurk in the Hearts of Many Galaxies,** John Noble Wilford, *New York Times,* January 14, 1997.   144
    John Wilford examines new evidence that has emerged to indicate that supermassive *black holes* are located at the core of nearly all *galaxies.*

E. THE MILKY WAY GALAXY

40. **The Milky Way,** Ken Croswell, *New Scientist,* May 25, 1996.   148
    By mapping the *Milky Way galaxy,* plotting the distribution of stars within the galaxy, and classifying the types of stars that occur within the galaxy, astronomers are attempting to learn the origin and evolution of the Milky Way.

F. DARK MATTER

41. **The Dark Side of the Galaxy,** Ken Croswell, *Astronomy,* October 1996.   152
    Astronomers have narrowed the field of candidates for *dark matter* to two broad possibilities: *MACHOs* (Massive Compact Halo Objects) and *WIMPs* (Weakly Interacting Massive Particles).

42. **Is the Dark Matter Mystery Solved?** James Glanz, *Science,* February 2, 1996.   157
    Recent work on the nature of *dark matter* has favored the idea that dark matter is composed of *MACHOs* (Massive Compact Halo Objects). However, critics point out that there simply are not enough observations yet to draw reasonable conclusions.

The concepts in bold italics are developed in the article. For further expansion please refer to the Topic Guide and the Index.

G. GALAXY STRUCTURE AND EVOLUTION

43. **The Evolution of Our Galaxy,** James Binney, *Sky & Telescope,* March 1995. — 158
    The study of the evolution of our *galaxy* is based both on observations within our galaxy and on studying other galaxies.

44. **Seeing How Galaxies Form,** Craig J. Hogan, *Nature,* October 19, 1995. — 163
    New observations of very different galaxies indicate two additional factors to be considered when studying *galaxy formation*.

45. **Before Galaxies Were Galaxies,** William Keel, *Astronomy,* July 1997. — 165
    A question that has plagued modern astrophysics is *How did galaxies form?* William Keel reviews some of the answers to this provocative question.

46. **What Makes Galaxies Change?** Marcia Bartusiak, *Astronomy,* January 1997. — 170
    Ideas about *galaxy evolution* are changing rapidly. Images from the Hubble Space Telescope have allowed astronomers to classify galaxies more accurately by size, type, and magnitude. A great understanding of what kinds of galaxies there are allows for more accurate models of galaxy evolution.

H. QUASARS

47. **Galactic Engines,** Neil de Grasse Tyson, *Natural History,* May 1997. — 178
    At the center of *quasars* are *black holes* that provide them with the energy to produce their light. In order for a quasar to stay healthy, its central black hole must "consume" stars to obtain energy.

48. **Beyond the Soapsuds Universe,** Gary Taubes, *Discover,* August 1997. — 181
    Gary Taubes considers the research of astronomer Margaret Geller, which offered compelling evidence that *galaxies are congregated on two-dimensional structures*.

UNIT 5

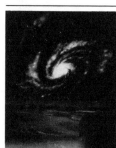

Ideas, Hypotheses, and Theories

The 12 articles in this section consider the concepts of space and time, extraterrestrial life, the Big Bang theory, and the age paradox.

Overview — 186

A. SPACE AND TIME

49. **The Nature of Space and Time,** Stephen W. Hawking and Roger Penrose, *Scientific American,* July 1996. — 188
    By discussing such concepts as quantum *black holes,* quantum theory, and quantum *cosmology,* two famous physicists define their own unique, and differing, views on quantum mechanics and the evolution of the universe.

B. EXTRATERRESTRIAL LIFE

50. **Searching for Alien Earth,** John Davies, *New Scientist,* May 13, 1995. — 193
    Search strategies for identifying a planet similar to Earth are being planned. Likely sources to have Earth-like planets include wobbly stars and *pulsars*.

51. **Is Anybody (Like Us) Out There?** Neil de Grasse Tyson, *Natural History,* September 1996. — 197
    Neil de Grasse Tyson presents a philosophical discussion on the questions arising from the notion of *extraterrestrial life*. The two most likely places to begin the search are *Mars* and Jupiter's moon *Europa*.

52. **When E.T. Calls Us,** Seth Shosak, *Astronomy,* September 1997. — 201
    The possibility that other civilizations dot our galaxy has recently been given a boost because of the discovery of several extrasolar planets. Although no life has been found, scientists now know that the galaxy teems with planets, some of which could harbor extraterrestrials.

C. THE BIG BANG THEORY

53. **The Best Cosmology There Is,** *Nature,* November 3, 1994. 206
    In this essay, the strengths and weaknesses of the *Big Bang theory* are examined. The abundance of deuterium (and other light nuclei) and the uniformity of the microwave background radiation are strong pieces of evidence for the *Big Bang.*

54. **Holes in the Big Bang,** *Nature,* November 3, 1994. 208
    While it does not suggest that the *Big Bang theory* is incorrect, this article does examine what needs to be done to complete the picture of the very early evolution of the universe.

55. **In Defense of the Big Bang,** Neil de Grasse Tyson, *Natural History,* December 1996/January 1997. 210
    The *Big Bang theory* is still regarded as a leading candidate for explaining the origin of the universe. The theory is explored by Neil de Grasse Tyson, and the criteria that theories must meet are also listed. For example, astronomers (and scientists) would insist that a theory make mathematical sense but not necessarily common sense.

56. **Everything You Wanted to Know about the Big Bang,** Richard Talcott, *Astronomy,* January 1994. 213
    Richard Talcott addresses the 10 most common questions about the *Big Bang theory.* Some of the questions that are considered are: Is space expanding or are galaxies moving apart? and, How do astronomers know how old and how big the universe is?

D. THE AGE PARADOX

57. **Conflict over the Age of the Universe,** M. Bolte and C. J. Hogan, *Nature,* August 3, 1995. 218
    The authors of this report provide an in-depth examination of the *age paradox.* They contend that "the ages of the oldest stars in our galaxy can be estimated by comparing stellar populations in global clusters to calibrated stellar models."

58. **Breakthroughs: Crisis Redux,** *Discover,* September 1995. 222
    Photos of *star clusters* taken by a telescope on a recent space shuttle indicate that the clusters are farther away than previously thought. This, in turn, affects the calculated ages of the stars within the clusters.

59. **Younger than They Look,** Cheryl Jones, *New Scientist,* September 7, 1996. 223
    Using new techniques, a group of Australian astronomers have measured the luminosity of a group of *pulsating stars.* Calculation of the ages of these stars was then determined to be between 9 and 12 billion years old, within other age estimates by standard cosmological models.

60. **Ages of the Oldest Clusters and the Age of the Universe,** Sidney van den Bergh, *Science,* December 22, 1995. 224
    A recalculation of the *Hubble parameter* is provided in this essay. The parameter is important because it gives the scale-size of the universe and provides constraints on the age of the universe.

Index	225
Article Review Form	228
Article Rating Form	229

The concepts in bold italics are developed in the article. For further expansion please refer to the Topic Guide and the Index.

Topic Guide

This topic guide suggests how the selections in this book relate to topics of traditional concern to students and professionals involved with the study of astronomy. It is useful for locating articles that relate to each other for reading and research. The guide is arranged alphabetically according to topic. Articles may, of course, treat topics that do not appear in the topic guide. In turn, entries in the topic guide do not necessarily constitute a comprehensive listing of all the contents of each selection. **In addition, relevant Web sites, which are annotated on the next two pages, are noted in bold italics under the topic articles.**

TOPIC AREA	TREATED IN	TOPIC AREA	TREATED IN
Age Paradox	57. Conflict Over the Age of the Universe (20, 21, 22, 23)	Galaxies (continued)	11. Radio Astronomy in the 21st Century 39. New Findings Suggest Massive Black Holes Lurk in the Hearts of Many Galaxies 43. Evolution of Our Galaxy 44. Seeing How Galaxies Form 45. Before Galaxies Were Galaxies 46. What Makes Galaxies Change? 48. Beyond the Soapsuds Universe (5, 6, 7, 8, 18, 19)
Asteroids	32. Day the Dinosaurs Died 33. Cosmic Collision (2, 14, 15, 17, 18, 25)		
Big Bang Theory	53. Best Cosmology There Is 54. Holes in the Big Bang 55. In Defense of the Big Bang 56. Everything You Wanted to Know about the Big Bang (2, 4, 22, 23)	Gamma Rays	7. In Line for a New Mission 8. Gamma-Ray Bursts (1, 2, 4)
		Hertzsprung-Russell Diagram	4. Spectral Types of Stars 36. Life and Times of a Star (2, 3, 4, 8, 20)
Black Holes	6. Recent Advances of X-Ray Astronomy 39. New Findings Suggest Massive Black Holes Lurk in the Hearts of Many Galaxies 47. Galactic Engines 49. Nature of Space and Time (2, 4, 13, 20, 21, 22, 23, 25)	Hubble Parameter	60. Ages of the Oldest Clusters and the Age of the Universe (20, 21, 22, 23)
		Hubble Space Telescope	1. In Golden Age of Discovery 2. Learning from Hubble's Deep Field 3. Beyond the Hubble Space Telescope (5, 6)
Calendar	16. Day in the Life of a Year		
Comets	14. How the Earth Got Its Atmosphere 15. What Makes a Planet a Friend for Life? 27. Kuiper Belt 28. Comets That Changed the World 29. Comet's Gift 33. Cosmic Collision (2, 3, 4, 9, 10, 15, 18)	Infrared Light Waves	9. Infrared View of Our Universe (7)
		Jupiter	21. Atmospheric Dynamics on the Outer Planets 22. Planet in a Bottle 23. Magnetic Fields on Distant Moons 24. Life in a Deep Freeze? 35. Strange New Planetary Zoo (15, 17, 18, 19, 22, 25, 26)
Cosmology	11. Radio Astronomy in the 21st Century 49. Nature of Space and Time (6, 7, 22, 23)		
Dark Matter	41. Dark Side of the Galaxy 42. Is the Dark Matter Mystery Solved? (20, 21, 22, 23)	Kuiper Belt	27. Kuiper Belt (15, 18)
		Life	15. What Makes a Planet a Friend for Life? 50. Searching for Alien Earth 51. Is Anybody (Like Us) Out There? 52. When E. T. Calls Us (22, 25, 26)
Earth	13. Measuring the Size of the Earth 14. How the Earth Got Its Atmosphere 16. Day in the Life of a Year (3, 4, 9, 10)		
Europa	24. Life in a Deep Freeze? 25. Water World 51. Is Anybody (Like Us) Out There? (12, 17, 18, 22, 25, 26)	MACHOs (Massive Compact Halo Objects)	41. Dark Side of the Galaxy 42. Is the Dark Matter Mystery Solved? (20, 21, 22, 23)
		Magnetic Fields	23. Magnetic Fields on Distant Moons (17, 18)
Exoplanets	1. In Golden Age of Discovery (5, 6)		
Galaxies	2. Learning from Hubble's Deep Field 9. Infrared View of Our Universe 10. Cool Gaze at Heartless Galaxies	Magnitude	34. Stellar Magnitude System (2, 20, 22)

TOPIC AREA	TREATED IN	TOPIC AREA	TREATED IN
Main-Sequence Stars	36. Life and Times of a Star (2, 3, 4, 8, 20)	Radio Astronomy	1. In Golden Age of Discovery 11. Radio Astronomy in the 21st Century 12. Radio Map of the Milky Way (5, 6, 7, 8)
Mars	22. Planet in a Bottle 26. Stars of Mars 30. Bits of Mars and Pieces of the Moon 31. Life from Ancient Mars? 51. Is Anybody (Like Us) Out There? (2, 3, 4, 12, 15, 16, 22, 23, 25, 26)	Saturn	21. Atmospheric Dynamics on the Outer Planets (15, 17)
Meteorites	30. Bits of Mars and Pieces of the Moon 31. Life from Ancient Mars? (2, 3, 4, 12, 15, 16)	Scientific Method	17. Moon Watching (11)
		Solar Flares	7. In Line for a New Mission (2, 4)
Milky Way Galaxy	12. Radio Map of the Milky Way 40. Milky Way 41. Dark Side of the Galaxy 43. Evolution of Our Galaxy (15, 17, 20, 21, 22)	Solar Wind	20. Unsolved Mysteries of the Sun—Part 2 (22, 23)
		Spectral Classes	4. Spectral Types of Stars (2, 3, 4, 8)
Moon	17. Moon Watching 18. Charting the Moon by Eye 23. Magnet Fields on Distant Moons 25. Water World 30. Bits of Mars and Pieces of the Moon (2, 3, 4, 11, 12, 22, 24)	Stars	10. Cool Gaze at Heartless Galaxies 11. Radio Astronomy in the 21st Century 34. Stellar Magnitude System 35. Strange New Planetary Zoo 36. Life and Times of a Star 37. Extreme Stars 38. Ka-Boom! How Stars Explode 48. Beyond the Soapsuds Universe 50. Searching for Alien Earth 58. Breakthroughs: Crisis Redux 59. Younger than They Look (5, 6, 7, 8, 20, 21, 22)
NASA (National Aeronautics & Space Administration)	3. Beyond the Hubble Space Telescope (4, 7, 9, 14, 19)		
Nebulae	9. Infrared View of Our Universe (21)		
Neptune	21. Atmospheric Dynamics on the Outer Planets 24. Life in a Deep Freeze? 27. Kuiper Belt (15, 17, 18, 19)	Stellar Magnitude System	34. Stellar Magnitude System (2, 20, 22)
		Sun	11. Radio Astronomy in the 21st Century 16. Day in the Life of a Year 19. Unsolved Mysteries of the Sun—Part 1 20. Unsolved Mysteries of the Sun—Part 2 (2, 4, 8, 13, 14, 15, 22)
Neutrinos	19. Unsolved Mysteries of the Sun—Part 1 (13, 14, 15, 18)		
Neutron Stars	5. Shrill Notes from the Stars 6. Recent Advances of X-Ray Astronomy (3, 4)	Supergiants	36. Life and Times of a Star (18, 20, 21, 22)
		Supernovae	6. Recent Advances of X-Ray Astronomy 7. In Line for a New Mission 38. Ka-Boom! How Stars Explode (5, 6, 7, 8, 20, 22, 23)
Planets	21. Atmospheric Dynamics on the Outer Planets 24. Life in a Deep Freeze? 26. Stars of Mars 35. Strange New Planetary Zoo (3, 4, 5, 15, 16, 17, 18, 19)		
		Uranus	21. Atmospheric Dynamics on the Outer Planets (17, 18, 19)
Pluto	27. Kuiper Belt (15, 18)	White Dwarfs	36. Life and Times of a Star (2, 3, 4, 8, 20)
Pulsars	6. Recent Advances of X-Ray Astronomy 11. Radio Astronomy in the 21st Century 50. Searching for Alien Earth 59. Younger than They Look (2, 4)	WIMPs (Weakly Interacting Massive Particles)	41. Dark Side of the Galaxy (20, 21, 22)
		X-Rays	5. Shrill Notes from the Stars 6. Recent Advances of X-Ray Astronomy (20, 22)
Quasars	47. Galactic Engines 50. Searching for Alien Earth (22, 23, 25, 26)		

Selected World Wide Web Sites for *Annual Editions: Astronomy*

All of these Web sites are hot-linked through the *Annual Editions* home page: http://www.dushkin.com/annualeditions (just click on a book). In addition, these sites are referenced by number and appear where relevant in the Topic Guide on the previous two pages.

Some Web sites are continually changing their structure and content, so the information listed may not always be available.

General Sites

1. American Astronomical Society Home Page—*http://www.aas.org/*—From this page you can reach the Astrophysical Journal's electronic edition, browse through the latest issues, and also access other astronomy links.

2. AstroWeb: Astronomy/Astrophysics on the Internet—*http://www.stsci.edu/astroweb/astronomy.html*—Basic site for all astronomical information, this is a collection of pointers to astronomy-related information available on the Internet. The AstroWeb Consortium, which maintains this database, also offers a search of its 2,501 distinct resource records.

3. The Space, Planetary, and Astronomical Cyber-Experience—*http://www.nss.org/space/home.html*—This site is presented by the National Space Society and features a Site of the Week as well as links to NASA, international space agencies, the International Space Station, and many other sites.

4. Yahoo!—Science: Astronomy—*http://www.yahoo.com/Science/Astronomy*—Basic search engine leads to much information and a multitude of sites in the field of astronomy.

Data Gathering Techniques

5. CADC HST Science Archive—*http://cadcwww.dao.nrc.ca/hst.html*—At this site you can *view* the HSA (HST Science Archive) data collected by the Hubble Space Telescope and maintained by the Canadian Astronomy Data Center. To *retrieve* the data, astronomers are requested to register.

6. Hubble Space Telescope—*http://ecf.hq.eso.org/HST.html*—The earth-orbiting Hubble Space Telescope is explained and explored at this site. Aso see the Guide Star Catalog, the world's largest astronomical star catalog, which contains coordinates and brightness for some 19 million objects. At *http://archive.eso.org/* you can now directly retrieve a list of HST Data Sets or search ESO (European Southern Observatory) and HST Databases.

7. SIRTF (Space Infrared Telescope Facility) Home Page—*http://sirtf.jpl.nasa.gov/sirtf/home.html*—Currently under design by NASA, this cryogenically cooled observatory to conduct infrared astronomy from space is planned to be launched December 12, 2001. SIRTF is expected to offer orders-of-magnitude improvements in sensitivity over previous IR missions. Read all about it at this site.

8. Telescopes—*http://www.stsci.edu/astroweb/yp_telescope.html*—This page will lead you to a long list of telescopes that can be accessed for retrieval data. Descriptions of projects are included.

The Earth and Moon

9. Observatory/NASA site—*http://observe.ivv.nasa.gov/*—NASA's Observatorium is a public access site for Earth and space data. Site uses satellite data (i.e., Landsat TD) and has excellent pictures of Earth, planets, stars, and other cool stuff, as well as the stories behind those images.

10. Welcome to Earth RISE—*http://earthrise.sdsc.edu/earthrise/main.html*—An easy graphical way to see 15 years of photos of Earth that were taken from space by astronauts out of the windows of the space shuttles. At Highlights, see the 500 best images.

11. Yahoo! Astronomy: Solar system: Moon, The—*http://www.yahoo.com/Science/Astronomy/Solar_System/Moon_The*—At this site, in addition to searching on your own, find a list of interesting material on the Moon, from the Clementine Lunar Image Browser, to a new theory on the Origin of the Moon, to many other Moon links.

The Solar System

12. Center for Mars Exploration—*http://cmex-www.arc.nasa.gov/*—This is the starting place for an exploration of the history of Mars, with links to the Whole Mars Catalog and Live from Mars information about Pathfinder and Global Surveyor.

13. Current Solar Images—*http://solar.uleth.ca/solar/www/images.html*—At this site you can click on recent solar imagery, hourly ionespheric maps, or a movie of Full-Disk Solar X-rays from the Yohkoh Satellite (October–November 1992). There is always a host of images to view.

14. Solar Data Analysis Center Home Page—*http://umbra.gsfc.nasa.gov/sdac.html#ECLIPSES*—Information collected by the SDAC at NASA's Goddard Space Flight Center includes frequent updates. See eclipses, the SUMER spectrum, the High Resolution UV Solar Atlas, and much more, plus links to other sources.

15. Solar System—*http://www.geocities.com/CapeCanaveral/ Lab/2683.index.html*—This excellent site enables you to navigate through the solar system to discover solar origins and planetary comparisons. You will be able to view all the major solar bodies. Click on Solar System, the Sun, or the Planets. You can also view the Mars Pathfinder from here.

16. The Mars Pathfinder Mission—*http://mpfwww.jpl.nasa.gov*—The site offers a comprehensive look at the entire Mars Mission and shows photos of what Pathfinder saw.

17. The Nine Planets—*http://www.seds.org/nineplanets/ nineplanets/*—William A. Arnett's multimedia tour of the solar system is an overview of the history, mythology, and current scientific knowledge of each of the planets and moons in our solar sytem. There are references to other related information. Besides pictures, find a glossary and many appendices. Site is constantly updated.

18. Views of the Solar System—*http://www.hawastsoc.org/ solar/homepage.htm*—This page, under the auspices of the Hawaian Astronomy Society, connects you to Calvin Hamilton's multimedia adventure that unfolds the splendor of the Sun, planets, moons, comets, asteroids, and more. Latiest scientific information, history of space exploration, rocketry, early astronauts, space missions, and spacecraft are all available through a vast archive of photographs, graphics, videos, and words.

19. Welcome to the Planets—*http://pds.jpl.nasa.gov/planets/*—Collection of many of the best images from NASA's planetary exploration program. Other items of interest at this site include Contacts and Related Pages, Glossary, and What's New. Mariner 10, Vikings 1 and 2, Voyagers 1 and 2, Galileo, Hubble, and Space Shuttle—the Explorers—are also shown with histories of their missions.

The Universe

20. 2MASS Home Page—*http://pegasus.phast.umass.edu/*—The Two Micron SII Sky Survey at the University of Massachusetts is shown and explained at this page. See Overview, Implementation, Images, and Status Report for more on this project, which will canvass the entire sky for stars and galaxies that are as much as 50,000 times fainter than the stars seen in the last survey, done 25 years ago.

21. Infrared Processing and Analysis Center Main Page—*http://www.ipac.caltech.edu/*—Explore IPAC's current projects and archives of images from this page. IRAS Galaxy Atlas is a useful data retrieval service. Projects include Midcourse Space Experiment, NASA/IPAC Extragalactic Database (NED), Wide-Field Infrared Explorer, and more.

Ideas, Hypotheses, Theories

22. The Astronomy Cafe—*http://www2.ari.net/home/odenwald/ cafe.html*—Click on Ask the Astronomer and Dr. Sten Odenwald will provide information on black holes, the Big Bang, Blue Moons, and many other topics. This site contains an archive of 3,001 questions with answers.

23. Cosmology and Astrophysics—*http://dept.physics.upenn. edu/~www/astro-cosmo*—From this page explore the work and thoughts of the University of Pennsylvania's Astrophysics and Cosmology Group. Links to other related sites.

24. LunaCorp—*http://www.lunacorp.com/*—In partnership with the Robotics Institute of Carnegie Mellon University, LunaCorp plans to launch rovers to the Moon and to allow "netizens" to navigate the Lunar Surface through a technology called telepresence—that is, navigation of the Moon's surface with mouse and keyboard. Learn about the project at this site.

25. P.E.R.M.A.N.E.N.T.—*http://www.permanent.com/*—Projects to Employ Resources of the Moon and Asteroids Near Term describes the plans to use materials in near-Earth space to establish settlements in orbit and beyond. Physicist Mark Prado acts as curator for the ideas of many engineers and scientists from around the world interested in the possibilities of the settlement of space.

26. SETI@home—*http://www.bigscience.com/setiathome.html*—This is the main page of Search for Extraterrestrial Intelligence, whose experiment is to harness the power of Internet-connected computers in the search for extraterrestrial intelligence (SETI). The experiment is to launch in Spring 1998.

We highly recommend that you review our Web site for expanded information and our other product lines. We are continually updating and adding links to our Web site in order to offer you the most usable and useful information that will support and expand the value of your *Annual Editions*. You can reach us at: *http:// www.dushkin.com/annualeditions/*.

Data-Gathering Techniques

Hubble Space Telescope (Articles 1–3)
Spectrum Analysis (Article 4)
X-Ray Astronomy (Articles 5 and 6)
Gamma-Ray Astronomy (Articles 7 and 8)
Infrared Astronomy (Articles 9 and 10)
Radio Astronomy (Articles 11 and 12)

Astronomy is unique in science in two different respects. First, it is one of the few sciences where the scientist has virtually no direct contact with the subject. (No one has traveled into space farther than our Moon.) Second, it is one of two sciences (paleontology is the other), where the amateur can still make important contributions.

Dealing with the first matter, astronomers are at a loss because they cannot view stars, black holes, or quasars close-up. Information about these phenomena arrive on Earth daily in the form of energy along the electromagnetic spectrum. This unit deals with this challenge of data gathering when the objects of interest may be at cosmological distances.

Visible light is one obvious source of data. Many powerful telescopes as well as more sensitive electronic detection instruments are sharpening our view of the cosmos. John Wilford examines astronomers' search of the heavens. The Hubble Space Telescope, a telescope in space and thus free from the interfering nature of our atmosphere, has opened up a whole new avenue of data collection. The article "Learning from Hubble's Deep Field," by Joshua Roth, discusses some of the advances made by images from the Hubble. The authors of the next essay, "Beyond the Hubble Space Telescope," offer NASA several recommendations for our space program after Hubble's mission is over.

A very important article in this unit is Alan MacRobert's "The Spectral Types of Stars." The spectral types of stars form the foundation of much of cosmology. Of particular importance is the Hertzsprung-Russell diagram, perhaps the most famous graph in astronomy. It shows how stars may be grouped into particular families.

The next two articles, by Guy Miller and Yasuo Tanaka, look at X-rays from the high end of the spectrum compared to visible light. X-ray emitters are numerous in the galaxy, and the most familiar examples include supernovae and black holes.

Gamma rays, also on the high end of the electromagnetic spectrum, allow astronomers to infer the composition of objects such as supernovae. Two great puzzles concerning gamma rays are their source and nature. Gerald Fishman and Dieter Hartmann, in "Gamma-Ray Bursts" discuss these two questions. Observations tend to confirm that gamma-ray bursts do not occur within our galaxy but at cosmological distances. However, what causes them still remains a mystery.

On the lower end of the spectrum, infrared light has been used by astronomers to "see" galaxies and nebulae in ways that are impossible with visible light. The reason for this is that infrared light can pass through the gas and dust of space that blocks visible light. Ian Gatley in "An Infrared View of Our Universe," provides a sample of the remarkable images that are achieved using infrared light.

Finally, radio waves have also been used to study celestial objects. The essay "Radio Astronomy in the 21st Century," by Kenneth Kellermann, gives a fairly complete view of the discipline, including suggestions for further advancements in the field. "A Radio Map of the Milky Way," by Paul Schuler III, was included specifically to show how amateur astronomers contribute to the field of astronomy.

Looking Ahead: Challenge Questions

What is the Hertzsprung-Russell diagram? What are its axes? What "fields" or regions occur on the diagram? Why is the diagram so useful to astronomers?

Why was X-ray astronomy developed only after the development of satellites?

What evidence is there that gamma rays originate outside our galaxy at cosmological distances? What other possibilities exist?

What type of information does infrared light provide astronomers that visible light does not?

What types of objects can be studied by radio waves?

UNIT 1

Article 1

In a Golden Age of Discovery, Faraway Worlds Beckon

JOHN NOBLE WILFORD

MOUNT HAMILTON, Calif.—Standing outside the dome of Lick Observatory on this lofty summit, two astronomers gazed beyond the foothills to the far horizon where California meets the Pacific Ocean. As the solid world at their feet rotated east, the great red sphere of glowing hydrogen seemed to sink perilously close to a doomsday collision, only to slip harmlessly out of sight in the west.

At the moment of sunset, birds somewhere in the trees broke into song, life sounding retreat at the loss of light. The astronomers turned back to the dome and the telescope within. Time to go to work, time to search the heavens for other stars not unlike the Sun and see whether some of them also have companion worlds—other places where night follows day, where there might be air and water, mountains and shore, even life and song.

The two astronomers, Dr. Geoffrey W. Marcy and Dr. R. Paul Butler of San Francisco State University, began another night of work at Lick Observatory, near San Jose, Calif., with the quiet confidence of professionals at the top of their game. In little more than a year, they and other teams in this country and

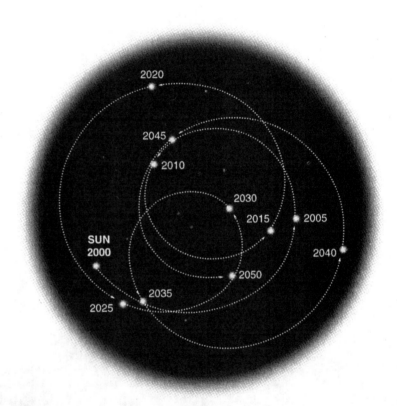

The Wobbling Sun

When a planet orbits a star, the planet's gravitational pull causes the star to wobble as the system moves through space. Astronomers search for these wobbling stars when trying to identify new, unseen planets. In fact, if alien astronomers began tracking the motion of the Sun in the year 2000, viewing it from its North Ecliptic Pole, this (above) is what they would see over the next 50 years, adjusted for the normal motions of the star and system.

Source: Sky and Telescope

N.Y. Times News Service

1. Golden Age of Discovery

Wobbling, Shifting Stars: Signs of New, Unseen Planets

Planets do not merely orbit stars. Instead, a star and its planet (or planets) tumble through space playing a virtual game of tug-of-war, both orbiting the point at which their gravitational pulls on each other equalize, called their center of mass. (Studying this behavior directly has led to one report of a new planet, or even two, as yet unconfirmed. It is expected to be a useful means of detection in the next few years.)

While tumbling through space, stars may appear to shift in color, periodically changing the wavelength of the light they emit. (Studying this pulsing has led to all eight of the confirmed extrasolar planet discoveries to date.)

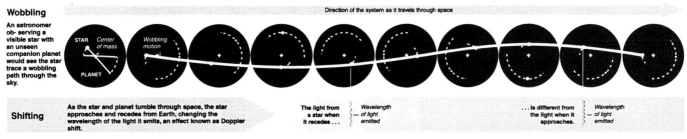

Sources: Sky & Telescope; San Francisco State University

The New York Times

Switzerland have for the first time detected planet-size objects—at least eight confirmed, and possibly two more—orbiting other stars like the Sun.

Even if a few skeptics still question whether these objects, called exoplanets, qualify as true planets, Dr. Alan P. Boss, a theoretical astrophysicist at the Carnegie Institution of Washington, expressed the prevailing view: "I truly believe we have indeed identified the first extrasolar planets."

All of a sudden, astronomers have turned a big corner and glimpsed in the dim light of distant lampposts a universe more wondrous than they had previously known. Other worlds are no longer the stuff of dreams and philosophic musings. They are out there, beckoning, with the potential to change forever humanity's perspective on its place in the universe.

Although no likely habitable planets beyond the solar system have been detected so far, the discoveries, coupled with fresh evidence of the possibility of early life on Mars, have already renewed enthusiasm in the search for extraterrestrial life. The first of a fleet of spacecraft and robotic landers took off late last year to resume Mars exploration. Space telescopes being planned for the next decade should be able to see planets as small as earth elsewhere and examine their atmospheres for signs of life.

More powerful telescopes on the ground and in space, especially the Hubble Space Telescope, and more sensitive electronic detection instruments are sharpening the view of the cosmos for other astronomers as well. Every few weeks brings more spectacular pictures from the depths of space, pictures of the moons of Jupiter, the nurseries of newborn stars and the galaxies taking shape when the universe was young. They embolden cosmologists, the historians of the universe, in the audacious belief that many answers to questions of cosmic origin and evolution may be within their grasp.

Little wonder scientists today feel justified in proclaiming this to be a new golden age of astronomy.

In that spirit, Dr. Marcy and Dr. Butler were now setting their sights on more planetary discoveries. Among the billions of galaxies, Earth's own Milky Way galaxy alone is populated with 100 billion stars, and a few hundred of these stars are close enough—less than 100 light-years away—to be in range of technology's new gifts of vision, making it possible to detect large planets there.

In the dark control room below the 120-inch Lick telescope, the two astronomers studied the light of target stars as they appeared, one by one, on a video screen. It is slow, tedious work. They were looking for faint variations, no more than 1 part in 100 million, in the frequencies of starlight, betraying the wobbling motions of a star caused by an unseen gravitational force nearby, something the size of a Jupiter or several Jupiters.

There could be no thought of an instant cry of "Eureka!" Hours of computer analysis and months of repeat observations precede any announcement of the discovery of one of these objects, which astronomers are calling extrasolar planets, or exoplanets. But in the wee hours, the two men reflected on the exhilaration of being young on a mountaintop on the only planet known to harbor life and having a part in discoveries with transforming implications.

Dr. Butler, who is 35, remembered the morning just over a year ago when he was "completely blown away" by the realization that a computer analysis showed that an object more than six times the mass of Jupiter was orbiting close to the star 70 Virginis, 80 light-years away. He could imagine the greats of astronomy nodding in awe. "I really felt

1. DATA-GATHERING TECHNIQUES: Hubble Space Telescope

the presence of Kepler standing at my shoulder," he said.

Professionally, Dr. Marcy, 42, mused, "for us this is the best that will be, at our young ages."

And the discoveries may be only beginning. One recent study suggested that planets might be lurking around half the Milky Way's stars. Astronomers have already seen enough to suspect that their definition of planets may have to be broadened considerably to encompass the new reality. As soon as they can detect several planets around a single star, they are almost resigned to finding that the Sun's family, previously their only example, is anything but typical among planetary systems.

The Quest
Seeking Worlds Around the Stars

The Epicurean philosophers of classical Greece would probably not have been surprised by such a turn of events. They believed that the chance conglomerations of infinite atoms in an infinite universe must form "innumerable worlds." Metrodorus of Chios, a disciple of Epicurus, wrote, "It would be strange if a single ear of corn grew in a large plain or were there only one world in the infinite."

When the Polish scholar Copernicus determined in the early 16th century that Earth orbited the Sun, a revolutionary idea began to take root in modern Western thinking: Earth might not be the center of the universe or even unique as an abode of life. Acceptance of that idea was grudging. Giordano Bruno was burned at the stake in 1600 for, among other things, the heresy of speculating about other inhabited worlds.

More recently, scientists have been encouraged in the search for other worlds by the recognition that the laws of physics appear to be universal and that life is a phenomenon based on natural chemical processes that need not be confined to one planet around one rather ordinary star. Since the 1960's, a few radio astronomers have been patiently listening to the heavens, seeking signal patterns that just might come from intelligent extraterrestrials.

Prospects for finding worlds around other stars improved substantially in the 1980's. Dr. Bradford A. Smith, of the University of Arizona in Tucson, and Dr. Richard J. Terrile, of the Jet Propulsion Laboratory in Pasadena, Calif., made infrared observations of a thick rotating disk of dust around the young star Beta Pictoris. This appeared to be a planetary system in the making and thus confirmation of the standard theory for the origin of such systems.

According to the theory, as a cloud of interstellar matter compresses and collapses, forming a new star, rotational forces cause the residual dust and gas to form a flat disk and then fragment and coalesce as planets. Similar disks were soon observed around other young stars. Some stellar clusters show signs of such disks around at least 60 percent of their stars. But detecting these broad disks has been easier than finding planets, relatively small objects obscured by the bright light of their parent stars.

The Methods
Planets' Effects Give Them Away

Planetary companions cannot be seen directly in visible light or other wavelenghts. Their presence must be inferred from their gravitational effects on the motions of their stars. The spectroscopic method, used by most astronomers, including Dr. Marcy and Dr. Butler, involves measurements of changes in the frequency of a star's light, the Doppler shift that reveals slight motions of the star toward or away from the observer. The simplest explanation for such radial motion is that a companion tugs at the star. The size of the frequency change in the starlight gives the amplitude of the star's motion, which in turn provides an estimate of the mass of the orbiting object.

Other scientists use the astrometric method, which requires years of tracking a star's course through space. Astronomers try to discern any speeding up or slowing down of a star, a sign that an orbiting object, like a planet, is alternatively holding back the star, then pulling it faster along its trajectory. Over the years, this technique has raised many false hopes, as when two companions were supposedly detected earlier in the century around Barnard's Star. They later proved to be artifacts of the telescope.

As it turned out, astronomers using radio-telescopes, not spectroscopy or astrometry, were the first to strike pay dirt in the search for planets, in 1994.

Dr. Alexander Wolszczan, a radio astronomer at Pennsylvania State University, reported detecting two and perhaps three planet-size objects orbiting a star in the Virgo constellation. He called this "a final proof that the first extrasolar planetary system has been unambiguously identified."

Although scientists accepted this assessment, they were nonetheless disappointed. Dr. Wolszczan seems to have found planets, but not planets around a normal star like the Sun. They are companions of a pulsar, a dense, rapidly spinning remnant of an exploded star, its thermonuclear furnace dead. The detections were made by observing regular fluctuations in the pulsar's rapid radio signals, indicating the planet's complex gravitational effects on the dead star.

But a pulsar is no place for supporting life. The environment there would lack warming starlight and be saturated with deadly radiation.

Meanwhile, other astronomers were busy studying the visible light of nearby stars for telltale variations

1. Golden Age of Discovery

of their radial velocity, with no success. Dr. Marcy and Dr. Butler had been at the task since 1987. A group at the University of British Columbia in Vancouver, led by Dr. Bruce Campbell and Dr. Gordon A. H. Walker, who pioneered the spectroscopic search technique, had been looking even longer. Discouraged, the British Columbia group abandoned its fruitless work in April 1995, just before two Swiss astronomers scored the breakthrough.

The Competition
The Unexpected
And a Jump-Start

On Oct. 6, 1995, Dr. Michel Mayor, of the Geneva Observatory in Switzerland, announced that he and a colleague, Dr. Didier Queloz, had discovered a planet orbiting a star similar to the Sun, 51 Pegasi, about 40 light-years away.

The planet has more than half the mass of Juputer, at least; since the radial-velocity technique can determine only an object's minimum size, the planet's mass could be some 10 times as great as that listed. And, to the surprise and puzzlement of astronomers, the planet is closer to its parent star than tiny Mercury is to the Sun.

At first, astronomers were wary, but the Swiss astronomers had checked their data with care. Dr. Mayor said they had repeated the observations and had also ruled out the possibility that the light variations were caused by the star's pulsations or eruptions like sunspots. When a theorist ran simulations showing that such a large planet could survive intact that near its star, Dr. Mayor decided to go public with the results.

Dr. Marcy and Dr. Butler raced to Lick Observatory to check it out. Yes, a planet was there. They had been scooped, Dr. Marcy said, in part because they had been looking in the wrong place, not expecting that such a large planet could be so close to its star. Their expectations were influenced by the one planetary system they knew well, in which the giant Jupiter, 317.8 times as massive as Earth, is half a billion miles out from its star.

"Mayor jump-started us," Dr. Butler remarked while beginning the night's work on the mountain. "His discovery brought a level of excitement to the field so that we were able to get more computational time and more telescope time. His discovery told us 'Jupiters' could orbit in close. We've each been able to confirm the other's results."

In astronomy, as in anything else, having a clearer idea of what to look for can improve the odds of success. Dr. Marcy and Dr. Butler took another look at their old data on 120 stars and conducted new observations. By a year ago, they had two more planets to report: the one around 70 Virginis, in the constellation Virgo, and another around 47 Ursae Majoris, in the Big Dipper.

More discoveries followed. They detected a planet around Tau Boötis and another around Rho Cancri. In October, Dr. William D. Cochran of the University of Texas at Austin and the Marcy-Butler team announced their independent discoveries of a planet traveling an eccentric orbit around the smaller of twin stars in the constellation Cygnus, designated 16 Cygni B.

Dr. George Gatewood, of the University of Pittsburgh's Allegheny Observatory, reported finding one and perhaps two planets orbiting Lalande 21185, the fourth-nearest star to the Sun. His report has yet to be confirmed by other observations. But if true, it is the only discovery so far to be made with the astrometric technique, which tracks stars' paths over years.

The possibility that this could be the first known extrasolar planetary system, not just a solitary planet detection, has generated excitement. "I can't see how it could not be a planetary system," Dr. Gatewood said. "But it is all somewhat mysterious."

The Discoveries
Better Instruments
For Better Data

So, by the latest count, if the two pulsar planets are included, Dr. Adam Burrows of the University of Arizona said, "We now know of more planets outside the solar system than within."

Astronomers attribute the sudden burst of discovery to significant advances in spectrometers, instruments like prisms that break up starlight into its component colors for detailed analysis; in electronic sensors, known as charge-coupled devices, that record the incoming starlight collected by telescope optics, and in computer software, which has been completely rewritten for discerning more reliably fluctuations in starlight that reveal the telltale motions.

By last year, the Marcy-Butler team had refined its techniques for measuring radial-velocity variations within an accuracy of 10 feet per second, which made the measurements five to seven times more accurate. They can now also compensate for imperfect telescope optics that cause a slight smearing of the light.

It also helped to have spent years gathering data on an increasing number of nearby stars. If the British Columbia team had examined at least 40 stars, not just 20, it might have made the first discoveries several years ago. "In retrospect, they were simply unlucky," said Dr. W. Latham, of the Harvard-Smithsonian Center for Astrophysics.

Still, astronomers have learned little about the nature of these exoplanets. Some are smaller than Jupiter, but most are larger. Some complete an orbit of their stars in months, while others take years. No one can tell if they are solid bodies, like Earth and Mars, or mainly gaseous spheres, like Jupiter and Saturn. No one can say if they have moons. So far, only one exoplanet, at 70 Virginis, seems to fall into the "habitable zone"—the region, governed by

1. DATA-GATHERING TECHNIQUES: Hubble Space Telescope

a planet's distance from its star, where water is liquid rather than solid or gaseous. Even so, the prospects for life there are far from encouraging.

"Theorists are now going a little bit crazy," Dr. Butler said. Observational astronomers seem to take a perverse delight in confounding their theoretically inclined colleagues, and the exoplanet findings have done that.

Debate continues over whether some or all of these objects are true planets or brown dwarfs. It is more than a question of their sizes and positions in relation to their parent stars; fundamentally, it is a matter of how they were formed, which cannot yet be determined.

Brown dwarfs form the same way a star does, by the collapse of a compressed interstellar cloud, but they never accumulate enough mass to support nuclear fusion in their interiors. It is conceivable that a brown dwarf may sometimes be as large as 80 Jupiter masses, but no one knows how small one can be. Only a few brown dwarf candidates have been identified, and only one confirmed.

"If a new object orbiting a star is a gas-giant like Jupiter," Dr. Boss, of the Carnegie Institution, wrote in a recent issue of Physics Today, "then in analogy with our own solar system, we would expect that Earthlike planets also formed around that star. However, if a new object is a brown dwarf star, then it is unclear whether or not Earthlike planets also formed—binary stars are thought to disrupt the planet formation process."

Dr. David C. Black, director of the Lunar and Planetary Institute in Houston, is especially outspoken in his belief that most of the objects will prove to be brown dwarfs. One discovery, first thought to be an exoplanet, has been revealed to be a brown dwarf companion to the star Gliese 229.

Dr. Black contends that it is "a bit hasty" to conclude that the objects around 51 Pegasi and 70 Virginis are planets and not brown dwarfs. If these are Jupiter-class planets, he said, they should be in more circular orbits farther out from their central stars, as is the rule in the solar system.

A study directed by Dr. Douglas N. C. Lin, of the University of California at Santa Cruz, yielded a possible explanation for finding actual planets in such cozy proximity to their stars: they could have formed farther out and then migrated inward to their observed positions. "The planet spiraled slowly but relentlessly toward the star," Dr. Lin said. "Finally, inward and outward forces on the planet's orbit canceled each other out just before the star would have consumed the planet."

Dr. Frederic A. Rasio, an astrophysicist at the Massachusetts Institute of Technology, approached the problem by assuming that relatively stable planetary systems like the Sun's, with its one dominant massive planet—Jupiter—may be extremely rare. Assume, instead, that many systems start with two Jupiter-size planets in fairly close proximity. In computer simulations, Dr. Rasio and a student, Eric B. Ford, showed that the strong gravitational interaction between the two planets could lead to one casting the other out of the entire system, while the survivor headed into a smaller orbit or sometimes crashed into the star.

One implication of this model, Dr. Rasio noted, is that any smaller Earthlike planets "are likely to be lost as a result of the instability." They either escape from the system or collide with the central star. Such a violent history could thus preclude the evolution of advanced life in such systems. By the same token, having only one Jupiter may have been a necessary condition making the solar system sufficiently stable for the evolution of intelligent life.

Dr. Black is skeptical of such explanations. "There is one other possibility, namely that planet hunters have discovered a new class of objects," he said.

Only when astronomers discover more than one planet candidate around a single star will they learn if what they are seeing are indeed planets and if other planetary systems bear much resemblance to the Sun's family. For all theorists know now, the solar system could be, as Dr. Marcy said, "the odd bird in the zoo."

The New Projects
'Goldilocks Orbits' Are Just Right

At a workshop where astronomers discussed new ideas for finding exoplanets, someone asked when would be the earliest anyone might begin detecting objects the size of Earth or Mars orbiting other stars at distances comparable to Earth's from the Sun. Scientists call this habitable zone the "Goldilocks orbit," where conditions should be neither too hot nor too cold but just right for life. Without hesitation, Dr. William J. Borucki, of the Ames Research Center in Mountain View, Calif., replied, "2001."

Dr. Borucki's optimism was based on a proposal by him and his colleagues for sending a small satellite into space to focus its telescopic electronic camera on thousands of stars considered to be prime candidates for planetary formation. The instrument should be able to detect a faint drop in the light intensity of a star, suggesting that a planet is passing across its face.

If the project wins Federal approval in the next few months, the spacecraft, called Kepler, could be launched in March 2001; within a few weeks, it could detect some 2,400 new planets, including perhaps 100 that might have a size and solid surface like Earth's.

A more ambitious concept is being developed by Dr. J. Roger P. Angel and Dr. Neville J. Woolf of the University of Arizona. They propose putting a large infrared telescope in deep space that would be capable of detecting the radiated heat of exoplanets. The emissions should also reveal the presence of any water, ozone or carbon dioxide on a planet, which could be evidence of life. But this project may have trouble winning support because of its complexity and a cost estimate of $2 billion.

At the Jet Propulsion Laboratory, engineers are drawing up plans for two space missions they expect will be centerpieces in the National Aeronautics and Space Administration's new Origins Program, a major goal of which is the search for Earthlike exoplanets. The missions are to apply a new technology called optical interferometry for the first time in space science. A variation on the idea that the whole can be greater than the sum of its parts, interferometry involves several small telescopes, separated but operated in unison, to make observations that are as sharp as those of a single telescope that would be so powerful and big, several hundred feet wide, that it would be impossible to build or deploy.

The first of these, the Space Interferometry Mission, which could be ready for launching in seven years, should give astronomers their most precise measure yet of the positions and motions of stars. Dr. Firouz Naderi, director of the program, put it this way: "If you were looking at the Moon with this system and there was an astronaut with a flashlight on the surface, you would be able to detect his passing the flashlight from one hand to the other."

The second step, to be taken in about 10 years, would be a mission called the Terrestrial Planet Finder. In a more elaborate application of interferometry, the spacecraft would operate four 60-inch telescopes placed along a 240-foot-long truss. The infrared light they collect would be combined in a way to eliminate light from the star but magnify any radiation from a nearby planet. Astronomers predict that the project should not only be able to discover Earth-size planets but also study their atmospheres for signs of life.

And that ultimately, is the goal.

"What we really want to find—let's face it—is an Earthlike planet, habitable and inhabited," Dr. Boss said.

"Scientists don't talk about it," he said, "but they want the same thing. It would be fantastic to happen in our lifetime, but the life we find may be only algae. Life may be close to us in space, but not in time."

The Prospects
Getting Ready to Cross a Line

Dr. Marcy and Dr. Butler know that exoplanets resembling Earth, if they exist, are probably beyond the sensitivity of their technology. Even so, they are excited to be in on the beginning of what could be discoveries of Copernican magnitude. They and other scientists struggle to find the words to convey their awakening sense of the possible intimations of new worlds.

After people accepted that the solar system is not the center of the cosmos, Dr. Naderi, of the Jet Propulsion Laboratory, said, "we drew a new line and said life is unique to us, at least intelligent life." He added: "I think we are beginning to chip away at that notion, too. Which would give us a warmer feeling? That we are something special in the universe? I would feel we are more special if we are part of a grander community."

In pressing for an accelerated search for exoplanets, Daniel S. Goldin, the NASA Administrator, said that "no human endeavor or thought would be unchanged" by the discovery of some form of life on other planets.

Dr. Boss predicted that the first images of another world like Earth, with the possibility of some form of life on it, "could shape what humans want to do for the next 1,000 years." People would want to explore it, at least with high-velocity robotic probes. They might send back close-up pictures to descendants living a millennium from now. "Hopefully, we won't burn anyone at the stake over it," he said.

By dawn at the Lick Observatory, at the sound of a bird-song reveille, Dr. Marcy and Dr. Butler were able to reflect on their night's work. Of 20 stars examined, at least 2 looked promising, but it could take another year of observations to be sure. Not a bad ratio, though too soon for observing astronomers to relax and leave their discoveries of other worlds to the meditations of philosophers.

"It's taken a long time getting good at this," Dr. Butler said. "Now I just want to find more planets."

Article 2
NEWS NOTES
Edited by Joshua Roth

Learning from Hubble's Deep Field

Today's 10- or 15-billion-year-old universe of stately, star-studded galaxies has been relatively well explored. Maps of the cosmic microwave background have even provided a fuzzy picture of the universe's infancy. Somewhere in between, galaxies came into being, though just how and when this happened remains a mystery.

A key piece in the puzzle of galaxy evolution has been the Hubble Deep Field (HDF), which scientists worldwide have had at their disposal for just over a year (*S&T:* May 1996, page 48). The result of 10 days' observing with the Hubble Space Telescope (HST), the HDF captured smudges down to 30th magnitude — several billion times fainter than the feeblest stars our unaided eyes can perceive. More than 3,000 extended objects within the HDF have been ranked by brightness, color, and shape.

To determine whether a given blob is a nearby dwarf or a distant monster, astronomers need each object's *redshift*, which tells how much space expanded while its light traveled to Earth. Most of

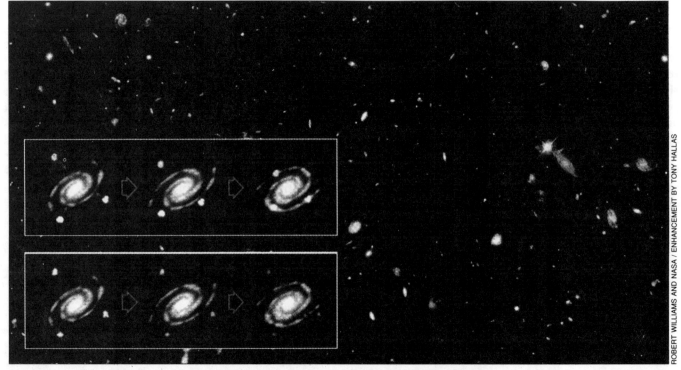

The deepest visible-light image of extragalactic space to date, the Hubble Deep Field has bolstered the notion that small, tightly packed galaxies dominated the cosmos when the universe was less than half its current age. However, it remains unclear whether such objects merged with their larger neighbors (upper panel at left) or simply faded from view (lower panel).

2. Learning from Hubble's Deep Field

the Deep Field's denizens are too faint for traditional redshift assessment by even the world's largest visible-light telescopes, the two Keck 10-meter reflectors. However, by February one of those leviathans had laboriously garnered spectra for nearly 100 Deep Field dwellers — enough to validate measures of age and distance that had been based on their color alone.

Emboldened by this confirmation, astronomers have gone on to ask how the intrinsic properties of galaxies have changed with time. As Henry C. Ferguson (Space Telescope Science Institute) recounted at January's meeting of the American Astronomical Society, most of the HDF's faintest blobs are just a fraction of the size of current galaxies like the Milky Way and may even be smaller than the Local Group's dwarf members. They also tend to outnumber today's large galaxies several times over, even after accounting for the expansion of space.

These demographics have led some astronomers to speculate that several precursors merged to form today's dominant ellipticals and spirals (*S&T:* November 1996, page 11). But Wesley Colley (Princeton University) thinks the HDF's smudges are simply giant star-forming regions similar to 30 Doradus in our neighboring Large Magellanic Cloud. If they were really dwarf systems that later merged, he says, the resulting galaxies would have more mass in their centers than do present-day galaxies.

Charles C. Steidel (Caltech) likewise doubts that HDF's overabundant fuzzies merged to form the galaxies we see today. His team's ground-based studies have found nearly enough star-forming galaxies in the universe's distant past to account for all the large ellipticals and spirals that exist now (*S&T:* May 1996, page 11). "I don't believe that [today's] big galaxies were in a hundred pieces" when the universe was already a few billion years old, he later told *Sky & Telescope*. He doesn't dispute the idea that the faint, blue denizens of deep-sky frames may have coalesced. But it seems just as plausible that they simply faded to invisibility once their modest gas reserves were exhausted or blown away.

Despite the unprecedented abundance of cutting-edge observations, Ferguson cautions, numerous selection effects will continue to bias censuses of the distant universe. For example, light arriving from high-redshift galaxies began its journey as ultraviolet radiation from hot, young stars, and seemingly normal galaxies can radically change their appearance when viewed through this spectral window (*S&T:* December 1995, page 12). The expansion of space renders low-surface-brightness galaxies all but invisible. And galaxies in clusters and groups may follow different life paths than do those in isolation. Researchers hope that HST's new infrared camera and a bevy of giant ground-based telescopes will lead to a more complete picture of the early universe and the evolution of its building blocks.

COMMENTARY

Beyond the Hubble Space Telescope

NASA needs to develop a new generation of space observatories embracing the theme of 'cosmic origins' if it is to maintain its lead in the exploration of space.

**Harley A. Thronson Jr,
Alan Dressler and
Douglas Richstone**

WHAT is past is prologue in astronomy: the content of the modern Universe—even the presence of life—is a product of events that took place billions of years ago. Fundamental questions about this epoch in cosmic evolution, about the origins of matter, structure and life in the cosmos, await a new generation of space observatories.

In 1993, the US National Aeronautics and Space Administration (NASA) and the Associated Universities for Research in Astronomy recognized that there were profound scientific questions that could not be answered with current space astronomy missions. Moreover, no new advanced space observatories were being considered for the period after the Hubble Space Telescope completes its primary science mission in 2005. In response, a committee of 18 astronomers from North America and Europe was established to identify core scientific questions about the cosmos that were unlikely to be answered with existing observatories and to recommend space missions capable of investigating them. This "HST and Beyond" committee, of which we are members, proposed "cosmic origins" as a unifying theme for a handful of fundamental problems involving the birth of normal galaxies, star formation and the possibility of Earth-like planets beyond our Solar System.

Fortunately, we can see the history of our universe directly: the farther we peer out into space, the further back we see into time. But the first few billion years of cosmic history are shrouded from the view of even the largest telescopes on Earth. The difficulty arises because the expansion of the Universe causes light from distant—and, consequently, young—galaxies to be shifted to longer wavelengths. Light from normal galaxies, such as our own Milky Way, is emitted by billions of stars at wavelengths mainly between 0.4 and 2 micrometres. But the large redshifts of young galaxies mean that this ancient light reaches Earth shifted to a wavelength longer than about 2.5 micrometres. At these wavelengths, Earth's atmosphere becomes opaque, and what light struggles through is overwhelmed by the thermal emission from both the atmosphere and the telescope. Only the most luminous, and most atypical, young galaxies can be detected by Earth-bound telescopes: the birth of the most common galaxies are accessible only from space.

As well as being interested in the birth of stars and galaxies, scientists and the public are intensely interested in whether Earth is alone in the cosmos in supporting life. A search for the origins of life, beginning with a search for Earth-like planets, requires a technology different from that used to study primaeval galaxies: spatial interferometry, where the optical system of many different telescopes can be combined destructively to cancel out the blinding light from a central star. This leaves behind the faint but detectable emission from surrounding planets. This residual light would be searched spectroscopically at infrared wavelengths for signatures of molecules believed to be necessary for life.

Important steps in understanding our cosmic origins are already being taken by both NASA and the European Space Agency. The latter is operating the Infrared Space Observatory, which is targeting luminous, very energetic galaxies. Similarly, NASA will launch its own powerful Space Infrared Telescope Facility, which will probe into the history of the Universe. Both observatories are essential milestones, but both are limited by modest apertures and lifetimes.

As a result of these considerations, our committee last week published three recommendations to NASA intended to guide the agency in space science at the start of the next century. First, NASA should develop an infrared-optimized space observatory with an aperture of more than 4 metres to study normal galaxies in the early Universe. The wavelength of optimum sensitivity of this observatory—about 1–5 micrometres—covers diagnostic spectral features from a wide range of other objects: the youngest and oldest stars, Solar System objects and galaxies throughout the local Universe. Furthermore, enlarging the wavelength coverage of the facility would greatly increase its usefulness to the broader astronomical community.

Second, NASA should develop the capability for space interferometry at optical and infrared wavelengths. The next major direction in space astronomy will be a notable increase in angular resolution: that is, we will be able to study very fine structure and determine positions and motions extremely accurately. A challenging application of interferometry in space will be the mission to search for Earth-like planets around neighbouring stars, and to explore their atmospheres for the molecules essential for life. The first main step in interferometry will be the Space Interferometry Mission, which will measure stellar positions to the limits of accuracy allowed by today's technology.

Third, the Hubble Space Telescope should continue to operate after 2005, when it is scheduled to complete its primary scientific mission, but in a 'no repair, no upgrade' mode. This would ensure the maintenance of a unique facility for ultraviolet astronomy and a continued return on the unprecedented US investment of thousands of man-years and billions of dollars in the observatory, whose success has dazzled people everywhere.

These three initiatives, along with a suite of smaller, more focused missions, will become the backbone of NASA's "Origins Program," which is intended to be a principal theme for future US research in space. In making these recommendations for the next generation of major ultraviolet–optical–infrared space astronomy missions, our committee was aware that NASA is likely to face severe constraints on its resources. So total costs cannot be comparable to those of the Hubble telescope. Alternatives to expanding costs range from applying innovative technologies, perhaps developed within defence-related industries, to exploring new management techniques, which would cut costs through accelerated development and international collaboration.

The initiatives recommended above would ensure NASA's continuing lead in the exploration of space well into the next century. By embracing our age-old human fascination with cosmic origins, such programmes will, we hope, excite humanity's imagination and encourage wider public participation in the exploration of the Universe. The recommendations do not chart an easy course for NASA, but the rewards of pursuing them are great.

Harley A. Thronson Jr is in the Department of Physics and Astronomy, University of Wyoming, PO Box 3905, Laramie, Wyoming 82071, USA. Alan Dressler is at the Carnegie Observatories, 813 Santa Barbara St, Pasadena, California 91101, USA. Douglas Richstone is in the Department of Astronomy, University of Michigan, Ann Arbor, Michigan 48109, USA.

Backyard Astronomy
Edited by Alan M. MacRobert

The Spectral Types of Stars

WHAT'S the most important thing to know about a star? Its apparent magnitude might top the list, but right behind would be its spectral type. Without it the star is a meaningless dot of light. Add a few letters and numbers, such as *G*2 V or *B*5 IV-Vshnne, and the star suddenly gains personality and character. To those who can read its meaning, the spectral code tells at a glance just what kind of object the star is — its color, size, and luminosity, its history and future, its peculiarities, and how it compares with the Sun and stars of all other types.

The modern spectral classification system is so successful that it has hardly been changed since 1943. It is based on just two physical properties that imprint themselves on the spectrum of a star's light: the star's temperature and atmospheric pressure. These reveal an abundance of information that paints the star's portrait and tells its life story.

The temperature sets the star's color and tells its surface brightness, how much light it gives off from each square meter of its surface. The pressure depends on the star's surface gravity and therefore, roughly, on its size — telling whether it is a giant, dwarf, or something in between. The size and surface brightness in turn yield the star's luminosity (its total light output, or absolute magnitude) and often its evolutionary status as young, middle-aged, or nearing death. The luminosity also gives a good idea of the star's distance. Appended to the basic spectral type may be letters for chemical peculiarities, an extended atmosphere, unusual surface activity, fast rotation, or other special characteristics.

Every starwatcher needs to have a feel for spectral types. Here are the most important things to know.

DISSECTING STARLIGHT

The story begins in 1802, when the English experimenter William Wollaston passed a beam of sunlight through a thin slit and then through a prism. The slit provided a sharp, high-resolution view of the familiar rainbow spectrum, with no colors overlapping. When seen this way, Wollaston noticed, the Sun's spectrum was marked by many narrow, black lines of various intensities. These dark lines stayed at exactly the same places in the colorful band from day to day and year to year. They were later measured and cataloged by Josef von Fraunhofer, for whom they are still called "Fraunhofer lines."

Similar spectral lines showed up in laboratory experiments. Using a slit and prism, physicists discovered that when a solid, a liquid, or a dense gas is heated to glow, it emits a smooth spectrum of light with no lines: a *continuum*. A rarefied hot gas, on the other hand, glows only in certain colors, or wavelengths: bright, narrow *emission lines* instead of a rainbow band. If a cooler sample of the same gas is placed in front of a glowing object, dark *absorption lines* appear at the wavelengths where the emission lines would be if the gas were hot.

By 1859 the situation was clear: we see the Sun's hot surface through its cooler atmosphere, which imposes the dark lines.

Every element, every chemical compound, shows its own set of spectral lines. They are as unique as fingerprints. They reveal not only which atoms and molecules are present but also many other physical conditions, starting with temperature. Here, scientists realized, was a way to bring the Sun down into the laboratory. When they put slit-and-prism devices (spectroscopes) on telescopes, they could even see spectral lines in the light of stars.

It was the 19th century's greatest astronomical breakthrough. Philosophers had cited the makeup of stars as something beyond all possible human knowing. Now finding the composition of the Sun and stars was just a matter of comparing spectral lines seen in a telescope to those in a laboratory. This wasn't always simple, but it gave birth to modern astrophysics — the treatment of stars as physical objects to be studied and understood, rather than as mere points of light on the sky to be measured.

SPECTRAL CLASSES

Today's classification scheme was born at Harvard College Observatory. Starting in 1886 under Edward C. Pickering, the staff photographed and classified thousands of stellar spectra. They assigned them letters from A through Q, generally in alphabetical order from the simplest-looking to the most complex. But soon a more natural system became clear. By rearranging and merging classifications, Antonia C. Maury and Annie J. Cannon found they could fit nearly all stars' spectra into one smooth, continuous sequence. The sequence matched the stars' color temperatures, from the hottest, blue-white stars to cool, orange-red ones.

But it was too late to reassign the letters. When the dust settled, the rearranged sequence ran *O B A F G K M* from hot to cool. Spectral types on the blue end were called "early" and those on the red end "late." These terms are still used today, though the incorrect idea of stellar evolution they embody — that stars simply cool with age — has been obsolete for generations.

The sequence could be cut even more

4. Spectral Types of Stars

finely. Cannon subdivided each letter from 0 to 9, so that a spectrum whose appearance placed it halfway between standard *G*0 and *K*0 stars was called *G*5. Using this scheme, Cannon led the classification at Harvard of 325,300 spectra recorded on wide-field photographs. The resulting *Henry Draper Catalogue* (HD) and *Henry Draper Extension* (HDE), published beginning in 1918, remain standard references today.

The time-honored mnemonic for remembering the spectral sequence, invented by Henry Norris Russell when astronomy's leadership was all male, is "Oh Be A Fine Girl Kiss Me." Last year *Mercury* magazine published a student's rejoinder: "Only Boys Accepting Feminism Get Kissed Meaningfully." Take your pick.

A few other spectral types don't fit the sequence but instead parallel it. Type *W* or Wolf-Rayet stars are as hot and blue as the hottest *O* stars but show strong emission lines, either of nitrogen (*WN*), carbon and oxygen (*WC*), or neither (*WR*). Emission lines indicate an especially thick shroud of hot gas surrounding a star.

Some giant stars at the cool end of the spectrum have an excess of carbon. These were originally called *R* and *N* but have been merged to form type *C*. "Carbon stars" can often be spotted at a glance in a telescope by their deep red color. A bright example in the autumn sky is 19 Piscium (TX Piscium) in the Circlet of Pisces, spectral type *C*5. Their distinctive absorption bands (masses of overlapping spectral lines) due to the carbon compounds C_2, CN, and CH darken or "blanket" the blue end of the spectrum. In other words, a carbon star's atmosphere is a red filter. When seen in emission instead of absorption, these same spectral bands glow blue; the same compounds that redden a carbon star in absorption gave Comet Hyakutake its blue-green tint in emission.

The rare type-*S* stars are also red giants. They parallel type *M* but show strong bands of zirconium oxide and lanthanum oxide instead of an *M* star's titanium oxide. We can imagine that planets of *S* stars, bathed in chemically peculiar stellar winds, might be encrusted with gems of cubic zirconia.

LUMINOSITY CLASSES

Even in stars of the same spectral type, the absorption lines don't always look alike. In some stars the lines are narrow and sharp; in others they are broadened

Different atoms and ions leave their fingerprints on starlight at different temperatures. Hydrogen, for instance, makes up the bulk of almost all stars, but it only shows itself strongly in spectra near type *A*. In cooler stars of type *K*, for example, the hydrogen lines are nearly lost among hundreds of others formed by heavier elements ("metals").

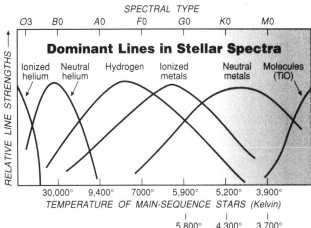

by various effects. Chief among these is atmospheric pressure, which also changes the intensity ratios of certain pressure-sensitive lines.

Astronomers quickly realized that atmospheric pressure tells a star's surface gravity and therefore suggests its size.

Narrow lines indicate an immense, bloated star with a weakly compressed atmosphere. In the *Henry Draper Catalogue,* spectral types were prefixed with *d* for dwarf, *sg* for subgiant, *g* for giant, and *c* for supergiant.

You'll still run across these letters

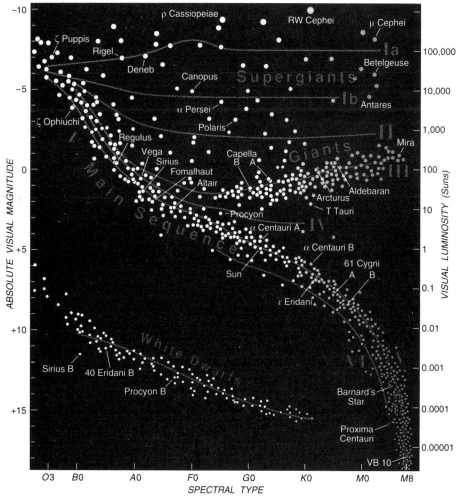

A Hertzsprung-Russell diagram plots stars' spectral types against their intrinsic luminosities (absolute magnitudes). The mix of real and simulated stars shown here gather on the main sequence and the giant branch; a scattering of bright giants and supergiants occurs at all spectral types. Based on a diagram by James Kaler.

1. DATA-GATHERING TECHNIQUES: Spectrum Analysis

from time to time, but beginning in 1941 they were replaced by a more detailed scheme first published by William W. Morgan and Philip C. Keenan. With only minor changes, this "MK" system of spectral classification remains the standard today. Stars are assigned to luminosity classes by Roman numerals: I for supergiants (often subdivided into classes Ia-0, Ia, Iab, and Ib in order of decreasing luminosity), II for bright giants, III for normal giants, IV for subgiants, V for dwarfs on the main sequence, and occasionally VI for subdwarfs.

Thus a designation such as *G*2 V, the Sun's spectral type, tells temperature and luminosity. When these are plotted against each other on a graph, the result is called a *Hertzsprung-Russell* or *H-R diagram*. This has been a fundamental astrophysical tool ever since it was invented around 1911. Most stars gather in certain narrow regions of the H-R diagram, as drawn at upper left, according to their masses and ages.

Stars arrive on the main sequence soon after they are born, and this is where they spend most of their lives. Massive stars blaze brightly on the hot, blue (*O* and *B*) end of the main sequence. They burn up their nuclear fuel in only millions or tens of millions of years. Stars with lower masses comprise the yellow, orange, and red dwarfs on the lower-right part of the main sequence, where they remain for billions of years.

As a star begins to exhaust the hydrogen fuel in its core, it evolves away from the main sequence toward the upper right and becomes a red giant or supergiant. Stars that began with more than six times the Sun's mass then evolve left and right through complicated loops on the H-R diagram as if in a frenzy to keep up their energy production, and finally they explode as supernovae. Less massive giants evolve to the left and then down to become white dwarfs; this is the track the Sun will trace through the H-R diagram 8 billion years from now (*S&T:* May 1994, page 12).

SOME SPECTRAL PECULIARITY CODES

Code	Meaning
comp	Composite spectrum; two spectral types are blended, indicating that the star is an unresolved binary
e	Emission lines (usually hydrogen)
m	Abnormally strong "metals" (elements other than hydrogen and helium); usually applied to *A* stars
n	Broad ("nebulous") absorption lines due to fast rotation
nn	Very broad lines due to very fast rotation
neb	A nebula's spectrum is mixed with the star's
p	Unspecified peculiarity, except when used with type *A*, where it denotes abnormally strong lines of "metals" (related to *A*m stars)
s	Very narrow ("sharp") lines
sh	Shell star (*B* to *F* main-sequence star with emission lines from a gas shell)
var	Varying spectral type
wl	Weak lines (suggesting an ancient, "metal"-poor star)

Symbols can be added for elements showing abnormally strong lines. For example, Epsilon Ursae Majoris in the Big Dipper is type *A*0p IV:(CrEu), indicating strong chromium and europium lines. The colon means uncertainty in the IV luminosity class.

ODDS AND ENDS

Spectra can reveal many other things about stars. Accordingly, lowercase letters are sometimes added to the end of a spectral type to indicate peculiarities. A list appears in the table below.

Certain spectral subtleties are not widely known among amateurs. Some visual observers pride themselves on being able to nail a star's type to the nearest letter by its color in the eyepiece. Color is indeed an excellent indicator of spectral type for stars earlier (hotter) than about *K*5, assuming no interstellar reddening is present. But the relationship often breaks down among the later *K* and *M* stars. Compare the tint of Betelgeuse, type *M*2 Iab, to that of Aldebaran, *K*5 III. Most people can't see a difference. At the same time, two red giants of the identical type may show different tints; compare Mu and Eta Geminorum, both cataloged as type *M*3 III.

In addition, dwarf *G*, *K*, and *M* stars are not as red as giants and supergiants of the same spectral types. The color difference is equivalent to one-half to one letter class.

Lastly, differences between spectra are far greater than differences in the actual compositions of stars. An *A* star might seem to be almost pure hydrogen, while a *K* star shows only trace evidence of hydrogen in a spectrum packed with lines of "metals" (the astronomer's term for all elements other than hydrogen and helium). But *A* and *K* stars are made of the same stuff. Different atoms and ions merely display their spectral lines at different temperatures. Even carbon stars are made mostly of hydrogen and helium. The true "abundances" of elements can indeed be measured in a star. But it's a tough job of comparing precise line strengths in a high-quality spectrum with those predicted by atomic theory or measured in the lab.

For much of the 20th century, the study of visible-light spectra practically *was* astronomy. In recent decades the opening of nonvisible wavelengths and other exciting advances have distracted attention from this field. Nevertheless it remains the bedrock on which modern astronomy rests.

— A. M.

FURTHER READING

Kaler, James B. **Stars and Their Spectra.** Cambridge University Press, 1989. Sections of this book were serialized in *Sky & Telescope* in 10 parts from February 1986 to May 1988.

X-RAY ASTRONOMY

Shrill notes from the stars

Guy Miller

ASTRONOMERS gather most of their information about the cosmos through spectral analysis and the interpretation of images. So it is notable that at a meeting last month in San Diego* the most exciting announcements involved neither images nor spectra. Instead, the highlight of the conference was the discovery of strange, extremely rapid variations in the brightnesses of several Galactic X-ray sources.

The sources in question are thought to be systems in which a weakly magnetized neutron star and a normal star orbit one another. The neutron star has a mass comparable to that of our Sun but is only a few tens of kilometres across, so the gravitational field near its surface is extremely strong. When gas lost by the companion star falls onto the neutron star, its tremendous gravitational potential energy is dissipated, and most of it goes into X-rays.

The gas has too much angular momentum to fall directly onto the neutron star, but instead spirals in slowly through a flattened accretion disk. If the magnetic field is very weak, the disk flow extends down to the stellar surface, ending in a hot, strongly sheared boundary layer. If the field is strong enough, the disk does not reach the stellar surface, but ends when magnetic stresses disrupt it at the magnetospheric boundary; material completes its journey to the stellar surface by moving along field lines to the star's magnetic poles.

The details of the flow structure and radiation mechanisms are unknown.

*Meeting of the AAS High Energy Astrophysics Division, San Diego, California, USA, 29 April–4 May 1996.

Also unknown is the interior structure of the neutron star, where matter is crushed to densities higher than are found anywhere else in the Universe (excluding, perhaps, the unobservable end state of material trapped by a black hole).

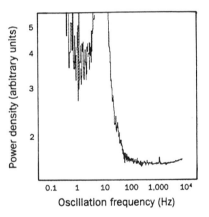

M. van der Klis, Univ. Amsterdam

Variability of the X-ray binary Scorpius X-1, over a wide range of frequencies. A startling quasi-periodic oscillation at 1,100 Hz (the small peak on the right) has been seen by NASA's new satellite, RXTE.

The Rossi X-ray Timing Explorer (RXTE) satellite, launched last December, has provided astronomers with a powerful tool for studying rapid X-ray variability, which should be an excellent indicator of conditions near neutron stars. A large proportional counter enables RXTE to collect and record X-ray photons at unprecedentedly high rates, so it can be used to study variations at higher frequencies and in greater detail than ever before. Not surprisingly, the first results from RXTE contained surprises.

Observations of the rapidly accreting source Scorpius X-1 (M. van der Klis, Univ. Amsterdam) show two new high-frequency quasi-periodic oscillations (QPO). A QPO is a narrow peak in the power spectrum of source variability—not a perfect 'tone' but a note spread over a small range of frequencies. Previous examples of QPO typically had frequencies lower than 100 HZ (ref. 1).

The new QPO are different. One is at about 800 Hz, and appears only at lower accretion rates; the other has a frequency centroid that rises from 1,060 to 1,130 Hz with increasing accretion rate. Both have amplitudes of less than about 1% of the total emission. Their reported widths, tens of Hz, are probably due to changes in frequency during the observations—the true widths are probably much smaller. Strangely, the frequency of the 1,100 Hz QPO seems to correlate linearly with the frequency of a previously known QPO that lies between 6 and 20 Hz.

The hoped-for periodic oscillation due to the rotation of the neutron star was nowhere to be seen, but evidence for stellar rotation may be present in the more slowly accreting source, 4U1728–34, which was observed during a recent series of X-ray outbursts (T. Strohmayer, NASA GSFC). Early in each outburst, as the luminosity of the source was still rising, a very strong QPO was seen at 363 Hz. In two of the outbursts it had a width of about 0.5 Hz. Later, as the X-ray luminosity declined, the width of the QPO peak became immeasurably small. Two other oscillations were present, at about 800 Hz and 1,100 Hz with a variable separation of roughly 363 Hz.

1. DATA-GATHERING TECHNIQUES: X-ray Astronomy

Kilohertz QPO have been discovered in two other sources (W. Zhang, NASA GSFC; I. Lapidus, Univ. Cambridge), but so far all these discoveries pose more questions than they answer. Weakly magnetized accreting neutron stars are expected to have rotation periods of 1 to 10 ms, and to be the progenitors of millisecond radio pulsars. But only in 4U1728–34 has a periodic oscillation been found that can be identified with stellar rotation. Even there, episodes occur during which the 363 Hz oscillation loses coherence and is broadened by 0.5 Hz. A repetitive modulation of the X-ray emission on a two-second timescale would provide the necessary broadening, but it is not evident why that should occur.

Even more mysterious are the variable-frequency QPO. Models of these must account simultaneously for excursions in frequency by 10% or more, and at the same time for the narrowness of the features. The oscillation mode must be sensitive enough to conditions in the accretion flow or in the outer layers of the neutron star to vary appreciably, and yet not be so sensitive that transient, localized irregularities alter its frequency too much and destroy its high-frequency coherence. The mode must also be able to produce intensity variations of 1 to 10%.

A number of candidate mechanisms wait in the wings. If the oscillation originates in the disk flow, it must come from near the neutron star, where the orbital and hydrodynamic timescales are short. The inner edge of the accretion disk, whether touching the stellar surface or kept away from it by the stellar magnetic field, is a natural site for the production of kilohertz variability via hydrodynamic oscillations. In sources such as Cygnus X-2, a class of QPO exists that is widely believed to arise from the interaction of the disk flow with the magnetosphere[2,3]. Searches for relations between this activity and kilohertz QPO will help in showing whether the latter have any connection with the inner edge of the disk.

Another possibility is that accretion or thermonuclear burning of the accreted material[4] excites oscillatory modes in the surface layers of the neutron star. Changes in the thermal state of the surface could be invoked to explain the frequency excursions of the observed oscillations. At present, however, it is not clear that stellar oscillations can achieve the necessary frequency variations or drive variations in the X-ray output as large as those observed.

A final proposal made at the conference (R. I. Klein, Univ. California, Berkeley) is that the accreting hot, magnetized plasma at the stellar surface is subject to a radiation-hydrodynamic instability that causes photons to collect in 'bubbles'. The X-ray intensity should rise briefly each time a photon bubble escapes from the accreting gas. Computer simulations show that under some conditions the radiative frothing of the accreting gas becomes organized into collective photon-bubble oscillations[5].

Although some mechanisms appear to have more difficulties than others, no leading candidate has emerged. It is not even certain whether there is a sole mechanism underlying the oscillations or a variety of causes. Nevertheless, the existence of extremely well-defined variabilities in the radiation from accreting neutron stars suggests that a powerful observational probe of these X-ray producing engines is at hand. It will depend on the development, in tandem, of a complete observational phenomenology and a secure theoretical framework. Clearly, many surprises lie ahead.

Guy Miller is at the Dearborn Observatory, Northwestern University, 2131 Sheridan Road, Evanston, Illinois 60218, USA.

Notes

1. van der Klis, M. A. *Rev. Astr. Astrophys.* **27**, 517–533 (1989).
2. Alpar, M. A. & Shaham, J. *Nature* **316**, 239–241 (1985).
3. Lamb, F. K., Shibazaki, N., Alpar, M. A. & Shaham, J. *Nature* **317**, 681–687 (1985).
4. Strohmayer, T. & Lee, U. *Astrophys. J.* (in the press).
5. Klein, R. I., Arons, J., Jernigan, G. & Hsu, J. J.-L. *Astrophys. J.* **457**, L85–L89 (1996).

Recent Advances of X-ray Astronomy

Yasuo Tanaka

Among the great astronomical discoveries of the 1960s was the totally unexpected finding by Giacconi et al. (1) of a bright x-ray star, now known as Scorpio X-1. Because cosmic x-rays are absorbed by the Earth's atmosphere, astronomers were effectively blind in the x-ray wavelength band until they were able to send instruments into space. The discovery of Sco X-1 was truly surprising because no bright x-ray sources other than the sun were considered to exist. Since then, x-ray astronomy has developed very rapidly, especially in the last 20 years, after satellite observations became possible.

X-rays are emitted either thermally, from hot plasmas with temperatures of millions to tens of millions of kelvins, or nonthermally through relativistic processes involving high-energy electrons. Therefore, many high-energy astrophysical processes are manifested most directly in the x-ray and gamma ray bands, and x-ray observations have become indispensable for the studies of high-energy astrophysics. At present, every class of astronomical object, from nearby stars through quasars at cosmological distances, is the subject of x-ray astronomy investigations.

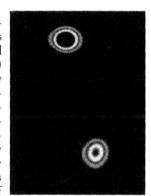

Fig. 1. X-ray image of M81 obtained with ASCA on 7 April 1993. The source near the top (north) is the nucleus. The bright source below is SN 1993J (3 arc min from the nucleus). Another source adjacent to the supernova is a probable x-ray binary in M81.

The first x-ray astronomy satellite, UHURU (1970), revealed a great many x-ray sources, galactic as well as extragalactic. One of UHURU's most important discoveries was that the brightest galactic x-ray sources were close binaries of which one member was a gravitationally collapsed, compact object—usually a neutron star, but possibly a black hole (2). The intense x-ray emission arises as matter dragged from the normal companion star falls into the extremely deep potential well of the compact object, releasing 100 MeV per nucleon of gravitational energy in the form of heat. This process, called mass accretion, produces a variety of phenomena in the x-ray band, allowing detailed investigation of the nature of these compact objects.

The richness of the x-ray sky was further disclosed as exploration progressed with succeeding x-ray satellites. The Einstein Observatory (1978) carried the first focusing x-ray telescope and had a sensitivity that was improved by several orders of magnitude over previous experiments. This enormously expanded the horizon of the x-ray sky and made x-ray studies of extragalactic sources an important branch of astronomy.

The Einstein Observatory revealed that many active galaxies, including quasars, were strong x-ray emitters (3). An active galaxy emits an immense quantity of electromagnetic radiation from a tiny region at the center of the host galaxy, called an active galactic nucleus (AGN). The emission is clearly of nonthermal origin, and the spectrum extends up to the gamma ray regime, indicating that complex relativistic processes are involved. From various arguments, an AGN is suspected to be a supermassive black hole, but the mechanism of the "central engine" of the AGNs still remains enigmatic.

Observations at x-ray wavelengths revealed that clusters of galaxies contain a large amount of hot (several tens of millions of kelvins) plasma (4). The mass of this intracluster gas is comparable with or more than the total mass of the constituent galaxies and accounts for a substantial fraction of the baryonic matter in the universe. The formation and evolution of clusters of galaxies have become increasingly important topics of x-ray astronomy.

More recently, the German x-ray satellite, the Roentgen Satellite (ROSAT), was launched in 1990. It was equipped with a larger x-ray telescope than that of the Einstein Observatory and, in the limited soft x-ray band of 0.3 to 2 keV, has the highest sensitivity of any soft x-ray telescope flown to date. Naturally, many discoveries are being made with ROSAT. The ROSAT all-sky survey vastly expanded the previous catalog to include nearly 100,000 x-ray sources.

The author is with the Institute of Space and Astronautical Science, 3-1-1 Yoshinodai, Sagamihara, Kanagawa-ken 229, Japan.

1. DATA-GATHERING TECHNIQUES: X-ray Astronomy

In the past, despite great scientific need, high-energy astronomy satellites have been scarce, too few compared to the ground-based observatories. Today, fortunately, several x-ray and gamma ray satellites with unique capabilities are simultaneously in operation: ROSAT (Germany), GRANAT (Russia), the Compton Gamma-Ray Observatory (CGRO) (United States), and the most recently launched Japanese satellite, ASCA. Significant advances in high-energy astrophysics can be expected.

In the remainder of this article, I shall focus on several topics in x-ray astronomy, using recent results obtained mainly with Ginga and ASCA. Ginga (1987–1991), the Japanese x-ray satellite before ASCA, had an x-ray collecting area of 4000 cm², the largest of its kind, covering a wide energy range between 1 and 40 keV. Launched in February 1993, ASCA is the first x-ray observatory capable of simultaneous imaging and spectroscopic observations over the range 0.5 to 10 keV. Although the angular resolution is modest (1 arc min), the x-ray charge-coupled device (CCD) cameras provide spectra of an unprecedented energy resolution.

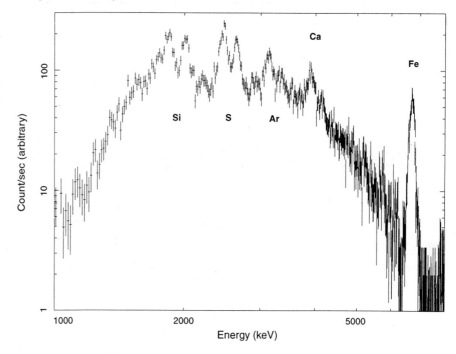

Fig. 2. The energy spectrum of the supernova remnant W49B obtained with the x-ray CCD camera of ASCA. Characteristic lines from silicon, sulfur, argon, calcium, and iron are clearly resolved.

There are about 200 x-ray binaries in our galaxy. About 30 of them include a strongly magnetized neutron star and appear as x-ray pulsars. Ginga observed the spectra of many binary x-ray pulsars and found cyclotron resonance features in 10 of them, which allows the direct measurement of the surface magnetic field of the neutron stars. These measured field strengths fall within a narrow range, 1×10^{12} to 4×10^{12} G (5), which casts doubt on the hypothesis that the magnetic field of neutron stars decays on a time scale of millions of years.

Most x-ray binaries, however, do not pulsate because the neutron stars in them have a much weaker magnetic field than those in the x-ray pulsars. Their companions are almost always low-mass stars, so the systems are called low-mass x-ray binaries. Sco X-1 is this kind of x-ray binary. How these low-mass x-ray binaries are formed is still an unresolved problem. The spectrum of a typical bright low-mass x-ray binary actually consists of two components: a soft component from the accretion disk formed around the compact object by the infalling matter, and a harder blackbody component from the neutron star surface.

About 20 bright x-ray binaries have been observed whose spectra are distinctly different from either the x-ray pulsars or the low-mass x-ray binaries (6). One group has a soft component characteristic of an accretion disk and is accompanied by a hard tail; the other group has a spectrum approximated by a single power law. The blackbody component expected from the neutron star surface is absent in both types of spectra. Optical observations of binary orbital motions allow us to estimate the mass of the compact object. For five of these binaries (Cygnus X-1 is a well-known example) for which such optical observations were made, the lower limit to the mass was found to exceed three solar masses. According to standard theory, a neutron star with a mass greater than three solar masses is unstable and collapses into a black hole, so the compact objects in these systems are most probably black holes. Even if optical measurements, and hence mass estimates, are not available, sources with similar x-ray spectra are also considered to be good candidates for black holes because of the unique spectral forms.

For some unknown reason, most of the candidate black hole binaries are transients. The records of previous observations and observations from Ginga, GRANAT, and CGRO indicate an occurrence rate for such transients of roughly one per year. These statistics lead to an estimate of more than several hundred black hole binaries in our galaxy (6).

Interestingly, the Ginga results show that there are striking similarities in the behavior of galactic black hole binaries and AGNs (6). Both systems exhibit rapid intensity fluctuations. This leads us to suspect that the fundamental physical processes are common to both systems, despite the difference of many orders of magnitude in the system scale and mass. However, the nature of such processes is not well understood at present. Multiwavelength studies of AGNs from radio waves through gamma rays are considered crucial for a better understanding of AGNs.

A supernova is a violent outburst of a massive star at the end of its evolution. Three weeks after Ginga was launched, a supernova (SN 1987A) occurred in the Large Magellanic Cloud. It did not emit x-rays at first, but after 5 months, Ginga and the Russian Mir-Kvant instrument detected x-rays from the radioactive decay of ^{56}Ni, synthesized during the supernova explosion and emerging from the expanding debris (7, 8). Five weeks after ASCA was launched, a supernova (SN 1993J) occurred in M81 (a nearby galaxy about 10 million light years away). This supernova emitted intense x-rays, which ROSAT and ASCA detected (Fig. 1) (9, 10). This was the first time that x-rays were observed from a supernova so soon (within 10 days) after outburst. The spectrum obtained by ASCA indicated that the temperature of the shock-heated plasma was initially greater than 10^8 K.

More than a hundred galactic supernova remnants (SNR) are known, and many are strong x-ray emitters. The best known is the Crab Nebula (SN 1054), which emits synchrotron radiation from radio through gamma rays generated by the neutron star. However, synchrotron nebulae are rather rare, and most SNRs emit thermal radiation from shock-heated plasmas. X-ray spectroscopy of SNRs, combined with detailed plasma diagnostics, enable us to study chemical abundances and dynamics of SNRs. This information is very important because all elements heavier than helium in the galaxy were synthesized in the interior of stars and spread over the interstellar space by supernovae. Spectra of SNR of unprecedented quality are being obtained with the ASCA x-ray CCD cameras, as the spectrum of the SNR W49B shows (Fig. 2). Individual characteristic lines from various elements up to iron are clearly resolved. A major advance in SNR physics is expected.

The interstellar space of our galaxy is complex. Most interstellar matter is in a cool phase, consisting of 100 K atomic hydrogen clouds and even colder molecular clouds and dust. However, x-ray observations have revealed that a large fraction of interstellar space is occupied by hot plasma at millions to tens of millions of kelvins. In fact, the solar system is within a huge volume filled with a tenuous 10^6 K plasma. The Ginga survey of the galactic plane showed that there exist regions that are much hotter, around several 10^7 K as characterized by emission lines from highly ionized iron, which are distributed all along the galactic plane (11). The origin of these hot plasmas is yet unknown. If they were of supernova origin, it would require a supernova rate of one in 10 years, 10 times higher than currently believed.

In addition, there is a strong peak of emission toward the galactic center (11). The amount of plasma responsible for this peak corresponds to thousands of supernovae. The center of our galaxy is not a bright x-ray source. Yet, this huge amount of hot plasma may indicate intermittent activity in the galactic nucleus. There is other observational evidence for a large mass concentration at the galactic center, suggestive of a massive black hole. Observations with ASCA have also revealed x-ray luminous nuclei in several spiral galaxies like ours (the nucleus of M81 in Fig. 1 is an example), indicating an activity similar to that in AGNs even in normal galaxies. These nuclei are less luminous than the previously known AGNs and hence may be called "mini AGNs."

ASCA is the first x-ray observatory able to provide spatially resolved spectra of the hot gas in clusters of galaxies. The results of such studies are important as they have significant implications for the cosmological evolution and the problem of dark matter which determines the gravitational potential of clusters of galaxies. Investigations of clusters of galaxies with ASCA have just started, and we anticipate significant advances.

Finally, the origin of the intense cosmic x-ray background (CXB), whose existence has been known since the birth of x-ray astronomy, is an important but still unresolved issue. As the sensitivity of observations has improved, the CXB has been increasingly resolved into discrete sources. The ROSAT deep survey has resolved more than 70% of the CXB below 2 keV into discrete sources, of which the majority are found to be AGNs, but there remains the puzzling "spectral paradox": The CXB has a slope (the exponent of a power law) of −1.4 in the photon number spectrum, whereas most AGNs have softer spectra, with an average slope of −1.7. Known AGNs therefore cannot account for the entire CXB, and a significant fraction of the CXB must be due to galaxies with flatter (harder) spectrum in order to account for the CXB spectrum. Capable of determining the spectral shape with its wide-band coverage, ASCA will be able to find the yet unidentified contributors to the CXB.

References

1. R. Giacconi et al., *Phys. Rev. Lett.* **9**, 439 (1962).
2. H. Tananbaum, *Astrophys. Space Sci. Libr.* **43**, 208 (1974).
3. T. Maccacaro et al., *Astrophys. J.* **253**, 504 (1982); R. Mushotzky, C. Done, K. Pounds, *Annu. Rev. Astron. Astrophys.*, in press.
4. W. Forman and C. Jones, *Annu. Rev. Astron. Astrophys.* **20**, 547 (1982).
5. K. Makishima, in *The Structure and Evolution of Neutron Stars*, D. Pines, R. Tamagaki, S. Tsuruta, Eds. (Addison-Wesley, New York, 1992), p. 86.
6. Y. Tanaka, in *Proceedings of the Ginga Memorial Symposium* (Institute of Space and Astronautical Science, Kanagawa, 1992), p. 19.
7. T. Dotani et al., *Nature* **330**, 230 (1987).
8. R. Sunyaev et al., *ibid.*, p. 227.
9. H. U. Zimmermann et al., *IAU Circular No. 5748* (1993).
10. Y. Tanaka and the ASCA Team, *IAU Circular No. 5753* (1993).
11. K. Koyama, *Nature* **339**, 603 (1989).

Article 7

GAMMA-RAY ASTRONOMY

In line for a new mission

Peter J. T. Leonard

THERE is a great wealth of astrophysical information to be gleaned from low-to-medium-energy γ-rays (between about 30 keV and 30 MeV), on a variety of phenomena including solar flares, supernovae and the mysterious cosmic γ-ray bursts. The latter emit most of their flux at these energies. These and other scientific topics were the subject of a meeting late last year*, as were the technologies from which to build telescopes sensitive enough to study them. I will limit myself to just one topic: nuclear emission lines.

The most direct way to measure the abundances of different elements and isotopes is through their nuclear emission lines. From a solar flare in 1991, for example, OSSE (the Oriented Scintillation Spectrometer Experiment) on board the Compton Gamma-Ray Observatory (CGRO) saw lines from ^{56}Fe, ^{24}Mg, ^{20}Ne, ^{28}Si, ^{12}C, ^{16}O and ^{15}N. These nuclei get excited by energetic protons and α-particles, accelerated to MeV energies by the flare; and they promptly de-excite by emitting γ-rays at characteristic energies. The relative strength of each line can be used to measure isotopic abundances in the Sun without resorting to assumptions about temperature or atomic physics.

The same MeV protons and α-particles should also create radioactive nuclei in the parts of the Sun near a flare, which will emit γ-rays as they decay. These emissions could be used to study mixing in the solar wind and in the Sun's outer convective layer (R. Ramaty, NASA Goddard Space Flight Center), much as studies of the radioactive fallout from a nuclear explosion can be used to study circulation in Earth's atmosphere. But at present there is no instrument sensitive enough and with enough angular resolution to study this phenomenon in and around the Sun.

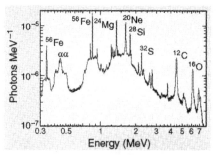

A theoretical nuclear de-excitation spectrum, showing both broad and narrow emission lines for material of solar composition. (Provided by Reuven Ramaty, NASA Goddard Space Flight Center.)

Supernovae produce detectable nuclear lines too. Types II and Ib supernovae (SNe II and Ib) occur when the cores of massive, evolved stars collapse into neutron stars, and the expanding remnants spread newly synthesized heavy elements into the interstellar medium. Shortly after the launch of CGRO, OSSE detected the ^{57}Co line at 122 keV in the four-year-old remnant of SN1987A. The abundance ratio of ^{57}Co/^{56}Co measured by OSSE is significantly smaller than that found by optical observations.

COMPTEL (the COMPton TELescope), also on CGRO, detected ^{44}Ti in the supernova remnant Cassiopeia A at 1.16 MeV. Cas A is a mystery, as the available modern evidence suggests that the explosion occurred in the second half of the seventeenth century, and it should have been easily seen at the time—but there is no undisputed record of it. Better γ-ray observations would help solve the mystery of what happened here. And γ-ray observations of a SN II or Ib as it happens should be a great help in understanding these explosions.

Type Ia supernovae (SNe Ia) are an order of magnitude more luminous than SNe II and Ib. They are probably explosions of relatively massive white dwarfs, triggered by accretion from binary companion stars. There is only a small dispersion in the peak luminosity of SNe Ia, which makes them useful extragalactic distance indicators.

COMPTEL barely detected two ^{56}Co lines from the type-Ia supernova, 1991T. These are the strongest of several predicted γ-ray lines, and it would be useful to have a much more sensitive instrument in place ready to observe the next bright SNe Ia. The different models for these explosions (detonation, delayed detonation, deflagration, deflagration plus mixing, and so on) make specific predictions for how the γ-ray spectrum evolves with time, and they can be distinguished with sensitive enough observations (P. Höflich, Univ. Texas, Austin).

In star-formation regions within our Galaxy, ^{26}Al is produced by supernovae and other evolved stars, and decays into ^{26}Mg with a half life of 7×10^5 years.

---

*Low/Medium Energy Gamma-Ray Astrophysics Mission Workshop, Lansdowne, Virginia, USA, 15–17 November 1996.

7. In Line for a New Mission

> **Compton telescopes**
>
> A COMPTON telescope consists of two separate layers of scintillator material. The upper layer is a converter and tracker made from a low-atomic-weight material; the lower is an imager and calorimeter made from a high-atomic-weight material.
>
> An incoming γ-ray Compton-scatters off one of the atomic electrons in the upper layer, changing its energy and direction. An energetic electron results from the process, and its energy is measured in the upper layer. The γ-ray goes on to the lower layer, where, ideally, it is absorbed, kicking out an even more energetic electron. The energy of the initial γ-ray is then the sum of the two electron energies, and their relative energies also constrain the source of the initial γ-ray to a circle on the sky. Measuring the track of the electron in the upper layer would reduce this circle to a point, tremendously improving the telescope's sensitivity and simplifying the overall analysis.
>
> The next generation of Compton telescopes should be at least an order of magnitude better in sensitivity, angular resolution and energy resolution than anything flown in space before.
>
> P.L.

COMPTEL has coarsely mapped this 1.81-MeV emission in the Galactic plane, and provided an upper limit of 350 parsecs on the distance to the Vela supernova remnant. But a more sensitive instrument could see groups of million-year-old supernova remnants and determine the Galactic supernova rate, which would constrain the formation rate of the most massive stars (M. Leising, Clemson Univ.). In addition, the rotation curve of the Galaxy can be measured by observing wavelength shifts in bright ^{26}Al sources.

One new satellite is already planned: INTEGRAL will be launched in the year 2001 on a two- to five-year mission (J. Matteson, Univ. California, San Diego). It will carry out high-resolution spectroscopy and imaging in the 15-keV to 10-MeV range. The coded-aperture-mask instruments on INTEGRAL are just about as sensitive as such instruments can be at these energies, but on the aforementioned questions many feel that they will only tease us.

At the meeting, there was a consensus that we should build a large Compton telescope (see box) at least ten times more sensitive than COMPTEL or INTEGRAL, and somehow squeeze it into a Medium-class Explorer (MIDEX) mission to be launched in the next five years. (NASA's goal is to launch roughly one MIDEX per year, each costing less than $100 million.) This is a challenge, but the new technologies presented at the meeting have the potential to meet it.

Peter J. T. Leonard is in the Compton Gamma-Ray Observatory Science Support Center, Goddard Space Flight Center, Greenbelt, Maryland 20771, USA.

Gamma-Ray Bursts

New observations illuminate the most powerful explosions in the universe

Gerald J. Fishman and
Dieter H. Hartmann

About three times a day our sky flashes with a powerful pulse of gamma rays, invisible to human eyes but not to astronomers' instruments. The sources of this intense radiation are likely to be emitting, within the span of seconds or minutes, more energy than the sun will in its entire 10 billion years of life. Where these bursts originate, and how they come to have such incredible energies, is a mystery that scientists have been attacking for three decades. The phenomenon has resisted study—the flashes come from random directions in space and vanish without trace—until very recently.

On February 28 of this year, we were lucky. One such burst hit the Italian-Dutch Beppo-SAX satellite for about 80 seconds. Its gamma-ray monitor established the position of the burst—prosaically labeled GRB 970228—to within a few arc minutes in the Orion constellation, about halfway between the stars Alpha Tauri and Gamma Orionis. Within eight hours, operators in Rome had turned the spacecraft around to look in the same region with an x-ray telescope. They found a source of x-rays (radiation of somewhat lower frequency than gamma rays) that was fading fast, and they fixed its location to within an arc minute.

Never before has a burst been pinpointed so accurately and so quickly, allowing powerful optical telescopes, which have narrow fields of view of a few arc minutes, to look for it. Astronomers on the Canary Islands, part of an international team led by Jan van Paradijs of the University of Amsterdam and the University of Alabama in Huntsville, learned of the finding by electronic mail. They had some time available on the 4.2-meter William Herschel Telescope, which they had been using to look for other bursts. They took a picture of the area 21 hours after GRB 970228. Eight days later they looked again and found that a spot of light seen in the earlier photograph had disappeared.

There is more. On March 13 the New Technology Telescope in La Silla, Chile, took a long, close look at those coordinates and discerned a diffuse, uneven glow. The Hubble Space Telescope later resolved it to be a bright point surrounded by a somewhat elongated background object. Many of us believe the latter to be a galaxy, but its true identity remains unknown as of this writing.

If indeed a galaxy—as current theories would have—it must be very far away, near the outer reaches of the observable universe. In that case, gamma-ray bursts must represent the most powerful explosions in the universe.

Confounding Expectations

For those of us studying gamma-ray bursts, this discovery salves two recent wounds. In November 1996 the High Energy Transient Explorer (HETE) spacecraft, equipped with very accurate instruments for locating gamma-ray bursts, failed to separate from its launch rocket. And in December the Russian Mars '96 spacecraft, with several gamma-ray detectors, fell into the Pacific Ocean after a rocket malfunction. These

8. Gamma-Ray Bursts

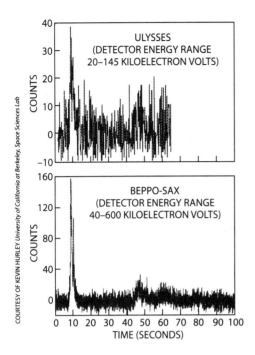

TIME PROFILE of GRB 970228 taken by the Ulysses spacecraft (*top*) and by Beppo-SAX (*bottom*) shows a brief, brilliant flash of gamma rays.

payloads were part of a carefully designed set for launching an attack on the origins of gamma-ray bursts. Of the newer satellites equipped with gamma-ray instruments, only Beppo-SAX—whose principal scientists include Luigi Piro, Enrico Costa and John Heise—made it into space on April 20, 1996.

Gamma-ray bursts were first discovered by accident, in the late 1960s, by the Vela series of spacecraft of the U.S. Department of Defense. These satellites were designed to ferret out the U.S.S.R.'s clandestine nuclear detonations in outer space—perhaps hidden behind the moon. Instead they came across spasms of radiation that did not originate from near the earth. In 1973 scientists concluded that a new astronomical phenomenon had been discovered.

These initial observations resulted in a flurry of speculation about the origins of gamma-ray bursts—involving black holes, supernovae or the dense, dark star remnants called neutron stars. There were, and still are, some critical unknowns. No one knew whether the bursts were coming from a mere 100 light-years away or a few billion. As a result, the energy of the original events could only be guessed at.

By the mid-1980s the consensus was that the bursts originated on nearby neutron stars in our galaxy. In particular, theorists were intrigued by dark lines in the spectra (component wavelengths spread out, as light is by a prism) of some bursts, which suggested the presence of intense magnetic fields. The gamma rays, they postulated, are emitted by electrons accelerated to relativistic speeds when magnetic-field lines from a neutron star reconnect. A similar phenomenon on the sun—but at far lower energies—leads to flares.

In April 1991 the space shuttle *Atlantis* launched the Compton Gamma Ray Observatory, a satellite that carried the Burst and Transient Source Experiment (BATSE). Within a year BATSE had confounded all expectations. The distribution of gamma-ray bursts did not trace out the Milky Way, nor were the bursts associated with nearby galaxies or clusters of galaxies. Instead they were distributed isotropically, with any direction in the sky having roughly the same number. Theorists soon refined the galactic model: the bursts were now said to come from neutron stars in an extended spherical halo surrounding the galaxy.

One problem with this scenario is that the earth lies in the suburbs of the Milky Way, about 30,000 light-years from the core. For us to find ourselves near the center of a galactic halo, the latter must be truly enormous, almost 600,000 light-years in outer radius. If

X-RAY IMAGE taken by Beppo-SAX on February 28 (*left image*) localized the burst to less than one arc minute, allowing ground-based telescopes to search for it. On March 3 the source was much fainter (*right image*).

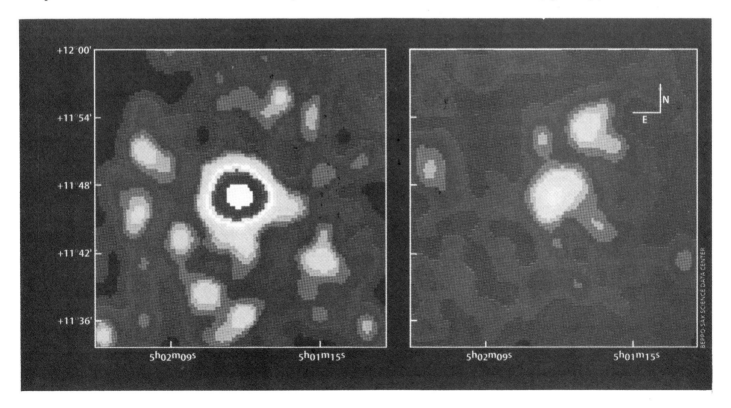

1. DATA-GATHERING TECHNIQUES: Gamma-Ray Astronomy

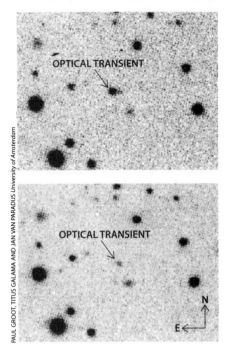

OPTICAL IMAGES of the region of the burst were taken by the William Herschel Telescope on the Canary Islands, on February 28 (*top*) and March 8 (*bottom*). A point of light in the first image has faded away in the second one, indicating a transient afterglow.

so, the halo of the neighboring Andromeda galaxy should be as extended and should start to appear in the distribution of gamma-ray bursts. But it does not. (Special models in which the neutron stars beam in the same direction as their motion can, however, overcome this objection.)

This uniformity has convinced most astrophysicists that the bursts come from cosmological distances, on the order of three billion to 10 billion light-years away. At such a distance, though, the bursts should show the effects of the expansion of the universe. Galaxies that are very distant are moving away from the earth at great speeds; we know this because the light they emit shifts to lower, or redder, frequencies. Likewise, gamma-ray bursts should also show a "redshift," as well as an increase in duration.

Unfortunately, BATSE does not see, in the spectrum of gamma rays, bright or dark lines characterizing specific elements whose displacements would betray a shift to the red. (Nor does it detect the dark lines found by earlier satellites.) In April astronomers using the Keck Telescope in Hawaii obtained an optical spectrum of the afterglow of GRB 970228. It is smooth and red, with no telltale lines. Still, Jay Norris of the National Aeronautics and Space Administration Goddard Space Flight Center and Robert Mallozzi of the University of Alabama in Huntsville have statistically analyzed the observed bursts and report that the weakest, and therefore the most distant, show both a time dilation and a redshift. There are, however, other (controversial) ways to interpret these findings.

A Cosmic Catastrophe

One feature that makes it difficult to explain the bursts is their great variety. A burst may last from about 30 milliseconds to almost 1,000 seconds—and in one case, 1.6 hours. Some bursts show spasms of intense radiation, with no detectable emission in between, whereas others are smooth. Also complicated are the spectra—essentially, the colors of the radiation, invisible though they are. The bulk of a burst's energy is in radiation of between 100,000 and one million electron volts, implying an exceedingly hot source. (The photons of optical light, the primary radiation from the sun, have energies of a few electron volts.) Some bursts evolve smoothly to lower frequencies such as x-rays as time passes. Although this x-ray tail has less energy, it contains many photons.

If originating at cosmological distances, the bursts must have energies of perhaps 10^{51} ergs. (About 1,000 ergs can lift a gram by one centimeter.) This energy must be emitted within seconds or less from a tiny region of space, a few tens of kilometers across. It would seem we are dealing with a fireball.

The first challenge is to conceive of circumstances that would create a sufficiently energetic fireball. Most theorists favor a scenario in which a binary neutron-star system collapses [see "Binary Neutron Stars," by Tsvi Piran; SCIENTIFIC AMERICAN, May 1995]. Such a pair gives off gravitational energy in the form of radiation. Consequently, the stars spiral in toward each other and may ultimately merge to form a black hole. Theoretical models estimate that one such event occurs every 10,000 to one million years in a galaxy. There are about 10 billion galaxies in the volume of space that BATSE observes; that yields up to 1,000 bursts a year in the sky, a number that fits the observations.

Variations on this scenario involve a neutron star, an ordinary star or a white dwarf colliding with a black hole. The details of such mergers are a focus of intense study. Nevertheless, theorists agree that before two neutron stars, say, collapse into a black hole, their death throes release as much as 10^{53} ergs. This energy emerges in the form of neutrinos and antineutrinos, which must somehow be converted into gamma rays. That requires a chain of events: neutrinos collide with antineutrinos to yield electrons and positrons, which then annihilate one another to yield photons. Unfortunately, this process is very inefficient, and recent simulations suggest it may not yield enough photons.

Worse, if too many heavy particles such as protons are in the fireball, they reduce the energy of the gamma rays. Such proton pollution is to be expected, because the collision of two neutron stars must yield a potpourri of particles. But then all the energy ends up in the

DEEP EXPOSURE of the optical remnant of GRB 970228 was taken by the Hubble Space Telescope. The afterglow (*near center of top image*), when seen in close-up (*bottom*), has a faint, elongated background glow that may correspond to a galaxy in which the burst occurred.

8. Gamma-Ray Bursts

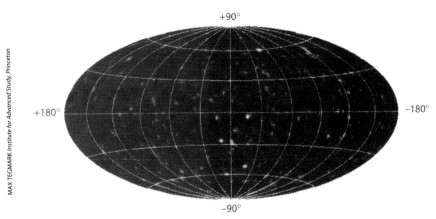
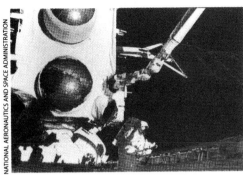

BURST DISTRIBUTION over the sky, measured by the Burst and Transient Source Experiment (BATSE), shows no clustering along the Milky Way (*along equatorial line*). BATSE is located on the Compton Observatory (*right*), here shown being deployed.

kinetic energy of the protons, leaving none for radiation. As a way out of this dilemma, Peter Mészáros of Pennsylvania State University and Martin J. Rees of the University of Cambridge have suggested that when the expanding fireball—essentially hot protons—hits surrounding gases, it produces a shock wave. Electrons accelerated by the intense electromagnetic fields in this wave then emit gamma rays.

A variation of this scenario involves internal shocks, which occur when different parts of the fireball hit one another at relativistic speeds, also generating gamma rays. Both the shock models imply that gamma-ray bursts should be followed by long afterglows of x-rays and visible light. In particular, Mario Vietri of the Astronomical Observatory of Rome has predicted detectable x-ray afterglows lasting for a month—and also noted that such afterglows do not occur in halo models. GRB 970228 provides the strongest evidence yet for such a tail. There are some problems, however: the binary collapse does not explain some long-lasting bursts. Last year, for instance, BATSE found a burst that endured for 1,100 seconds and possibly repeated two days later.

There are other ways of generating the required gamma rays. Nir Shaviv and Arnon Dar of the Israel Institute of Technology in Haifa start with a fireball of unknown origin that is rich in heavy metals. Hot ions of iron or nickel could then interact with radiation from nearby stars to give off gamma rays. Simulations show that the time profiles of the resulting bursts are quite close to observations, but a fireball consisting entirely of heavy metals seems unrealistic.

Another popular mechanism invokes immensely powerful magnetic engines, similar to the dynamos that churn in the cores of galaxies. Theorists envision that instead of a fireball, a merger of two stars—of whatever kind—could yield a black hole surrounded by a thick, rotating disk of debris. Such a disk would be very short-lived, but the magnetic fields inside it would be astounding, some 10^{15} times those on the earth. Much as an ordinary dynamo does, the fields would extract rotational energy from the system, channeling it into two jets bursting out along the rotation axis.

The cores of these jets—the regions closest to the axis—would be free of proton pollution. Relativistic electrons inside them can then generate an intense, focused pulse of gamma rays. Although quite a few of the details remain to be worked out, many such scenarios ensure that mergers are the leading contenders for explaining bursts.

Still, gamma-ray bursts have been the subject of more than 2,500 papers—about one publication per recorded burst. Their transience has made them difficult to observe with a variety of instruments, and the resulting paucity of data has allowed for a proliferation of theories.

If one of the satellites detects a lensed burst, astronomers would know for sure that bursts occur at cosmological distances. Such an event might occur if an intervening galaxy or other massive object serves as a gravitational lens to bend the rays from a gamma-ray burst toward the earth. When optical light from a distant star is focused in this manner, it appears as multiple images of the original star, arranged in arcs around the lens. Gamma rays cannot be pinpointed with such accuracy; instead they are currently detected by instruments that have poor directional resolution.

Moreover, bursts are not steady sources like stars. A lensed gamma-ray burst would therefore show up as two bursts coming from roughly the same direction, having identical spectra and time profiles but different intensities and arrival times. The time difference would come from the rays' traversing curved paths of different lengths through the lens.

To further nail down the origins of the underlying explosion, we need data on other kinds of radiation that might accompany a burst. Even better would be to identify the source. Until the fortuitous observation of GRB 970228—we are astonished that its afterglow lasted long enough to be seen—such "counterparts" had proved exceedingly elusive. To find others, we will need to locate the bursts very precisely.

Watching and Waiting

Since the early 1970s, Kevin Hurley of the University of California at

ROBOTIC TELESCOPES on a hill near Lawrence Livermore National Laboratory search for a burst within seconds of obtaining its location from BATSE.

1. DATA-GATHERING TECHNIQUES: Gamma-Ray Astronomy

  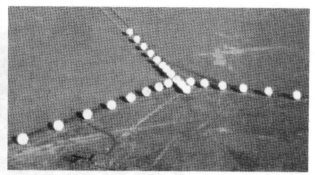

left to right: BEPPO-SAX SCIENCE DATA CENTER; ROGER RESSMEYER *Corbis;* NRAO

VARIETY OF INSTRUMENTS contribute to the study of gamma-ray bursts. The Beppo-SAX satellite (*left*), shown in the process of assembly, has gamma-ray and x-ray detectors that have proved crucial to locating recent bursts. The William Herschel Telescope (*center*) on the Canary Islands photographed the optical transient of GRB 970228. In May the Very Large Array (*right*) of radio telescopes in Socorro, N.M., found radio waves from a burst—for the first time.

Berkeley and Thomas Cline of the NASA Goddard Space Flight Center have worked to establish "interplanetary networks" of burst instruments. They try to put a gamma-ray detector on any spacecraft available or to send aloft dedicated devices. The motive is to derive a location to within arc minutes, by comparing the times at which a burst arrives at spacecraft separated by large distances.

From year to year, the network varies greatly in efficacy, depending on the number of participating instruments and their separation. At present, there are five components: BATSE, Beppo-SAX and the military satellite DMSP, all near the earth; Ulysses, far above the plane of the solar system; and the spacecraft Wind, orbiting the sun. The data from Beppo-SAX, Ulysses and Wind were used to triangulate GRB 970228. (BATSE was in the earth's shadow at the time.) The process, unfortunately, is slow—eight hours at best.

Time is of the essence if we are to direct diverse detectors at a burst while it is glowing. Scott Barthelmy of the Universities Space Research Association at the NASA Goddard Space Flight Center has developed a system called BACODINE (BAtse COordinates DIstribution NEtwork) to transmit within seconds BATSE data on burst locations to ground-based telescopes.

BATSE consists of eight gamma-ray detectors pointing in different directions from eight corners of the Compton satellite; comparing the intensity of a burst at these detectors provides its location to roughly a few degrees but within several seconds. Often BACODINE can locate the burst even while it is in progress. The location is transmitted over the Internet to several dozen sites worldwide. In five more seconds, robotically controlled telescopes at Lawrence Livermore National Laboratory, among others, slew to the location for a look.

Unfortunately, only the fast-moving smaller telescopes, which would miss a faint image, can contribute to the effort. The Livermore devices, for instance, could not have seen the afterglow of GRB 970228 (unless the optical emission immediately after the burst is many times brighter, as some theories suggest). Telescopes that are 100 times more sensitive are required. These mid-size telescopes would also need to be robotically controlled so they can slew very fast, and they must be capable of searching reasonably large regions. If they do find a transient afterglow, they will determine its location rather well, allowing much larger telescopes such as Hubble and Keck to look for a counterpart.

The long-lasting, faint afterglow following GRB 970228 gives new hope for this strategy. The HETE mission, directed by George Ricker of the Massachusetts Institute of Technology, is to be rebuilt and launched in about two years. It will survey the full sky with x-ray detectors that can localize bursts to within several arc minutes. A network of ground-based optical telescopes will receive these locations immediately and start searching for transients.

Of course, we do not know what fraction of bursts actually exhibit a detectable afterglow; GRB 970228 could be a rare and fortuitous exception. Moreover, even an observation field as small as arc minutes contains too many faint objects to make a search for counterparts easy. It would be marvelous if we could derive accurate locations within fractions of a second from the gamma rays themselves. Astronomers have proposed new kinds of gamma-ray telescopes that can instantly derive the position of a burst to within arc seconds.

To further constrain the models, we will need to look at radiation of both higher and lower frequency than that currently observed. The Energetic Gamma Ray Experiment Telescope (EGRET), which is also on the Compton satellite, has seen a handful of bursts that emit radiation of up to 10 billion electron volts, sometimes lasting for hours. Better data in this regime, from the Gamma Ray Large Area Space Telescope (GLAST), a satellite being developed by an international team of scientists, will greatly aid theorists. And photons of even higher energy—of about a trillion electron volts—might be captured by special ground-based gamma-ray telescopes. At the other end of the spectrum, soft x-rays, which have energies of up to roughly one kiloelectron volt (keV), are helpful for testing models of bursts and also for getting better fixes on position. In the range of 0.1 to 10 keV, there is a good chance of discovering absorption or emission lines that would tell volumes about the underlying fireball and its magnetic fields. Such lines might also yield a direct measurement of the redshift and, hence, the distance. Sensitive instruments for detecting soft x-rays are being built in various institutions around the world.

Even as we finish this article, we have just learned of another coup. On the night of May 8, Beppo-SAX operators located a 15-second burst. Soon after, Howard E. Bond of the Space Telescope Science Institute in Baltimore photographed the region with the 0.9-meter optical telescope at Kitt Peak; the next night a point of light in the field had actually brightened. Other telescopes confirm that after becoming most brilliant on May 10, the source began to fade. This is the first time that a burst has been observed reaching its optical peak—which, astonishingly, lagged its gamma-ray peak by a few days.

Also for the first time, on May 13 the Very Large Array of radio telescopes in New Mexico detected radio emissions from the burst remnant. Even more exciting, the primarily blue spectrum of this burst, taken on May 11 with the Keck II telescope on Hawaii, showed a few dark lines, apparently caused by iron and magnesium in an intervening cloud. Astronomers at the California Institute of Technology find that the displacement of these absorption lines indicates a distance of more than seven billion light-years. If this interpretation holds up, it will establish once and for all that bursts occur at cosmological distances.

In that case, it may not be too long before we know what catastrophic event was responsible for that burst—and for one that might be flooding the skies even as you read.

The Authors

GERALD J. FISHMAN and DIETER H. HARTMANN bring complementary skills to the study of gamma-ray bursts. Fishman is an experimenter—the principal investigator for BATSE and a senior astrophysicist at the National Aeronautics and Space Administration Marshall Space Flight Center in Huntsville, Ala. He has received the NASA Medal for Exceptional Scientific Achievement three times and in 1994 was awarded the Bruno Rossi Prize of the American Astronomical Society. Hartmann is a theoretical astrophysicist at Clemson University in South Carolina; he obtained his Ph.D. in 1989 from the University of California, Santa Cruz. Apart from gamma-ray astronomy, his primary interests are the chemical dynamics and evolution of galaxies and stars.

Further Reading

THE GAMMA-RAY UNIVERSE. D. Kniffen in *American Scientist,* Vol. 81, No. 4, pages 342–350; July 1993.

THE COMPTON GAMMA RAY OBSERVATORY. Neil Gehrels, Carl E. Fichtel, Gerald J. Fishman, James D. Kurfess and Volker Schönfelder in *Scientific American,* Vol. 269, No. 6, pages 68–77; December 1993.

THE GAMMA-RAY BURST MYSTERY. D. H. Hartmann in *The Lives of Neutron Stars.* Edited by A. Alpar, Ü. Kiziloglu and J. van Paradijs. NATO Advanced Studies Institute, Kluwer Academic Publishers, 1994.

GAMMA RAY BURSTS. G. J. Fishman and C. A. Meegan in *Annual Review of Astronomy and Astrophysics,* Vol. 33, pages 415–458; 1995.

Beppo-SAX Mission home page is available on the World Wide Web at http://www.sdc.asi.it/

An Infrared View of Our Universe

New detectors are bringing us fresh, detailed views of nebulae and galaxies.

Ian Gatley

Ian Gatley is an infrared astronomer at the National Optical Astronomy Observatories in Tucson, Arizona.

Ten years ago you could easily tell which were the infrared images at an astronomy conference: They were the crude ones, compared with their optical counterparts. An astronomer "imaging" an object spent all night scanning the telescope's light beam back-and-forth over a single, relatively insensitive detector. The researcher then plotted these measurements one by one to create a primitive map of infrared brightness. Even early infrared cameras provided too small a field of view. Researchers had to tape together lots of exposures of small regions to make an image of the entire object. As recently as five years ago, pictures made in this crude way caused loud gasps at astronomy meetings as everyone admired these wonderful images.

But technology has progressed. Researchers have developed new detectors with greater sensitivity to infrared radiation and recently have produced large arrays that make imaging a simple task instead of a time-consuming chore. These arrays let astronomers probe regions of star formation, view the complex shells of gas cast off by dying stars, and more easily see structures in galaxies.

Why worry about imaging in the infrared anyway? Because infrared light waves have longer wavelengths than visible light, which enables them to pass through gas and dust in space that blocks visible light. And many molecules emit energy only in the infrared part of the spectrum.

To record the low energy of infrared waves, scientists make their detectors out of hybrid combinations of exotic chemicals such as tellurium, cadmium, indium, and antimony instead of the silicon found in CCD and video cameras, which is insensitive to infrared light. The new arrays are huge by old infrared detector standards: they have 256 by 256 elements.

Actually, though, it's not the sensitivity and size of the arrays that make the new infrared detectors exciting to astronomers. What excites them is what they are seeing with the new arrays.

For example, consider the Omega Nebula (M17) in Sagittarius, which is familiar to many backyard astronomers. Visible light images (right) show that this is a place where massive stars are being born. The stars formed in a cool cloud containing molecular hydrogen gas. But the radiation from the young, hot stars heats up the leftover gas and makes it shine, producing the beautiful emission nebula we see.

Yet this view is deceiving. The brightest part of the nebula doesn't appear in visible light photographs because dust in the densest part of the gas cloud hides it from view. Infrared radiation, however, penetrates the dust, making infrared images a perfect means to see [a] rich stellar nursery in this part of the nebula.

Of course our eyes can't see infrared. But to reap the information contained in infrared "color" images, . . . colors [are assigned] to different infrared wavelengths. . . . These false-color images show us what our eyes would see if they could detect infrared light.

Besides penetrating dust, infrared imaging has another advantage for studying nebulae. Molecular hydrogen doesn't emit at all in the visible part of the spectrum, making infrared images important for studying the distribution of this gas through the nebula.

Because both atomic and molecular hydrogen gas emit at different wavelengths in the infrared, we can use infrared images to isolate different features of the gas cloud. Choosing the proper wavelengths allows us to view just the molecular hydrogen gas or just the atomic hydrogen gas. Subtracting another image made at a nearby wavelength removes all the stars so just the gas appears. This star-subtraction technique has revealed jets shooting from the molecular cloud located north of the Orion Nebula (M42).

Reversing the star-subtraction method takes away the gas and highlights the stars. The visible light and infrared images opposite reveal, as did the M17 images, many more stars in the infrared image. These stars lie buried in the molecular cloud—another star factory—behind the Orion Nebula.

Other images show that the Orion Nebula emits bright emission lines of molecular hydrogen. These lines indicate that the wind blowing off the surface of a young star in the molecular cloud collides with gas in the nebula, heating it to about 2000 kelvins. Other emission lines suggest that ultraviolet radiation from young, hot stars is lighting up a wispy infrared halo around the Orion Nebula. Yet other observations reveal that ultraviolet radiation from young stars strips electrons from the hydrogen gas to ionize it. This ionized gas forms a cavity within the molecular cloud. So infrared observations reveal a wealth of features unseen in visible light.

INFRARED CUTS THROUGH the gas and dust of M17 seen in visible light (below) to show the heart of the nebula (bottom).

Michael Stecker

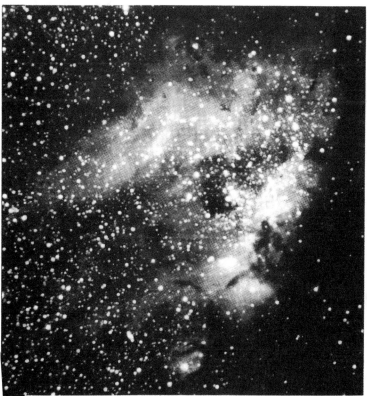

Ian Gatley, NOAO

1. DATA-GATHERING TECHNIQUES: Infrared Astronomy

INFRARED IMAGES SHOW A BAR in the companion galaxy to the spiral galaxy M51 (bottom) that is unseen in this visible light image (below).

U. S. Naval Observatory

Ian Gatley, NOAO

But more important, they show how the structure of the Orion Nebula is intimately tied to the cycle of star birth inside the gas cloud.

Another marvelous discovery came out of images of another famous object, the Ring Nebula (M57) in Lyra (opposite page). Previous low-sensitivity detectors couldn't see the faint, diffuse halo of filaments. The halo appears in some visible light images, but it appears prominently in recent infrared images. Astronomers had always interpreted this ring as a spherical shell of gas ejected by the central star during its final death throes. At the nebula's center we look easily through the shell's thin gas. But at the edge we look along the shell and thus see the emission of much more gas.

But the new images suggest something. Perhaps the structure isn't spherical after all. The ring could be a cylinder that we are peering down upon. Other planetary nebulae (so-named because they appear like the disk of a planet when observed through a small telescope) also suggest that these stellar eruptions occur as cylindrical or hourglass-shaped, not spherical, ejections of gas.

One example is NGC 2346 in the constellation Monoceros. We see this cylindrical planetary nebula nearly edge-on. If we could look down into this cyl-

Ian Gatley, NOAO

Bruce Balick, U. Washington/NOAO

THE PLANETARY NEBULA NGC 2346 looks like a disk with wispy wings in visible light images (right) but shows clearly an hourglass shape in the infrared (left). This indicates that the dying star at the center of the nebula didn't blow off a spherical shell of gas but rather ejected a doughnut-shaped shell. The star's wind interacts with this shell to form a cylinder of material, which seen at an angle forms the hourglass shape we see.

9. Infrared View of Our Universe

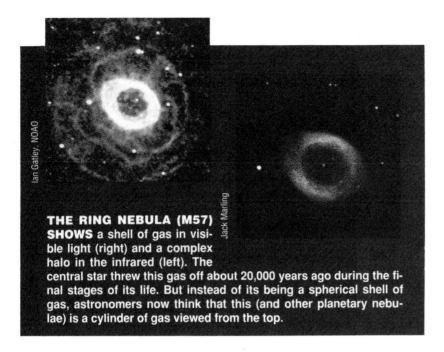

THE RING NEBULA (M57) SHOWS a shell of gas in visible light (right) and a complex halo in the infrared (left). The central star threw this gas off about 20,000 years ago during the final stages of its life. But instead of its being a spherical shell of gas, astronomers now think that this (and other planetary nebulae) is a cylinder of gas viewed from the top.

STAR FACTORY EXTRAORDINAIRE: the Great Orion Nebula (M42) shows a wealth of newly formed stars in the infrared (top) that the nebula's gas hides in visible light images (above).

inder, we might see something like the Ring Nebula. From our side view, the inner ring appears as the bright part near the center, while the halo of filaments shows up as distinct X-shaped extensions of the nebula.

Because of its ability to penetrate gas and dust, infrared imaging also has provided new detailed views of galaxies. For example, visible light images of the companion to the spiral galaxy M51 in Canes Venatici have suggested the existence of a bar. But infrared images of the companion reveal the bar easily. This galaxy (NGC 5195) has collided with M51, disrupting its original shape. Perhaps the bar is a result of this collision.

These infrared images are just a few that are helping astronomers reshape their views of nebulae and galaxies. But that's not all. The next big breakthrough is imaging spectroscopy. The same device that holds the infrared camera at the telescope can also hold an infrared spectrometer. This instrument not only breaks the infrared light into its component wavelengths but does so at many different points of an object simultaneously. Think of it as taking an infrared image at many wavelengths all at once. Such an instrument lets researchers image an object in the infrared and analyze the object's chemical make-up, motion, and temperature. This is a giant step forward in technology that will change infrared astronomers' ability to study celestial objects as dramatically as the rapid advance in detectors over the last decade has.

INFRARED ASTRONOMY

Cool gaze at heartless galaxies

Gerry Gilmore

Gerry Gilmore is at the Institute of Astronomy, University of Cambridge, Madingley Road, Cambridge CB3 0HA, UK.

STAR formation is ubiquitous, but almost entirely mysterious. Some galaxies completed their star formation long ago, whereas many form stars sporadically, punctuated by long, pregnant pauses. Most spiral galaxies, like the Milky Way, seem to have formed stars at a fairly constant, low rate for the whole lifetime of their disk. Other galaxies again, especially those in the throes of major mergers, are forming stars so rapidly that they will exhaust their gas reservoirs very soon. How do stars form, and why do they form at such different rates? Are the most luminous galaxies powered by extreme rates of star formation, or do they harbour black-hole-powered active nuclei as well? A clear answer to this last question, and partial answers to the others, are found in a special issue of *Astronomy and Astrophysics* this month[1-4], which reports the first scientific results from the Infrared Space Observatory (ISO)*.

ISO is the first infrared astronomy satellite to have high spatial and spectral resolution, good sensitivity and wide wave length range. Its 60-cm telescope supplies light to four instruments—two spectrographs, a long-wavelength photometer and a camera operating at wavelengths of 2 to 20 micrometres—all cooled to two kelvin by 2,500 litres of superfluid helium. Targets range from the Solar System to the most distant proto-galaxies, but perhaps the most substantial advances reported to date are in two related fields: star formation in our Galaxy, and the nature of the most luminous known galaxies, in which extreme rates of star formation are perhaps complemented by massive black holes in active nuclei.

The best models of these luminous and ultra-luminous infrared galaxies have treated them as mergers of two large, gas-rich galaxies, with very rapid star formation triggered by the interaction of the huge gas masses, and with an active galactic nucleus (AGN) fed by gas flow into the central regions but hidden by dust. The most infrared-luminous galaxies in particular seemed not to be forming many of the most massive stars, suggesting both the presence of an AGN to provide the enormous infrared power, and that the mass distribution of stars formed in a starburst is different from that produced under more sedate circumstances.

Starbursts have softer radiation spectra than AGN, and so they excite different levels of ionization in the galactic gas which result in different emission-line ratios. It is these ratios that we use to discriminate between the two types of source, in effect determining the temperature of the radiation field. At optical and near-infrared wavelengths, however, very large corrections must be made for colour-dependent absorption by dust: a typical model requires a factor of 10^4 reduction in short-wavelength intensity, which introduces the possibility of systematic uncertainties. But ISO can see directly the mid- and far-infrared emission lines, which suffer very little absorption and are sensitive to a very wide range of densities and ionization levels.

Already, ISO has been used to study three ultra-luminous infrared galaxies, and a sample of less extreme star-forming galaxies[1]. The remarkable result is that they do not harbour AGN—they are powered solely by rapid star formation, and the stars form with the same mass distribution as that seen in the Milky Way. In one lovely example, of two intersecting disk galaxies known as the Antennae, it is even possible to resolve the spot where the two disks currently cross, and to see the progression of star formation across the

---

*ISO is a collaboration between the European Space Agency, ISAS of Japan and NASA.

disk as the two galaxies orbit through each other[2].

At the other end of the distance scale, in regions of star formation in our Galaxy, one sees a lot of detail: dark, cold cores; bright, ionized filaments; newly formed stars busily heating, evaporating and illuminating the region; and complex molecules cooling the region, being created and destroyed in the process. To understand star formation, one must understand the relationships between all these phenomena. One long-suspected chemical change in star-forming regions is the formation of hot water—long suspected, but not observed (except as a microwave maser line in special circumstances) because water in the Earth's atmosphere blocks the important wavelengths. But ISO, being above the atmosphere, has already seen hot water, cold water, ice, carbon monoxide, carbon dioxide, polycyclic aromatic hydrocarbons, silicate, methane... and so the list goes on. The study of the molecular environment of star-forming cores is set to become an observational science *par excellence*[3].

A critical issue in understanding star formation is the density structure of the cold, dense cores of molecular clouds, which are believed to be the precursors of star-forming regions. The low temperatures of these clouds and their extremely high optical obscuration have hampered observations, because we have been limited by the low spatial resolution of ground-based, sub-millimetre telescopes, and by the restriction to densities and temperatures probed by molecular transitions that happen to be observable through the Earth's atmosphere. One of the nearest star-forming regions—rho Ophiuchi, some 160 parsecs away—has now been analysed by ISO. Dense cores are indeed seen, the distribution of ionizing radiation can be mapped out, and detailed studies have begun of the evolution of a region in which new hot stars compete with far-infrared cooling and complex chemical and molecular evolution to stimulate or prevent further star formation[4].

In combination, these first detailed and unobscured studies of nearby and distant star formation are a firm empirical basis to build on. The early ISO results are extremely encouraging in showing that star formation differs more in rate than in kind; that reliable dust-absorption-free data can be obtained; and that the Universe is kind, at least sometimes, in providing clean examples of star-forming galaxies uncontaminated by AGN.

Notes

1. Lutz, D. *et al. Astron. Astrophys.* **315**, L137-L140 (1996). Refs 1–4 also available at http://isowww.estec.esa.nl:80/ISO/AandA
2. Vigroux, L. *et al. Astron. Astrophys.* **315**, L93–L96 (1996).
3. Helmich, F. P. *et al. Astron. Astrophys.* **315**, L173–L176 (1996).
4. Abergel, A. *et al. Astron. Astrophys.* **315**, L329–L332 (1996).
5. Sanders, D. & Mirabel, I. *Annu. Rev. Astron. Astrophys.* (in the press).

RADIO ASTRONOMY IN THE 21st CENTURY

Astronomers are arming themselves to make the most of radio waves from space.

KENNETH I. KELLERMANN

Kellermann is Chief Scientist at the National Radio Astronomy Observatory's headquarters in Charlottesville, Virginia.

SINCE KARL JANSKY and Grote Reber first detected cosmic radio waves in the 1930s, the sensitivity and resolution of radio telescopes have improved dramatically. Meanwhile, the wavelengths that radio telescopes can focus and amplify have been pushed down by a factor of more than one thousand from meter to submillimeter wavelengths. A new generation of powerful radio telescopes is now poised for a new era of discovery, especially in the relatively unexplored millimeter and submillimeter regions of the electromagnetic spectrum. What opportunities will these innovative instruments offer the astronomers of the 21st century?

A UNIQUE WINDOW ON SPACE

Visible-light astronomers look at the familiar and largely unchanging sky of planets, stars, nebulae, and galaxies. At radio wavelengths, we see an entirely different sky: one including star-forming regions whose visible light is blocked by dust; one containing the cold hydrogen gas found both within and between galaxies; one seeded with abundant material at temperatures too low to give off visible light. Radio astronomers also study vast, energetic phenomena, such as cosmic masers, beams of relativistic gas, pulsars, and radio bursts from the Sun and planets.

Aside from the narrow visible-light window between 4000 and 7000 angstroms and the radio band between 1 millimeter and 10 meters, the Earth's atmosphere is nearly opaque to electromagnetic radiation. The dramatic discoveries of ultraviolet, infrared, X-ray, and gamma-ray astronomy have been made by telescopes placed above the Earth's obscuring atmosphere. Even at visual wavelengths, where the atmosphere is relatively transparent, image distortion from atmospheric turbulence ("seeing") has driven visible-light instruments such as the Hubble Space Telescope into orbit.

In contrast, radio waves pass through the Earth's atmosphere relatively unhindered. Thus, there has been little need to put radio telescopes in space. As a result, even the most powerful and complex ground-based radio telescopes are much cheaper to build and operate than are space observatories. With relative ease and at modest cost they can be repaired or upgraded to maintain pace with rapidly changing technology. Thus, unlike many other areas of astronomy, which are primarily supported by organizations such as NASA and the European Space Agency, radio astronomy gets funding principally from bodies like the National Science Foundation (NSF) and universities as well as state and national governments.

INCREASING SENSITIVITY

The first large fully steerable radio telescope built in the United States was the 140-foot (42-meter) telescope in Green Bank, West Virginia (the first of the U.S. National Radio Astronomy Observatory's several current sites). Completed in 1965, the 140-foot has a sufficiently precise reflecting surface to work at wavelengths as short as 1 centimeter. Still used today, the 140-foot is the largest equatorially mounted telescope ever constructed. Most of its contemporaries — including the fully steerable instruments at Jodrell Bank, England (75 meters in diameter), Parkes, Australia (63 meters), and Effelsberg, Germany (100 meters) — use lighter, lower-cost, altazimuth designs.

One exceptional early telescope was the 300-foot (90-meter) transit radio telescope, which NRAO designed and built in less than two years. This instrument, which began operating in 1962, traded full steering capability to gain what was then the largest movable telescope "mirror" in the world. Originally intended to have a scientific lifetime of only about five years, the economical instrument (built for less than one million dollars) enabled more than a thousand astronomers to observe radio sources ranging from solar-system objects to galaxies and quasars billions of light-years away. Eventually, however, the 300-foot collapsed from metal fatigue on the fateful evening of November 15, 1988, while completing a 6-centimeter survey of the northern sky (*S&T:* March 1989, page 252).

When the 300-foot died, the U.S. was

11. Radio Astronomy in the 21st Century

Rapidly developing technology has enabled astronomers to study an ever-increasing range of physical phenomena.

left without a large, steerable radio telescope. (The 1,000-foot Arecibo reflector in Puerto Rico, while the world's biggest single radio dish, is able to scan only a 40°-wide band that misses the Milky Way's energetic center and many other interesting parts of the sky.) Since the 1960s, essentially every review of U.S. radio astronomy facilities has recommended the construction of a fully steerable, 100-meter radio telescope. But money remained unavailable until, in the wake of the 300-foot telescope's demise, the NSF and Congress included funds for a modern new radio telescope at Green Bank.

Now nearing completion, the new Green Bank Telescope (GBT) will study celestial radio waves with an actively controlled, 100-meter-by-110-meter surface that will be continuously adjusted to stay within a few millimeters of a paraboloidal shape despite constantly changing gravitational and thermal distortions and wind.

The GBT will hold either a secondary reflector or its radio-wave detectors on an offset "feed" arm located outside the path of incoming radio signals. Conventional radio telescopes support their detectors (or secondary reflectors) with three or four symmetrical legs. While giving good mechanical stability, such structures block part of the aperture from incoming radio waves and slightly reduce sensitivity. Worse, by scattering radio waves from off-axis directions into the detectors, obstructing structures also introduce "sidelobes," patches of sensitivity to radio sources (both natural and artificial) located far away from the telescope's intended target.

Unblocked-aperture antennas are familiar to us all as some of the newer home satellite-TV receivers. This type of construction, however, has never previously been attempted on such a scale as GBT's. The new telescope's feed arm will tower more than 200 feet — almost the height of the Statue of Liberty — above the telescope's reflector.

The GBT will give an order-of-magnitude improvement in sensitivity over the 140-foot telescope at short centimeter wavelengths. As a result, when the GBT goes into operation in 1998, the venerable 140-foot will be used by the privately funded SETI Institute to search for extraterrestrial intelligence.

Even after the GBT is built, the giant 1,000-foot reflector at Arecibo Observatory in Puerto Rico will remain by far the largest and most sensitive single-dish radio telescope in the world. Although this instrument was originally built in the early 1960s for radar studies of the planets and the Earth's ionosphere, its vast collecting area has been used for a wide range of astronomical applications. However, the telescope's limited tracking ability, relatively coarse surface, and spherical shape (which focuses radiation along a line, rather than at a point) restricted early observations.

In 1974 the original wire-mesh surface was replaced by 38,778 aluminum panels to give the improved surface accuracy needed for operation at centimeter wavelengths. More recently, the Arecibo dish has undergone major renovations that will greatly improve its performance for radio and radar astronomy programs requiring the ultimate in sensitivity.

In the first phase of this upgrade, a 50-foot-high screen was erected around the reflector to reduce the amount of thermal radiation picked up from the ground. In the second phase, now nearing completion, the cumbersome line feeds (the instrument's actual "antennas") have been replaced by a dual-reflector Gregorian assembly that corrects for the reflector's spherical aberration. This 75-ton set of corrective optics will allow Arecibo to operate at wavelengths from a few centimeters to about one meter with up

Left: A behemoth being built. The Green Bank Telescope as it appeared last July. When completed in 1998, it will be the largest fully steerable telescope in the world, and its 8,000-ton surface will point to targets with arcsecond precision. *Above:* This artist's conception shows the 140-meter-tall telescope dwarfing its companions at NRAO's original site in Green Bank, West Virginia. Courtesy NRAO.

1. DATA-GATHERING TECHNIQUES: Radio Astronomy

to 10 times better sensitivity than before. Radar studies of the solar system will enjoy an even greater improvement (up to 40 times) in sensitivity.

New drive and control systems, receivers, and a 1-megawatt transmitter for planetary radar will complete the facility's face-lift. The first observations with the new Gregorian system are expected early in 1997, and radio astronomers throughout the world are eagerly awaiting their turns at Arecibo.

TOWARD SHORTER WAVELENGTHS

The first radio astronomy observations were made at meter wavelengths. But rapidly developing technology at decimeter, centimeter, and most recently at millimeter and submillimeter wavelengths has enabled astronomers to study an ever-increasing range of physical conditions and phenomena. The development of sensitive amplifiers in the 1960s led to the discovery of an unexpected and remarkably large number of molecules in interstellar space. Observations of these molecules' millimeter and submillimeter spectral lines help us to understand the chemistry of the interstellar medium and the formation of stars.

However, at millimeter and especially submillimeter wavelengths, water vapor in the Earth's lower atmosphere limits the sensitivity of astronomical observations. Thus most millimeter-wave telescopes are sited at high, relatively dry mountaintop locations. For example, the Caltech Sub-Millimeter Telescope on the summit of Mauna Kea operates in the newly opened spectral region between the radio and infrared bands.

The 1,000-foot Arecibo radio telescope — the world's largest single "dish" — boasts several new features that will enhance its utility. A 50-foot screen rimming the reflector will cut down on unwanted radiation, while Gregorian optics (hidden within the radome at upper center) will correct for spherical aberration. Courtesy Cornell University.

The world's largest millimeter-wave telescopes are currently the French-German 30-meter instrument in the mountains near Granada, Spain, and the 45-meter dish near Nobeyama, Japan. These will soon be joined by the 50-meter Large Millimeter Telescope (LMT) now being designed by the University of Massachusetts and Mexico's National Institute of Astrophysics, Optics, and Electronics. The LMT will work at wavelengths as short as 1 mm and will be located on a high, dry site in Mexico, where it will have an exceptional view of the galactic-center region. The LMT will use active optics and receiver arrays to form images of molecular clouds and newborn stars.

ENHANCING ANGULAR RESOLUTION

All telescopes are limited by diffraction to an angular resolution (in radians) given roughly by the wavelength of observation divided by the instrument's aperture. For many years it was naturally thought that radio telescopes would always rank a distant second in resolution when compared to visible-light telescopes, since visible-light waves are about 10,000 times shorter than radio waves. Curiously, however, in practice the re-

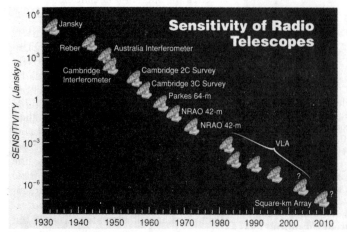
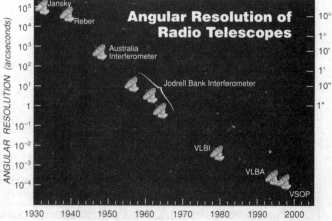

The sensitivity and resolution of radio telescopes have both improved by many orders of magnitude since Karl Jansky and Grote Reber built their groundbreaking instruments more than a half century ago. (Note that, at 10,000 janskys or more, a cellular telephone on the Moon would appear on Earth as one of the sky's brightest radio "stars.") Courtesy the author.

11. Radio Astronomy in the 21st Century

Interferometer par excellence, the Very Large Array in New Mexico moves 27 antennas along three railroad tracks that provide baselines as large as 35 kilometers. Radio images are formed by correlating the signals garnered by each antenna. This photograph shows the array in its most compact configuration, wherein the antennas lie no farther than 600 meters from the center. Courtesy NRAO.

verse has been true, as the resolution of conventional ground-based optical telescopes is generally limited by atmospheric "seeing" and not by diffraction. Furthermore, to operate at its theoretical diffraction limit, a telescope must be built with a precision that is significantly finer than the wavelength of the electromagnetic radiation it observes. Thus, while visible-light telescopes need to be figured to better-than-micron accuracy, radio telescopes operating at centimeter wavelengths need precision good to only a millimeter or so. This relatively modest requirement can be met over essentially unlimited distances.

In the 1940s and 1950s, radio astronomers in England and Australia built a series of interferometers, which obtained angular resolution by combining the signals from separate, individual telescopes. The angular resolution of an interferometer is determined by the distance *between* antennas, rather than by the size of the individual elements. By the 1960s, a series of multielement arrays was constructed, primarily under the leadership of Martin Ryle at Cambridge University in England. These early radio arrays gave the first crude images of distant radio galaxies as well as of supernova remnants and other sources of cosmic radio emission within our Milky Way.

Inspired by the successes of the pioneering interferometer systems in England, the Netherlands, and California, NRAO then built the Very Large Array (VLA) radio telescope in the high desert of central New Mexico. Completed in 1981, the VLA is now being used by some 600 astronomers each year. Its extraordinary speed, sensitivity, resolution, and image quality permit a wide variety of research programs. Powerful image-processing techniques are used to remove any residual effects of atmospheric distortion to give essentially diffraction-limited images with resolutions comparable to that of the largest visible-light telescopes at the best mountain sites. At 0.7-cm and 1.3-cm wavelengths, the VLA's resolution rivals that of the orbiting Hubble Space Telescope.

The VLA is prevented by its midnorthern latitude from observing the southern sky, where the Magellanic Clouds and important parts of the Milky Way are found. The six-element, 6-km-long core of the Australia Telescope (AT), located in northern New South Wales, was built

Even an Earth-size baseline may not be large enough to resolve many of the most interesting radio sources.

1. DATA-GATHERING TECHNIQUES: Radio Astronomy

in part to fill this astronomical gap; it went into operation in 1990.

The most recent addition to the world's arsenal of radio arrays is the Giant Metrewave Radio Telescope now approaching "first light" near Poona, India. The GMRT's near-equatorial latitude of 18° north will enable its users to see almost all of the celestial sphere.

The VLA remains by far the most powerful and productive radio telescope in the world, but it still uses most of the original instrumentation, which was designed and built in the 1970s. By replacing the obsolete equipment, the sensitivity of the VLA will be dramatically improved. And radio astronomers are already looking beyond these enhancements to a Square Kilometer Array — a radio telescope with a collecting area of one million square meters that would give a 100-times improvement over existing radio telescopes. An international working group has already begun to investigate design concepts for this next-generation instrument.

Although physically linked arrays such as the Dutch Westerbork Array, the VLA, the AT, and the GMRT will continue to be powerful tools for radio astronomy well into the next century, even they do not have sufficient angular resolution to study some of the smallest sources associated with quasars, active galactic nuclei, and cosmic masers. This has motivated astronomers to build interferometers whose telescopes are separated by hundreds or even thousands of miles.

The MERLIN array in England uses radio links to connect as many as seven antennas with spacings up to 230 km. In the 1960s U.S. and Canadian radio astronomers developed interferometer systems with no direct connections between the elements. These very long baseline interferometer (VLBI) systems use atomic clocks to provide precise time references, allowing the signals from widely separated antennas to be recorded with high-speed tape recorders and later merged at one of the central processing facilities in Germany or New Mexico. As many as 23 radio telescopes located in 10 separate countries around the globe have been simultaneously joined in this way to produce milliarcsecond images of sources like radio galaxies and quasars.

VLBI research routinely involves many radio observatories in the U.S., most western European countries, and Poland, Russia, China, Australia, South Africa, and Japan. Four times each year, many of the world's largest radio telescopes cooperate in joint VLBI observations. With a single application, radio astronomers throughout the world can propose to use these global arrays.

Recently, NRAO completed a continent-wide array of 10 identical 25-meter antennas. This Very Long Baseline Array (VLBA) stretches from the Caribbean to Hawaii. Magnetic tapes recorded at each site are sent to a correlator facility in New Mexico, where radio images are constructed at an angular resolution better than 0.001 arcsecond — more than 100 times sharper than images obtained with the Hubble Space Telescope at visible wavelengths. With this extraordinary angular resolution, it is possible to resolve the gaseous clouds surrounding stars and to directly observe the outflow of relativistic plasma from deep within the nuclei of quasars and active galaxies billions of light-years from Earth.

But even an Earth-size baseline is not large enough to resolve many of the smallest and most interesting sources. This year Japan's Institute of Space and Astronautical Science will launch an 8-meter antenna into Earth orbit as part of its VLBI Space Observatory Programme (VSOP). The orbiting VSOP antenna will be linked to ground-based telescopes worldwide to obtain radio images with better angular resolution than is possible from the surface of the Earth. Even higher resolution is eventually expected from Russia's Radioastron satellite, which is designed to go into a highly elliptical orbit that will yield baselines as large as 77,000 km.

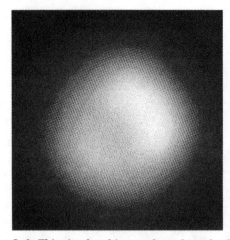 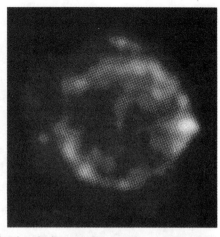

Left: This simulated image shows how the Cassiopeia A source (a supernova remnant in our own galaxy) would would appear to today's Very Large Array if it were located in the Pinwheel Galaxy (M33), some 3 million light-years away. *Right:* Adding signals from six more antennas scattered across New Mexico (two of which already exist) would enable the interferometer to capture nearly 10 times more detail. Courtesy NRAO.

MILLIMETER-WAVE ARRAYS

Millimeter-wave radio telescopes have opened up a new frontier, detecting the spectral signatures of circumstellar disks, rich molecular clouds, and copious stellar nurseries. Yet single-dish, millimeter-wave telescopes do not have adequate resolution to actually make clear images of most of their targets. So astronomers have married the push for higher resolution to the push toward shorter wavelengths by building millimeter arrays. Four such arrays are currently in operation, one each in France and Japan and two in California (Caltech's Owens Valley Radio Observatory, or OVRO, and the Berkeley-Illinois-Maryland Association, or BIMA, array at Hat Creek).

However, despite numerous recent improvements, the imaging capability of these arrays is still limited by their small numbers of antennas and, especially, by their relatively low-altitude sites, which lack the exceptionally dry conditions needed to exploit the full scientific potential of millimeter-wave astronomy. To overcome these limitations, American, European, and Japanese astronomers are looking at high desert sites on the Altiplano of northern Chile for the next generation of millimeter arrays.

The U.S. Millimeter Array (MMA) is planned for a site more than 4,875 meters (16,000 feet) high near the remote Atacama Desert — one of the driest places on Earth. It will operate at wavelengths as short as 0.8 mm with a sensitivity equivalent to that provided by a 50-meter antenna and a resolution (bet-

11. Radio Astronomy in the 21st Century

The real excitement in these facilities lies not in the old problems that will be solved but in the new ones that will be uncovered.

ter than 0.1 arcsecond) equivalent to that of an antenna 3 km in size. Because the MMA will serve astronomers throughout the world, the U.S. is seeking international partners to contribute to the design and to share costs with NRAO.

For high-resolution images at even shorter wavelengths, astronomers will use the Sub-Millimeter Array (SMA) now under construction near the summit of Mauna Kea. A joint project of the Harvard-Smithsonian Center for Astrophysics and Taiwan's Institute of Astronomy and Astrophysics, the SMA will contain eight 6-meter antennas operating at wavelengths as short as 0.35 mm. Astronomical observations with the SMA's first six elements are scheduled to begin in 1998. Prime targets for this instrument will include bodies within our solar system, star-forming regions in the galaxy, dust and molecular-gas structures in galaxies, and distant galaxies whose infrared spectral lines are redshifted into the submillimeter band.

RADIO ASTRONOMY IN THE 21st CENTURY

The instruments described above — both new and reborn — will open up a new era of astronomical discovery within every arena of our multifaceted discipline. Promising targets for these impressive tools will include:

Solar-system objects. New studies of the molecular contents of comets will give clues to the formation of our own solar system and, possibly, to the origin of life on Earth. All of the planets will be imaged with the Millimeter Array, and features such as volcanoes on Jupiter's satellite Io will be studied. The powerful new radar system at Arecibo will open up a new era in planetary mapping.

Stars. Radio astronomers look forward to new opportunities to direct the upgraded VLA to unusual, active objects like flaring stars, X-ray binaries, and novae; in addition, the Millimeter Array will measure the "blackbody" emission from thousands of stars throughout the Hertzsprung-Russell diagram. The VLA and the MMA will image binary star systems, stellar winds, nova outbursts, protoplanetary disks, and accretion disks and collapsing clouds around newly forming stars.

The interstellar medium. The new generation of millimeter and submillimeter telescopes will continue to explore the interstellar chemistry of organic molecules and trace the evolution of chemical elements in the Milky Way. They will also be used to investigate possible relationships between large organic molecules in space and the origin of life.

Pulsars. In addition to providing unique insights into solid-state physics and stellar evolution, pulsars provide a laboratory that radio astronomers can use to study subtle but important gravitational effects. The first observational evidence for the existence of extrasolar planets came from precise Arecibo pulsar observations by Aleksander Wolszczan (Pennsylvania State University), and future observations at Arecibo and Green Bank and in India are sure to un-

India's Giant Metrewave Radio Telescope spans a 25-kilometer-wide stretch along the plains near Poona. The array's 30 dishes, each 45 meters in diameter, will form the world's most powerful radio telescope at meter wavelengths. GMRT offers astronomers one of their best chances to detect the primordial hydrogen clouds out of which galaxy superclusters later grew. *Sky & Telescope* photograph by Leif J. Robinson.

1. DATA GATHERING TECHNIQUES: Radio Astronomy

cover many more planets that orbit around pulsars. Pulsar astronomers will also be in a position to look for signs of gravitational radiation. The only experimental evidence so far for gravitational radiation — a key prediction of Einstein's general theory of relativity — has come from Arecibo observations of the pulsar 1913+16 on the Aquila-Sagitta border. For this finding, Russell Hulse and Joseph Taylor (now at Princeton University) shared the 1993 Nobel Prize in physics.

Galaxies. Molecular spectroscopy will shed light on the nuclear regions of active galaxies, as will observations of gas and dust in those systems. Highly redshifted spectral lines of atoms and molecules will reveal products of nuclear fusion in distant quasars and galaxies. At millimeter and submillimeter wavelengths, it will be possible to image redshifted infrared emission from very distant galaxies.

The three-dimensional distribution of galaxies in space will be extended by hundreds of millions of light-years, and galaxies that are hidden from visible-light telescopes by the Milky Way's obscuring dust will be uncovered at radio wavelengths by means of 21-cm radiation from their cold atomic hydrogen gas.

Cosmology. Spectral observations of absorption lines due to matter in front of radio galaxies and quasars will extend the study of intergalactic gas clouds to the most distant parts of the universe. Of special interest is the search for massive hydrogen "pancakes," thought by many cosmologists to be the progenitors of galaxy superclusters. India's GMRT, in particular, will actively seek such precursors to the largest coherent structures known in the universe.

While it is exciting to think about the many observational puzzles that can be addressed by the new generation of radio telescopes, the real excitement of these facilities lies not in the old problems that will be solved, but in the new

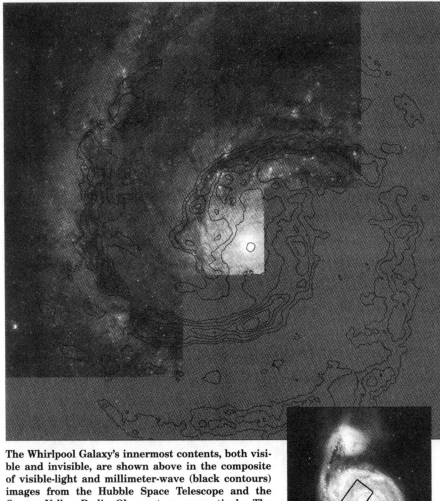

The Whirlpool Galaxy's innermost contents, both visible and invisible, are shown above in the composite of visible-light and millimeter-wave (black contours) images from the Hubble Space Telescope and the Owens Valley Radio Observatory, respectively. The juxtaposition clearly shows how carbon-monoxide gas is associated with the dust lanes that lie along the edges of the galaxy's spiral arms. Individual molecular-cloud associations, roughly 500 light-years across, are clearly resolved. The blue-light photograph at right was taken with the 5-meter reflector on Palomar Mountain. Courtesy Susanne Aalto and Caltech.

ones that will be uncovered. Starting with the pioneering work of Jansky and Reber, radio astronomy has led to a series of remarkable and unexpected finds that raise a whole new set of questions never thought of before. Radio galaxies, pulsars, cosmic masers, and the cosmic microwave background radiation were all discoveries that changed our view of the cosmos in a fundamental way. If history is a guide, we cannot begin to speculate on the discoveries that will be made in radio astronomy in the next decade and beyond.

Telescope Making
Edited by Roger W. Sinnott

A Radio Map of the Milky Way

I FIRST BECAME involved with radio astronomy eight years ago, recognizing that I was about to explore an area not frequented by many amateurs. As with the early days of observing with optical telescopes, I knew that if I wanted to get involved in this field I would have to design and build my own instruments. I also learned that the construction of an electronic device such as a radio telescope requires experience in such varied disciplines as radio-frequency electronics, analog and digital design, and computer programming. And familiarity with radio astronomy's history didn't hurt.

Not long after Heinrich Hertz discovered low-frequency radiation in the late 19th century, astronomers began to suspect that celestial objects emit radio waves. But early attempts to detect the Sun's radio emissions met with no success. It wasn't until 1932 that a Bell Laboratories engineer named Karl Jansky detected radio waves emanating from a celestial source, the Milky Way, while doing a radio background survey. Jansky's accidental discovery went essentially unnoticed until 1938, when Grote Reber, an electrical engineer and radio buff, tried his hand at detecting these celestial radio waves. Reber built a 31-foot paraboloidal dish antenna in his backyard and, after several attempts, detected radio emissions from the Milky Way and subsequently from the Sun.

Paul W. Schuler of Burlington, Connecticut, adapted a 12-foot satellite-TV "dish" for study of our galaxy's radio emissions from neutral hydrogen. Schuler provided all illustrations for this article.

1. DATA-GATHERING TECHNIQUES: Radio Astronomy

It was after this deliberate attempt to detect celestial radio emissions that astronomers began to take notice. But it should be noted that Jansky had indeed recognized the radio waves he detected as extraterrestrial in origin, and consequently he is considered the father of radio astronomy. The radio flux unit of measure is named after him.

I entered the field of amateur radio astronomy with a long-term objective: to construct an instrument that would allow me to observe the feeble radio emissions from neutral atomic hydrogen at a wavelength of 21 centimeters. There were three reasons for my choice: this wavelength is reserved by the Federal Communications Commission and is supposedly free of interference, it is detectable with equipment of modest cost, and it provides much opportunity for experimentation.

Radio telescopes are very similar in design to optical telescopes, but they are usually a lot bigger. The reason is simple enough, as radio waves are much larger than their visible-light counterparts and a telescope's resolution depends on its size relative to the wavelength it is receiving. Furthermore, since radio telescopes are exposed to the elements and can be operated continuously (radio energy is unaffected by daylight or most inclement weather), they must be rugged and weather resistant.

Unlike optical telescopes a single radio instrument does not create images, again due to the long wavelengths, but instead produces intensity measurements. However, the results of many scans at different declinations can be built up until an "image" is formed. The radio scans themselves can be produced in a number of ways. Most amateur observations and many professional ones employ the meridian-drift scanning technique. Here the antenna remains fixed with respect to the Earth and pointed at the local meridian. The Earth's rotation causes the dish to scan a band of declination in the sky, making a full sweep in 24 sidereal hours. Instruments that employ this method are generally steerable in declination only, as is the radio telescope I eventually built.

I began the project by building a small instrument that employs homemade 6-foot reflectors designed to operate at the 45-cm wavelength. I used this radio telescope for some time, gradually refining it until my scans of the Sun proved satisfactory.

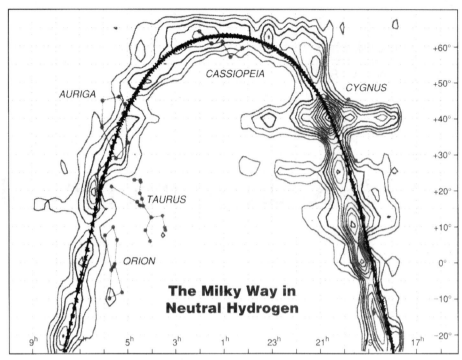

Schuler produced this map of the Milky Way from declination –25° to +70°. The thick black line representing the galactic equator and certain constellations have been added for reference. The strong emissions in Cygnus are believed to be due, in part, to ground noise entering the feed rather than a celestial source.

Later I acquired a bigger dish and after a few false starts got it operating at the 21-cm neutral-hydrogen spectral line. It was during this period of "learning the ropes" that I discovered the difficulty of building a sensitive instrument required to remain outside. I encountered problems of nearly every kind, including getting sunburned while working at the focus, birds nesting in the dish trusses every spring, accumulated rainwater dousing my head while I was adjusting the declination, and having to shovel out the dish after every snowstorm. At times I found myself muttering, "And I do this for *fun?*"

The reflector is a large (3.66-meter diameter) paraboloid that was originally designed for satellite-television reception. I adapted it for my purposes by designing electronics and a prime-focus feed to receive the 21-centimeter line. The dish and feed, akin to the mirror and eyepiece in an optical reflecting telescope, ride on an altazimuth mount fixed toward due south. The antenna's altitude is then adjusted to the declination of interest. The rest of the system is a simple "total power" radio telescope.

I first attempted to detect some of the more powerful discrete radio sources, such as Cygnus A and Cassiopeia A, but to no avail. So I decided to concentrate on the Milky Way, which my instrument *could* receive. A series of experiments verified that I was indeed detecting the Milky Way solely in hydrogen emissions. With this confirmed I began to map the galaxy.

My mapping procedure was quite simple. Starting at declination –25° the telescope observed continuously until I had

Radio signals collected, amplified, and converted by Schuler's telescope are fed to this receiving and signal processing equipment in his basement, the actual center of operations.

48

obtained three high-quality 24-hour scans. I then moved the antenna up 5° (the antenna records 4°-wide swaths of sky at a time), acquired three more scans, and continued the process until I reached declination +70°, well beyond the Milky Way's 62° northern limit.

As with all first attempts, I encountered difficulties. For instance, I had to aim the antenna between two houses and through the tops of a stand of trees to observe the galactic center, resulting in less emission at 21 cm than expected. For future endeavors I want to increase the positional accuracy of my observations, improve my data-handling capabilities, and refine my calibration techniques.

Problems notwithstanding, the results of this project were very gratifying. I am continuing my Milky Way research using a multichannel signal processor that will allow me to measure the radial velocities of neutral-hydrogen clouds by means of their Doppler shifts.

Amateur radio astronomy is an interesting avenue for backyard exploration, offering a look at the universe in a new "light" invisible to the human eye. For those interested in finding out more, the Society of Amateur Radio Astronomers (SARA), a worldwide organization, can be contacted through Vincent Caracci, 247 N. Linden St., Massapequa, NY 11758. I would like to thank Sam Palmer (Harvard-Smithsonian Center for Astrophysics) and Gerrit Verschuur (Rhodes College) for their encouragement, the members of SARA and the staff of the National Radio Astronomy Observatories for their help and comradeship, and my wife, Ruth Ann, for her patience and support.

PAUL W. SCHULER III
31 Nelson Drive
Burlington, CT 06013

12. Radio Map of the Milky Way

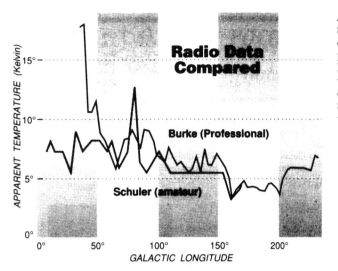

A comparison of Schuler's data with that obtained in the 1950s by professional astronomer Bernard F. Burke reveals some discrepancies. Overall, however, Schuler is pleased with the match.

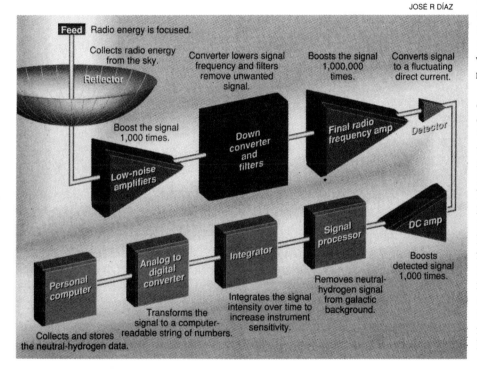

The signal collected by Schuler's radio telescope must undergo considerable processing before it can be analyzed.

The Earth and the Moon

The Earth (Articles 13–16)
The Moon (Articles 17 and 18)

After considering how astronomers acquire data, the next step is to consider what we know as a result of studying the data. And it makes sense to begin closest to home with Earth and the Moon.

In "Measuring the Size of the Earth" Ferrando Espinoza describes Eratosthenes' method of measuring Earth—a goal of Greek astronomers. Incredibly, with two long sticks, a few hours of driving time, and careful measuring, one can measure the size of the planet to within a few percent of the actual value.

Tabias Owen's article on how Earth got its atmosphere is related to historical geology as well as astronomy. While other planets in our solar system have atmospheres, ours is unique in composition. Understanding how Earth's atmosphere evolved and is maintained is important when considering planetary evolution and the evolution of life.

As far as we know, we are alone in the universe. However, considering life and other life forms is both intriguing for scientists and entertaining for others. The essay "What Makes a Planet a Friend for Life?" considers fundamental questions about the nature of life and the possibility of life elsewhere. The articles in Unit 3 on Jupiter's moon Europa (see "Magnetic Fields on Distant Moons Hint at Hidden Life," "Life in a Deep Freeze," and "Water World") and the Martian meteorite (see "Life from Ancient Mars?") could certainly be read in conjunction with this one.

One aspect of the Earth-Sun relationship that has immediate practical importance is the keeping of time, that is, calendars. Ian Stewart, in "A Day in the Life of a Year," examines the problems of keeping an accurate calendar and the complexities that are involved with the bodies of the solar system.

The origin of the Moon has long puzzled astronomers. The problem of the Moon's origin became murkier after Moon samples that were retrieved by Apollo astronauts were analyzed by scientists. The data showed that the Moon was like Earth in certain ways, but unlike it in others. There are no positive explanations that seem to explain all the variations. Lately, however, one hypothesis has emerged as favored—the impact hypothesis.

Two articles: "Moon Watching: An Experiment in Scientific Observation," by William Lovegrove, and "Charting the Moon by Eye," by Edmund Fortier, deal with viewing the Moon by eye alone. Fortier lists "Pickering's Dozen," the 12 objects that are visible on the Moon's surface with the naked eye.

Looking Ahead: Challenge Questions

Describe Eratosthenes' method of measuring the size of Earth.

What was/is the source of the abundant oxygen in our atmosphere?

Describe the problems that arise by using our current calendar to keep track of days and years.

List five features on the Moon's surface that are visible to the naked eye.

UNIT 2

MEASURING THE SIZE OF THE EARTH

Some physical science concepts are easier for students to understand if they have a visual aid that provides a hands-on introduction to the concept.

One such concept is measuring the size of the Earth. When learning how the Earth's size and shape have been determined throughout history, students usually understand the use of ancient evidence for the curvature of the planet (a ship's mast disappearing over the horizon, for example). However, the method used to measure its size appears abstract and esoteric.

The best-known method is attributed to Eratosthenes of Alexandria (250 B.C.). He used the length of an object's shadow when the Sun is at its highest point in the sky (solar noon). He also used a geometric technique that projects the angle subtended by the shadow to the angle formed at the Earth's center (Figure 1). Although several versions of the experiment appear in texts, students often have difficulty understanding the concepts involved and applying the technique to questions about other planets or satellites.

I have developed a technique that involves a globe of the Earth, an overhead projector to represent the Sun, and two pins stuck into the globe at Alexandria and Aswan, the two locations said to have been used by Eratosthenes in his experiments.

The pin at Aswan projects "no shadow." In other words, there is a shadow, but it lies exactly along the length of the pin. The other pin shows the shadow at Alexandria displaced from the line of the pin. Measuring the distance from the line of the pin to the displaced shadow of the pin-

FIGURE 1.
Angle subtended by the shadow.

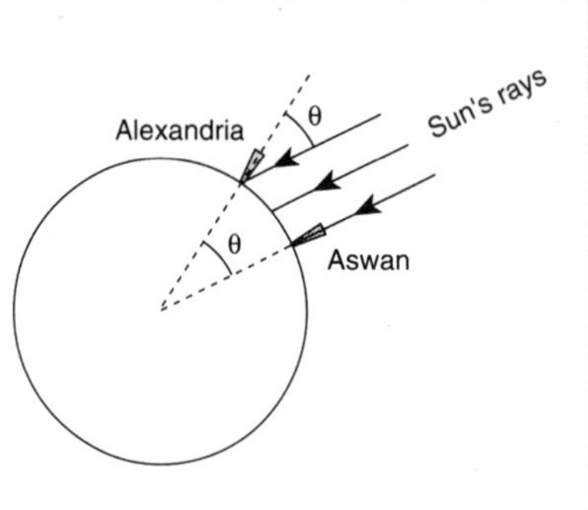

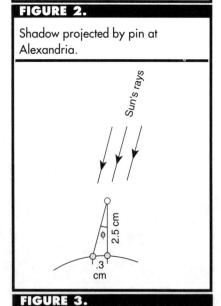

FIGURE 2.
Shadow projected by pin at Alexandria.

FIGURE 3.
Ratio of total angle of circle to angle at the center.

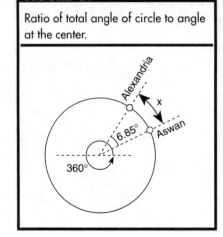

head, along with the length of the pin, gives the angle subtended by the pin's shadow (Figure 2).

Use a trigonometric relation

$$\tan\theta = 0.3\,\text{cm}/2.5\,\text{cm} = 0.12 \Rightarrow \theta \approx 6.85$$

to get the angle subtended by the shadow.

Then use the following proportion.

$$360°/6.85° = x/2\,\text{cm} \Rightarrow x = 105.11\,\text{cm}$$

to get the circumference.

The ratio of the total angle of a circle to the angle of the shadow is the same as the ratio of the Earth's circumference to the distance between the two locations. In other words, the angle is about 1/50 of 360°, so x is approximately 50 times the distance between the two locations (Figure 3).

Since one degree of latitude is about 112 km on Earth, and the globe being used has a scale of 1° = 3 cm, one degree of latitude on the globe is about 0.3 cm. The two scales combine to yield

112 km = 0.3 cm or 1 km = 373.3 cm from the globe.

The value obtained for the circumference, $x = 105.11$ cm can be converted to kilometers on the Earth scale by multiplying 105.11 by 373.3 to give 39 238 km, the value for the circumference of the Earth. Comparing this result with the known circumference of the Earth, 40 074 km, shows that this activity is only about 2 percent off the value.

Fernando Espinoza, Archbishop Molloy High School, 83-53 Manton St., Briarwood, NY 11435.

Article 14

How The Earth Got Its Atmosphere

By TOBIAS OWEN

WHY IS THERE AIR? Mercury and the Moon have essentially no atmospheres at all, while Jupiter and the other giant planets have atmospheres more massive than the entire Earth. Our own atmosphere is mostly nitrogen, whereas our nearest neighbors, Mars and Venus, both have atmospheres dominated by carbon dioxide.

The thin atmosphere of Mars produces a surface pressure that is only 1/150 the sea level pressure on Earth, while the atmospheric pressure on Venus is 90 times the terrestrial value. In the outer solar system, hydrogen and helium are the most abundant gases, except on Triton, Titan and Pluto, where once again nitrogen rules the roost. What accounts for these differences? Where do all these gases come from? What does this have to do with ice?

The ability to maintain an atmosphere over the lifetime of the solar system depends on a planet's gravitational field, the composition of the atmosphere and the temperature of the "exosphere"—the outermost layer of the atmosphere from which gases can escape into space. The density of our atmosphere steadily decreases with altitude until we reach a level about 500 kilometers above the ground at which a molecule heading upward will no longer encounter another molecule. This is the base of the exosphere.

All that is necessary for escape is that the kinetic energy (or energy of motion) of a moving particle, be it a gas molecule or a space ship, is equal to its potential energy (the energy it owes to its position in the Earth's gravitational field). The lightest molecules move the fastest, and only the fastest molecules can escape. On our own planet, hydrogen and helium can flee into space with ease, while heavier gases such as neon, nitrogen and oxygen remain in the atmosphere essentially undiminished over the lifetime of the solar system.

In other words, the Earth retains its atmosphere because its mass is sufficient to produce a gravitational field that prevents most gases from escaping. Giant Jupiter has 318 times our mass and can keep even hydrogen and helium. Our Moon, with 1/80 the Earth's mass, cannot keep any atmosphere except an extremely tenuous, flickering envelope that is constantly resupplied by gases leaking from the lunar interior and others carried by the solar wind.

We might expect Venus, Earth and Mars to have very similar atmospheres, yet, as already indicated, carbon dioxide composes over 90% of the atmospheres of both Mars and Venus whereas this gas accounts for only 0.03% of our atmosphere. Why are even these neighboring planets so different?

Earth is the only planet in the solar system with abundant molecular oxygen. This life-giving gas composes 21% of our air, compared with nitrogen's 78%. When we emphasize oxygen's vital role for life on Earth we are being rather chauvinistic—it is animal life that depends on oxygen. For the green plants that produce it, this gas is simply a waste product. No green plants, no oxygen; no oxygen, no animals. It's a very simple equation. If the green plants were wiped out by some giant catastrophe today, the oxygen in our atmosphere would disappear in less than 40 years, as would all animal life.

2. THE EARTH AND THE MOON: The Earth

In this picture of Comet Humason (1961e), background stars appear as trails because the telescope was set to guide on the moving comet. The bright, white coma of dust and gas and a tenuous plasma tail (see arrow) are visible. The tail's color comes from ionized carbon monoxide but ions of nitrogen, water vapor, carbon dioxide and other molecules are also present.

So that is why only Earth has oxygen; ours is the only planet with abundant life. What about the other gases in our atmosphere? For example, where does the nitrogen come from? You might suspect the Earth's volcanoes. A major difficulty with this idea emerges as soon as one tries to identify the minerals from which the nitrogen would come. It turns out that there is no rich source of nitrogen in rocks, and almost no nitrogen is actually present in volcanic gases. The nitrates that do exist are overwhelmingly biological in origin. Nitrates in any form are so rare that production of artificial fertilizers containing nitrogen is an important industry on Earth.

We find a similar story when we investigate the origin of Earth's carbon dioxide. Almost no carbon minerals occur in nature, diamonds and graphite being notable exceptions. Instead, we find deposits of oil, natural gas, coal, and huge amounts of carbonate rocks (such as the White Cliffs of Dover) all of which were created by living organisms. This explains the great differences between the Earth and both Mars and Venus. In the absence of life (and liquid water, which can also form carbonates), the Earth would have a carbon dioxide–dominated atmosphere almost as thick as that of Venus.

Evidently we need an outside source for these volatile elements but what is that source? The tradition has been to assume that it was the meteorites, particularly the carbonaceous chondrites, as they are especially rich in volatiles. There has always been a problem with this idea, however. In the meteorites, the abundances of krypton and xenon are approximately equal, whereas in the Earth's atmosphere, xenon is only one tenth as abundant as krypton. For many years, scientists assumed that this "missing" xenon must be trapped on or below the Earth's surface in rocks such as shales, or possibly in ice. However, attempts to find this hidden reservoir have failed; the xenon simply isn't there. Thus, the Earth's inventory of volatiles differs in this essential respect from the inventory that meteorites could supply.

Why is this so important? After all, xenon is only a tiny

trace constituent, less than one part per million of the gases we find in our atmosphere today. The reason we care about this tiny trace is that xenon, like all of the noble gases, is chemically inert. It is also very heavy, with three times the molecular weight of carbon dioxide. Thus, it will simply accumulate in a planetary atmosphere, neither escaping into space nor combining with other elements to form compounds (like the carbonates) on a planet's surface. To be believable, any model for the origin and evolution of a planetary atmosphere must satisfy the constraints imposed by observed abundances and isotope ratios of the noble gases.

If meteorites cannot supply the right mixture of these critical gases, how about the comets? Comets, or icy planetesimals as they are more generally called, are the low-temperature equivalent of the asteroid and meteorite debris left over from the formation of the planets and satellites. All of this debris has left a record in the form of countless impact craters on solid surfaces throughout the solar system. Almost all of these craters were formed as the system "cleaned itself out" during the first 700 million years of its history, with a few random impacts occurring even today.

What we know about the composition of comets is certainly consistent with the idea that they, rather than the meteorites, could have been the major source of the atmospheres of inner planets. Like the Earth's inventory of gases, comets appear to be deficient in nitrogen when compared to solar abundances. Thus comets would supply the carbon and nitrogen we find on Earth today in the right proportions.

Unfortunately, we know nothing about noble gases in comets, despite the great success of the missions to Halley's comet in 1986. That is not surprising because we expect that the abundances of these important gases in a cometary nucleus would be so low that the instruments on the spacecraft could not have detected them. Without direct observations, we are forced to rely on laboratory simulations, attempts to demonstrate what gases an icy comet nucleus is likely to contain, assuming that it forms at some specified temperature—hence distance from the Sun—in the outer solar nebula (the cloud of gas and dust from which the entire solar system formed).

Results from work carried out by Professor Akiva Bar-Nun and colleagues at the University of Tel-Aviv in Israel support the idea that comets must have been an important source of the volatiles we now find on the inner planets. At a temperature of 30 degrees Kelvin (water freezes at 273 on the Kelvin scale and boils at 373 Kelvin), argon, krypton and xenon are trapped exactly in their original proportions.

At a temperature of 50 Kelvin, argon is trapped much more poorly than krypton and xenon, such that the proportions of these three gases in the ice are entirely consistent with the abundances we find in the atmospheres of Mars and Earth. In other words, if the krypton and xenon in the Earth's atmosphere were brought to our planet by ice that formed in the solar nebula at a temperature near 50 Kelvin, we would find the ratio of the abundances of these two gases to be very similar to the ratio we find in our atmosphere today, rather than the nearly equal abundances carried by the meteorites.

The temperature of 50 Kelvin is significant because this is the average temperature estimated for the region of the solar nebula in which Uranus and Neptune formed. All four giant planets are mixtures of materials contributed by icy planetesimals and gases and dust from the solar nebula. As these giants grew, their powerful gravitational fields strongly perturbed the motion of the residual icy planetesimals in their vicinities. The resulting random orbits of these planetesimals allowed many of them to enter the inner solar system where they could collide with the forming inner planets.

The laboratory experiments suggest that icy planetesimals formed in the Uranus-Neptune region carried noble gases in the right proportions to produce the mixture we find on Mars and Earth today. To duplicate the gas mixture on Venus requires an additional contribution from one or more comets formed in the Kuiper Belt, at a temperature near 30 Kelvin. Icy planetesimals formed in the Jupiter-Saturn region would contribute most of the water, nitrogen and carbon in these atmospheres.

The comets did not contribute all the gases we find on the inner planets, however. There is evidence in both the martian and terrestrial atmospheres for another source, perhaps a late-accreting veneer of meteorites, perhaps the rocks that constitute the bulk of these planets. Furthermore, there were some losses, particularly in the case of Mars whose present atmosphere is so terribly thin. Calculations by Jay Melosh and Ann Vickery show that at least one hundred times the present mass of the martian atmosphere must have been removed by the same early bombardment that brought in volatiles and left a record of impact craters.

Mars was particularly vulnerable to this process of impact erosion of its atmosphere because its mass is only 1/11 the mass of the Earth. Poor Mars! Its fate was sealed by its small size. Yet it may have enjoyed episodes of high surface pressure during the first 800 million years of its history. The constantly shifting balance between delivery and removal of volatiles as the ferocious early bombardment ran its course would have included intervals in which the atmosphere was much denser than it is today. These episodes would have allowed the formation of the famous martian channels and other features caused by aqueous erosion.

So the answer to the question "Why is there air?" is thus "because there are comets!" We like to call this the "icy-impact" model for the origin and early evolution of inner planet atmospheres. It seems to be generally self-consistent.

A mixture of comets found in different regions of the solar nebula can deliver most of the carbon, nitrogen, water and noble gases we find on the inner planets. Nitrogen dominates the atmospheres of Titan, Triton and Pluto because the abundant carbon-containing gases are all unstable on these low-temperature objects: carbon monoxide is converted to carbon dioxide, which freezes out on their surfaces and methane is destroyed by photochemical processes.

2. THE EARTH AND THE MOON: The Earth

The arid surface of Mars serves as a reminder that the early bombardment by icy and rocky planetesimals could bring both life and death. Once water raged across the surface of this planet but repeated impacts depleted the atmosphere so severely that no water runs on Mars today.

However, there are many more details to be worked out before this model can be considered as a comprehensive theory and we are still lacking some essential data. At the top of the list are the abundances and isotope ratios of noble gases in icy planetesimals, so we don't have to rely on the laboratory simulations. We may learn some of these when the *Galileo* atmospheric probe sends back data from Jupiter in December. We will learn more from ground-based and near-Earth observations of the next bright comet (which may be Comet Hale-Bopp, scheduled to reach maximum brightness in 1997).

The coup de grace should come in 2012 with the return of data from the ESA Rosetta Mission. This highly promising mission to the nucleus of Comet Wirtanen would become even more valuable from the volatile-delivery perspective if the nucleus could be broken apart near the end of the mission to expose the deep interior. This would be our best chance of finding pristine ices that are still saturated with the gases they trapped from the outer solar nebula, 4.5 billion years ago.

Tobias Owen is Professor of Astronomy at the Institute for Astronomy of the University of Hawaii. He is studying planets, satellites and comets with the telescopes on Mauna Kea and by means of deep space exploration. Having participated in the Viking *and* Voyager *missions, Owen is presently an Interdisciplinary Scientist and a member of the Probe Mass Spectrometer Teams on both the* Galileo *and* Cassini-Huygens *missions.*

What Makes a Planet
A FRIEND FOR LIFE?

Julie Paque

Julie Paque is an associate research scientist at the SETI Institute in Moffett Field, California.

By all rights, Earth should be lifeless. As dead as a doornail. As dead, in fact, as the Moon, which it largely resembled four billion years ago.

How life began on Earth is at present a mystery, although scientists have made a bit of progress charting its first steps.

Life's beginnings can be traced back to primitive organic molecules that formed on Earth and learned how to copy themselves. How the organic stuff that made self-replicating molecules got on Earth, exactly how it combined, and what that process says about the existence of life elsewhere in the universe are still largely open questions. Many scientists believe the molecules arose in the soupy oceans of Earth, while others think they were deposited on Earth from space. Understanding which idea is correct will permit astronomers to better estimate the relative number of living planets, which they have already done in rough fashion using a formula called the Drake equation. It uses a number of variables about stars, planetary systems, and the chemistry of life to estimate the possible number of living planets in the universe.

It's ALIVE!

To search for life's origins astronomers must first define life. Living things are organisms as we know them on Earth, organisms based primarily on carbon but also on hydrogen, oxygen, and nitrogen. Imagination, with generous assistance from television dramas, suggests that other forms of life, perhaps based primarily on some element other than carbon, may exist elsewhere in the universe. But as yet scientists have no evidence to support the possibility of life based on other elements, so this discussion will focus on life as it exists on Earth.

Once upon a time humans considered the constituents of life to be the "elements" fire, water, air, and earth. Now of course scientists know the periodic table of the elements and have a detailed knowledge of organisms on the molecular and cellular level.

All living beings are composed of cells, and within each cell are proteins, which regulate carbohydrates and fats. Proteins in turn are composed of amino acids. The nucleic acids RNA (ribonucleic acid) and DNA (dioxyribonucleic acid) within the cells orchestrate how these amino acids link into proteins.

All living things metabolize energy from the environment and pass their genes to succeeding generations.

Besides being composed of cells, living beings share other characteristics. They metabolize energy from the environment (by eating), and they reproduce and pass along their genes to succeeding generations. Mutation, which introduces random changes in cells, is also a factor in the evolution of life, although only a moderate one. Such changes sometimes create stronger, more efficient individuals that compete more successfully for the available food and have a better chance of surviving to pass along their genes than average beings. Additionally, some living beings are sentient, meaning they have the power of perception by sense. Animals and plants share these traits.

The Cosmic Seeds of Life

Until recently, science continued to ask how the nucleic and amino acids arose. They have long believed compounds as complex as amino acids formed from methane, water, and ammonia. To do this on early Earth, two other things were needed, an oxygen-poor environment (like that on early Earth) and energy such as lightning, volcanic eruptions, or simply the Sun's radiation. The amino acids then formed proteins and nucleic acids. DNA and RNA are the blueprints for proteins, but they also need proteins as catalysts. So molecular biologists have something of a chicken and egg problem. Which came first—proteins or nucleic acids? Current research leans towards the RNA-first solution.

Whichever chemicals battled their way out of the soup first, life's stage was set as long as four billion years ago, when the number of objects impacting onto our planet started to slow down. Periodic impacts of large comets and asteroids undoubtedly changed the variety and quantity of organic material available to make more complex molecules. The first fossil evidence of life contains bacterialike beings entombed in rocks about 3.5 billion years old.

The early oceans provided a habitat where organic molecules could combine into amino acids.

In 1953, chemists Stanley L. Miller and Harold C. Urey produced amino acids in a laboratory from methane, ammonia, water, and hydrogen. If Miller and Urey had left this simple brew of chemicals on a lab bench, in a freezer, or in a warm bowl, nothing would have happened. Instead, they carefully recreated an oxygen-poor atmosphere like that of the early Earth by adding excess hydrogen, methane, carbon monoxide, and ammonia, and sparked the mixture with energy to simulate that contributed by the lightning common on early Earth. These classic experiments showed that the chemicals and conditions on early Earth could have produced amino acids in a very short time.

Since the Miller-Urey experiment, scientists have investigated where such chemical activity might have occurred on early Earth. Two locales may have acted as the mixing bowl: Earth's atmosphere or its oceans. The viability of each area depends on the nature of Earth's early atmosphere, and scientists hold a range of opinions on that. Many believe it was highly reducing, or oxygen-poor, while others disagree. Compounds important in forming life, like formaldehyde and hydrogen cyanide, are easiest to produce experimentally under oxygen-poor conditions. Earth's early ocean was almost certainly oxygen-poor, illustrated by Precambrian banded iron formations that precipitated from early seas. If the early atmosphere didn't, the early oceans almost certainly provided a habitat where organic molecules could combine into amino acids.

There's another possibility. Sherwood Chang, a planetary biologist at NASA Ames Research Center, points out that the ocean surface could have been a favorable setting. A complex set of physical and chemical processes operate here. Gases, aerosols, and dust from the atmosphere were continually mixed by wind and waves with the contents of the early ocean. Compounds were able to exist in gas, liquid, and solid states and ideal conditions existed for organic molecules to form.

So the first amino acids and proteins may have formed above the ocean or at the meeting place of ocean and volcano, which would have provided energy as well, and spread from there. Anyone who has seen dramatic images of an undersea vent or volcano splattering lava into the ocean can appreciate the unusual conditions. Life may have taken a foothold around a "black smoker" pouring energy out deep within the ocean.

Other locales may have existed, too. Verne Oberbeck, also at Ames, has proposed that raindrops may have hosted organic chemical reactions on early Earth.

Organic Stuff from Space

On the other hand, comets and asteroids dumped a great amount of material onto early Earth. Anyone who followed last summer's impact of Comet Shoemaker-Levy 9 into Jupiter can appreciate how a comet could dump material onto a planet. Several billion years ago vastly more such objects were orbiting and impacting bodies in the inner solar system. Analysis of Comet Halley indicates that approximately one-third of its mass consists of organic molecules. If Halley is like other comets, then, comets contain substantial amounts of organic molecules.

Comets contain organic molecules. Could the organics have survived a planetary strike?

Organics may have come from asteroids striking Earth, too. Carbonaceous chondrites, as their name implies, contain carbon and as much as 5 percent organic matter. The carbonaceous chondrite called Murchison, which fell in Australia in 1969, contains a variety of amino acids. Astronomers believe, then, that in some cases organic molecules from an asteroid or comet have survived the incredible crushing pressures and superhot temperatures of a planetary strike.

Not all of the organic material that struck Earth hit with the dramatic force of asteroids and comets, however. Just as significantly, a continuous silent rain of interplanetary dust particles falls slowly to Earth even today. These particles are microscopic, highly porous aggregates of mineral and organic material thought to have originated in comets and asteroids. Their carbon content can be as high as 40 percent. Three hundred tons of this stuff falls gracefully to Earth every year, gently depositing particles in our backyards and offices and on our houses and cars.

Some planetary scientists approach the extraterrestrial delivery of organics more creatively. Ted Bunch of NASA's Ames Research Center and Jeff Bada and Luann Becker of the Scripps Institution of Oceanography have found fullerenes in the Sudbury impact crater in Ontario, Canada. These organic compounds, also called buckyballs, probably formed during the Sudbury impact from carbonlike compounds in the impacting body. Bunch and his colleagues are now experimentally testing the survivability of organic material that could remain viable on Earth or create new compounds like buckyballs in an impact. Early test results are encouraging: Bunch's group has found that complex organics called polycyclic aromatic hydrocarbons are created during the impact of a carbonaceous chondrite into aluminum at velocities of up to 6.5 kilometers per second.

Survival of organics on a comet or asteroid remains an open question. Three years ago Christopher Chyba and his colleagues at Cornell University estimated the survivability of organic material as it passes through the atmosphere and impacts Earth. Large asteroids vaporize rock and water, destroying organic material, and generally devastate the environment. The large impact that produced the Chicxulub Crater on the Yucatan peninsula wiped out the dinosaurs. Such events were likely more common earlier in Earth's history and probably more devastating to the limited range of life forms present then. If Earth's early atmosphere had been denser, comets could have been slowed enough during entry for organic material to survive. Chyba's research shows that smaller, less dense asteroids or comets probably could contribute organic material to Earth, and certainly interplanetary dust particles could. (See "The Cosmic Origins of Life on Earth" by Christopher Chyba, *Astronomy,* November 1992.)

Even if asteroids and comets delivered life's building blocks, the terrific power of volcanoes along with the ongoing bombardment, may have snuffed out life many times in its infancy. Early Earth was a rather inhospitable place. That's fortunate for humans, though. Life might be very different in one of the earlier attempts had taken. Wouldn't it be ironic if a comet carried the building blocks of life and another comet wiped out the dinosaurs? Perhaps a comet will strike Earth again, rendering humans extinct, and leaving the planet open for a third master species—maybe ferrets or 13-lined ground squirrels.

Is Anybody Home?

Whether or not comets or asteroids helped life arise on Earth, our home planet and its inhabitants imply that life might be common on other planets in the Milky Way Galaxy and in other galaxies scattered across the universe. Frank Drake, president of the SETI (Search for Extraterrestrial Intelligence) Institute, has devised an equation to estimate the likelihood that technological civilizations exist in the universe (N):

$$N = R^* \bullet f_p \bullet n_e \bullet f_l \bullet f_i \bullet f_c \bullet L.$$

R^* is the rate of formation of stars suitable for development of intelligent life, and f_p is the fraction of stars with planetary systems. Of the planets in a planetary system, only a certain proportion have the environmental conditions necessary to sustain life (n_e), and only some of those planets will actually have life (f_l). Intelligent life will emerge on only a fraction of the planets (f_i) that contain life of any sort, and only some will be technologically advanced enough for their existence to be detectable from space (f_c) The factor L in the Drake equation reflects the length of time detectable signals are released into space. If all global lifespans of technological species survive only thousands of years our chance of detecting them is lower than if they survive for millions of years.

Sentient life has the power of sense, intelligent life the power of creativity. How much of it lies out among the stars?

Placing conservative estimates for each of these factors into the equation leads to the conclusion that the chance for intelligent life elsewhere in our universe is indeed good. The chance that life of any sort exists elsewhere is even better. Astronomers know the physical laws and chemistry of the universe hold constant everywhere astronomers observe them, even in the hearts of quasars in the early days of the universe some 15 billion years ago. Our Milky Way Galaxy contains as many as 400 billion stars. Astronomers estimate that at least 100 billion galaxies exist. You figure it out: Even with ridiculously conservative estimates, the potential number of stars with planets that have life is astonishingly high. That does not mean that astronomers can find life with ease, however.

Living on the only planet known to harbor life, astronomers have thus far concentrated their efforts on identifying star systems with Earth-like planets. Several problems complicate searches for planetary systems around other stars, mainly the difficulty in resolving a small, distant dark object near a star that is orders of magnitude brighter. Rather than observing in optical light, however, astronomers can observe in the infrared part of the spectrum to decrease this contrast, or they can indirectly detect planetary systems by measuring the gentle wobble in a star's path across the sky caused by the mass of orbiting planets.

Currently forming planetary systems are easier to observe, as the disks of gas and dust that surround their young suns absorb and emit lots more radiation than planets do.

A planet that supports carbon-based life such as ours must also have liquid water, and this considerably restricts the range of temperatures on the planet and requires an atmosphere so the water won't escape into space. Many other factors make life possible for humanlike beings, including visible light, gravity, a breathable atmosphere of oxygen and water vapor, and the right range of temperatures.

Beginning to comprehend how life on Earth rose out of a cosmic stew of organic molecules certainly gives us fresh respect for our fragile planet. Knowing what makes Earth habitable will help humans maintain their home and perhaps one day in the distant future allow our descendants to leave the planet, when our Sun ages, to find shelter among the stars. Perhaps in the interim humans will discover sentient life in the universe and come to learn the perspective of this life on our own civilization. If astronomers do discover the range of life that probably exists out there, it will certainly bring human beings a little closer in our common journey through space and time.

A DAY IN THE LIFE OF A YEAR

Has your year got off to a bad start? Not to worry, there are at least 14 other new years around the world left to choose from, says *Ian Stewart*

NOW that you've recovered from the new year bash, ask yourself a simple question. Why should it all happen on 1 January? Until AD 800, France preferred 1 March. For nearly two centuries after that it was 25 March, and from AD 996 until 1051 New Year's Day coincided with Easter. The English had completely different ideas. Between the 7th century and 1338 they considered Christmas Day to be the start of the new year. But starting in 1339 New Year's Day was moved to 25 March for civil purposes and to Easter for religious ones.

> 'The history of the calendar is a long-running, planet-wide soap opera'

The history of the calendar is a long-running, planet-wide soap opera—a stream of brave attempts to put the seasons in their rightful places, accompanied by just as many chronological blunders. It is a wonderful example of one of humanity's most endearing and infuriating traits: the inability to get the simplest and most basic things right, or even consistent.

We have still not sorted it out. Only in those countries that have adopted the Gregorian calendar does 1 January count as New Year's Day. According to the Chinese calendar, it will be on 19 February this year, for the Burmese 15 April, in the Islamic world 19 May, and for the Jews it will not arrive until 14 September. And every calendar still needs to have a fudge factor thrown in—a few seconds or perhaps a day here and there—to stop day turning into night or summer into winter. In humanity's defence, it must be said that the blame for our chronic calendric confusion lies ultimately with the bodies of the Solar System. Before you can even start the tricky task of predicting the date of the new year, you have to remember some of the fundamental facts about the heavens as perceived from the Earth.

Our planet rotates on its axis once every 23.9345 hours, or 23 hours 56 minutes and 4 seconds. This is the time it takes for one rotation relative to the "fixed" stars. It is less than the 24 hours we know as a day, the interval between successive occasions when the Sun is overhead—that is, when it crosses a chosen meridian. While the Earth is rotating, it is also revolving around the Sun, and it takes that extra four minutes or so for the rotation to catch up with the Sun's apparent slippage back across the sky. Even the 24-hour figure is only an average: the actual length varies, mainly because the Earth's orbit is an ellipse rather than a perfect circle.

To ancient humanity, the next most obvious celestial cycle was the repeating sequence of phases of the Moon. Relative to the fixed stars, the Moon revolves around the Earth once every 27.32166 days. However, its phases are governed by the relative positions of the Sun and the Moon as observed from the Earth. Once more there is some slippage in the Sun's position which has to be made up, leading to an average "synodic lunar month" of 29.53058 days.

Tropical trouble

Finally, there is the year, the time that it takes the Earth to travel once round the Sun. This is the "sidereal year" of 365.25636 days. For calendars, however, the more important period is the time between the start of a season in one year and its start in the next: this is called the "tropical year". The Earth's axis is tilted at an angle of 23.5° relative to the ecliptic, the plane of the Earth's orbit. This causes variations throughout the year in the day/night division of the 24-hour day, giving long nights in winter and short ones in summer. There are two "equinoxes" at which the division is 50:50; these fall, near enough, on the first day of spring and the first day of autumn. But because the Earth bulges at the equator, its axis of spin slowly precesses like that of a spinning top: the direction in which the North or South Pole points revolves round a complete circle in the sky once every 25 800 years. One consequence of this is that the equinoxes come a little earlier each year and the tropical year is 365.24219 days long, slightly shorter than the sidereal year. Calendars have yet to deal with this effect, because it is smaller than other in-built errors.

The important point here as far as calendars are concerned is that neither the tropical year, nor the lunar month, are simple multiples of a day. The multiples are instead irrational numbers, which, like $\sqrt{2}$ or π, cannot be represented as exact fractions. If the solar year were a rational multiple of a day then after some integral number of days, the year and the day would be back in step at exactly the same point. For example, if the tropical year were exactly

16. Day in the Life of a Year

365.25 days, a ratio of 1461/4, then 4 years would be exactly equal to 1491 days, so that all the astronomical events related to the apparent position of the Sun would repeat precisely over that period. But both the tropical year and the lunar month contain an irrational number of days. Worse, the tropical year also contains an irrational number of lunar months. So nothing ever repeats exactly.

Calendars therefore have to make compromises, and it is the history of those compromises in different cultures that has led to a plethora of calendar systems. Depending on which one you choose, there will be at least 26 different New Year's Days in any given year—an average of one a fortnight. Admittedly, some of these have been obsolete for a few millennia, but there are around 16 still in widespread use. These fall into two basic types: lunar and solar. In one, primary attention is paid to the apparent motion of the Moon; in the other the Sun has pride of place. Nevertheless, most lunar calendars include clever solar-related jiggles to keep them roughly in tune with the seasons, and most solar calendars at least pay lip service to the movement of the Moon.

Our own calendar is solar, and goes back to ancient Rome: In 46 BC Julius Caesar reformed the previously erratic Roman calendar, a lunar one. On the advice of the astronomer Sosigenes he took the length of the tropical year to be 365.25 days, and set up a cycle of 1461 days consisting of one leap year of 366 days and three common years of 365. He added 90 days to the year 46

New Year's Days occurring in 1996

1	January	Gregorian
19	February	Chinese
26	February	Akbar
20	March	Saka
21	March	Iranian solar
24	March	Jelali
15	April	Burmese
19	May	Islamic
6	June	Soor San
23	July	Yezdezred
28	July	Zoroastrian
3	August	Fasli Deccan
11	September	Coptic, Ethiopian
14	September	Jewish
17	September	Parasuram

BC to get spring back to its traditional date of mid-March, and he also decided that the year would start on 1 January, which proves what a sensible chap he was.

Julius decreed that from that point on there would be 12 months, relics of the lunar system but no longer linked to the lunar month. Their lengths were not quite those we have now; in particular February had 29 days in a common year and 30 in a leap year. Our current lengths seem to have been introduced by Julius's successor Augustus Caesar, and an apocryphal story has it that he pinched a day from February to make August (renamed after him from its original, Sextilis) exactly the same length as July (previously renamed from Quintilis in honour of Julius).

As a result of a monumental cockup by Roman priests, the calendar had to be reformed again around 10 BC. The priests had been instructed to insert a leap year every four years. But to count the years they used the common Roman procedure of beginning a new count with the end of the previous one. This meant that a leap year occurred every third year instead of every fourth. So to get the calendar back on track, leap years were omitted until AD 4.

The Julian year exceeds the true tropical year by 0.00781 days, so by 1582 the spring equinox had slipped back from 21 March to 11 March. To prevent further slippage, Pope Gregory XIII reformed the calendar once more. Leap years thenceforth were to be omitted in years ending 00, unless that year happened to be a multiple of 400. This resulted in the omission of 3 days from every 400-year cycle, and reduced the "theoretical" length assumed by the structure of the calendar to 365.2425 days, much closer to the true value of 365.24219. To bring the equinoxes back into alignment, ten days of 1582 were removed, 5 October becoming 15 October. The new year, as in the Julian calendar, began on 1 January.

Taxing changes

The adoption of the Gregorian calendar was rapid in Catholic countries or provinces, but slower in Protestant ones. Italy and France made the switch in 1582, the Germans followed at various times between 1583 and 1700, while Finland left it as late as 1918. In England the Gregorian calendar was adopted in 1750 and put into practice in 1752, by which time the year 1600 had come and gone, so equinoxes had slipped by a further day. This meant catching up 11 days, and 3 September to 14 September were the ones that were omitted. A consequence of this move that still haunts us is the date of 6 April as the start of the financial year. It was originally 25 March—an inaccurate approximation to the spring equinox, but one consistent with three-monthly accounting periods, including the all-important 25 December. With the addition of 11 days it became 5 April. In 1800 it became 6 April because a Julian leap year was omitted, but this anachronistic modification was not applied again in 1900.

A good example of a lunar calendar tuned to accommodate the solar year is the Jewish one. The calendar uses a lunar cycle of 19 years combined with a solar cycle of 28 years. To approximate the lunar month, the 12 Jewish calendar months contain either 29 days (when they are called "defective") or 30 ("full"). Most months have fixed lengths but two are variable, depending on the solar cycle. There is also a 30-day intercalary month, which occurs only in years 3, 6, 8, 11, 14, 17 and 19 of the lunar cycle. The structure of the Jewish calendar is complicated by the need to avoid certain events falling on certain days of the week. For instance, the year cannot begin on a Sunday, Wednesday or Friday. New year wanders erratically through September and early October in our Gregorian calendar. For anyone who feels like an extra celebration, the next five Jewish new years fall on the following dates: 14 September 1996, 2 October 1997, 21 September 1998, 11 September 1999, and 30 September 2000.

The Muslim calendar is unique in being totally lunar. The year consists of 12 lunar months or 354 days, so significant dates and festivals drift relative to the seasons. The calendar starts counting from the Hijrah, the prophet Mohammed's flight from Mecca to Medina to escape religious persecution. Most Muslims consider his arrival time in Medina to be sunset on 16 July 622, but a few who count days from midnight to midnight (rather than the normal sunset to sunset) employ 15 July as their starting date. The calendar has 12 months, which are alternately 30 and 29 days long. However, the 12th month, which is usu-

2. THE EARTH AND THE MOON: The Earth

Historic New Year's Days in Europe, extrapolated to 1996

1	January	Roman Empire
1	March	France until AD 800
25	March	France AD 801 to AD 995
5	April	(Good Friday) Delft, 10th century
6	April	(Easter eve) Péronne
7	April	(Easter day) Council of Tours, AD 775
16	April	(Easter, previous year) France AD 996 to 1051
1	May	Annalis Pitaviennes
1	July	Sicily until 17th century
12	August	Parts of Denmark
1	November	Celts until 1179
25	December	Mayence (Mainz) until 15th century

For any party animals who are interested, the two tables give a selection of new years to celebrate. The first shows new years during 1996 for calendars that are still in use. For those with a historical turn of mind, the second table lists the different New Year's Days from past times, extrapolated into 1996—for instance, by choice of Easter where appropriate. The table lists only the first recorded occurrence of each date for the new year. In total, the tables give 27 excuses to celebrate.

ally 29 days long, acquires an extra day in intercalary years 2, 5, 7, 10, 13, 16, 18, 21, 24, 26 and 29 of a repetitive 30-year cycle. Those who start the year on 15 July add the extra day in the 16th year, not the 15th.

> **'Whatever calendar you settle on, the butterfly effect will cause an unpredictable drift'**

For the Chinese, our brand-new 1996 will be the Year of the Rat (prophesying a British general election, perhaps). The Chinese calendar, known as the *yin-yang-li*, goes back to 2953 BC, making it older than any other in current use—although since 1911 China has followed the Gregorian calendar for official business. However, it has seldom been the only calendar in use in China: one authority estimates that at least 102 different types of calendar system have been used there at one time or another. The Japanese and the Koreans also use the *yin-yang-li*, with some minor modifications. It is based on a 60-year cycle which combines a cycle of 10 constellations with a zodiacal cycle of 12 animals, and it allocates the timing of the new year by the phase of the Moon.

The constellations (for which there is no sensible translation) are *kiah, yih, ping, ting, wu, ki, kang, sin, jin* and *kwei*. The animals are *tse* (rat), *chau* (ox), *yin* (tiger), *mau* (hare), *shin* (dragon), *se* (snake), *wu* (horse), *wi* (sheep), *shin* (monkey), *yu* (rooster), *siuh* (dog) and *hai* (pig). The combined cycle starts with *kiah-tse* and then moves one step along each list to give *yih-chau, ping-yin,* and so on. Each list wraps round to its start, so that after *kwei-yu* comes *kiah-siuh*. Exactly half of the 120 possible combinations occur, because 10 and 12 have the common factor 2. The new year is defined independently of this cycle: it falls on the new Moon nearest to the time when the Sun is at a certain fixed point in the constellation Aquarius. This always turns out to be within 15 days of 4 February. For example in 1996 (*ping-tse* in the cycle) it will fall on 19 February; in the following years it will be on 7 February 1997 (*ting-chau*), 28 January 1998 (*wu-yin*), 15 February 1999 (*ki-mau*), and 4 February 2000 (*kang-shin*).

Calendar chaos

What of the future? It's certainly not getting any simpler. All of the various astronomical cycles are slowly changing their lengths: the day, the tropical year and the lunar month are all lengthening because of tidal gravitational forces. There are other difficulties in the pipeline, such as irregularities in the precession of the equinoxes. These are caused by occasional glitches, called Milankovitch shifts, in the tilt of the Earth's axis. And thanks to the work of astronomers such as Jack Wisdom of Massachusetts Institute of Technology in Cambridge and Jacques Laskar of the Bureau des Longitudes in Paris, we have known since 1993 that the motion of objects in the Solar System is chaotic. No matter what scheme you settle on and how carefully you have accounted for all the variables, the "butterfly effect" will cause an unpredictable drift away from whatever calendar you calculated in advance.

So it would be best to institute an interactive method of tinkering with the calendar, as is already done to keep the length of the year in tune with a gradually slowing Earth. However, this means that science fiction titles such as A. E. van Vogt's *200 000 000 AD* need to be taken with a pinch of salt—especially if you want to see in the new year on the right day. Look far enough ahead, and even that is not predictable.

Moon Watching: An Experiment in Scientific Observation

William P. Lovegrove, *Department of Physics, Bob Jones University, Greenville, SC 29614*

Because we live so much of our lives indoors, most of us have little personal experience with observing the motion of the Moon and the stars. Moreover, many of our laboratory exercises are rigidly structured. Students are given detailed step-by-step procedures that are sure to produce reasonably good results. Often this is unavoidable because of time, personnel, and class size constraints. However, this type of exercise bears little resemblance to the exploratory type of research actually carried out by scientists. Experiments in which the objective is to verify some already-known behavior have their place as a pedagogical tool but fail to convey the excitement of discovery.

"Moon watching" is an exercise that can be carried out by any number of students at any time without any equipment or supervision. The motion and phases of the Moon are complex enough that careful observation and careful thought are required to fully understand them. I've discovered that few if any students, including the ones who have studied lunar motion in previous science classes, have even a basic understanding of the motion of the Moon prior to performing this experiment.

My students go Moon watching prior to any discussion in class about the Moon. I give students the following introduction:

This lab is intended not to teach you a group of facts about the Moon and stars, but to help you make some observations of your own and draw your own conclusions. You will act like an amateur astronomer and see what you can discover. You are not to consult reference books. Everything must be based on your own personal observations.

They are told to watch for the Moon for a period of several weeks. Each time they see it, they are to "make an observation" of the Moon. This involves writing down the date and time, noting the location of the Moon in the sky (by whatever means they can think of), and describing its appearance. Instructions are deliberately vague so that students are forced to develop their own method of recording data. They are told:

Do not worry about using "standard" terminology or units. Do whatever seems to work best.

They are told that at the end of the observation period, they will be given a quiz in which they will have to answer questions and make predictions about the Moon based only on their observations. The contents of this quiz are deliberately left unspecified so that students must decide how much and what kinds of data to collect.

As the weeks pass, I informally monitor how the students are progressing and drop a few hints. Some students will not think to look for the Moon during daylight hours. Some will not realize that multiple observations at different times during the same night can be helpful. As they struggle to describe where the Moon is and what it looks like, they gain an appreciation for how the language of science develops.

When the day of the quiz comes, the class is divided into small groups (typically 4 or 5 persons). They are asked to discuss the assigned questions (listed here in the Appendix) and determine the answers as a group. They turn in one set of answers and all receive the same grade. A lively discussion results. I have found that a full 50-minute class period is required for this discussion. I provide Earth and Moon globes and an overhead projector to use for a "Sun." The students don't have to be told what to do with them. They naturally pick them up and try to figure things out as their discussion progresses.

It soon becomes clear to the students that merely observing the Moon is not sufficient for real learning to take place. Only when they are confronted with difficult questions and forced to discuss the answers do their misunderstandings surface.

2. THE EARTH AND THE MOON: The Moon

Most of the questions are conceptual. For some of the questions an exact answer is difficult. I expect to receive only a general answer that demonstrates correct understanding.

When I hear students comment that they would be able to do a much better job of taking data now that they know what to look for, a discussion naturally follows of the problems real scientists face in collecting data for previously unexplored phenomena.

My students are amazed (and so am I, sometimes) at how well they are able to answer the difficult questions. The students enjoy this activity, and I observe that real learning takes place.

In summary, this experiment is highly successful in teaching my students what real scientific observation is about. It is suitable for large or small classes in physics, physical science, or astronomy. It requires minimal equipment and is performed largely outside of class.

Appendix

Answer each of the following questions. Give data to support your answer, if possible. Turn in your data with your answers. Turn in one set of answers per group, please.

View A   View B

Fig. 1. Which view of the Moon is possible?

1. The Moon is full and it is at its highest point in the sky. Approximately what time is it?

2. I see the Moon both in the evening right after sunset and in the morning just before sunrise. What is the phase of the Moon?

3. If I looked for the Moon on February 14, when and where should I have looked and what would I have seen?

4. We know that the Moon's light is actually reflected light from the Sun, but *how* do we know? Can you prove it from your data?

5. The Sun and the Moon both seem to move across the sky. Does the Moon appear to move more quickly or more slowly than the Sun?

6. I am looking at the Moon shortly after sunset. Which of the views shown in Fig. 1 will I never see? Why not?

7. Approximately how long is it from one Full Moon to the next?

8. If the Moon is directly overhead at 9:00 tonight, where will it be at 9:00 tomorrow night?

9. The spinning of the Earth is partly responsible for the apparent motion of the Moon, as is the actual motion of the Moon. Is the Moon moving around the Earth in the same direction the Earth is spinning or in the opposite direction? How do you know?

10. If I told you that a solar eclipse would occur within the next month, on what day would you predict that it would occur?

Charting the Moon by Eye

Stalking Luna's surface features with the naked eye requires timing, precision, and skill.

by Edmund A. Fortier

Imagine a crisp, clear morning just before dawn. The waning gibbous moon, scarcely past full, still rides high in the western sky. Bright enough to cast shadows through most of the night, the moon now is a pale disk, its surface textured with soft, delicate detail. You think you recognize some of the more prominent features and wonder how many others might be visible to a sharp-eyed observer.

The answer, surprisingly, is quite a few. The moon's diameter (2,160 miles) and average distance from Earth (239,004 miles) combine to yield an image that spans half a degree of sky. Although this is large compared to the angular width of other solar system objects, it's smaller than many people realize — about the size of an aspirin held at arm's length. Yet in terms of discernible surface features, the moon offers about as much detail to the unaided eye as does the planet Mars at opposition when viewed through a medium-size telescope.

Success in identifying lunar surface features, particularly subtle ones, doesn't depend entirely on good observing conditions and visual acuity. You must also know what to look for and where to find it. That's an advantage early moon watchers didn't have, and it affected the accuracy of their observations. The only known moon map that predates the telescope was made by the English scientist William Gilbert. Although Gilbert was a careful observer, his placement of the lunar maria only roughly approximates their actual location. Early telescopic drawings of the moon tend to be equally imprecise — or worse. None of the features recorded by Galileo, for example, correspond to any known lunar formations.

Therefore, in order to detect faint surface features, you must first be thoroughly familiar with the moon and its topography. The most brilliant region of the lunar disk is the southern highlands, a crater-strewn area that is virtually featureless to the unaided eye. In contrast, the northern and western hemispheres are dominated by a number of well-defined dusky markings. These dark areas are the maria, vast basaltic plains that formed about 3.8 billion years ago when lava from the moon's interior oozed to the surface and flooded gigantic impact basins. The most elusive details are generally found along the boundaries of these two regions, where tones and shadings are enhanced by contrast.

The maria are the moon's largest surface features and consequently the easiest to see. Close to the eastern limb, at about two o'clock, is the small, detached spot that is Mare Crisium. Just to its west lies Mare Tranquillitatis and above that, Mare Serenitatis. Together, these two lava plains appear as an elongated dusky patch. Two smaller "seas" — Mare Fecunditatis and Mare Nectaris — extend from either side of Tranquillitatis's southern rim and give it a forked appearance. The Fecunditatis branch is the more prominent of the two, but you should be able to detect the Nectaris branch as well.

West of Serenitatis is Mare Imbrium, the largest impact structure on the moon. Clearly defined, this feature is separated from its neighbor by the thin curve of the Apennine and Caucasus Mountains. Its nearly circular shape and 900-mile diameter make it visually distinctive. South of Imbrium, and extending along much of the western side of the moon, is the vast expanse of Oceanus Procellarum. At its most southern extent you'll find two small dusky lobes, Mare Humorum and Mare Nibium. Skirting the northern edge on the moon is a narrow strip of Mare Frigoris. Despite its faintness, this feature is visible even to casual moon watchers as an indistinct line just above the Imbrium basin.

Now that you're familiar with the moon's overall appearance, it's time to put your observing skills to the test. The accompanying map plots 12 naked-eye objects in increasing order of difficulty. The

2. THE EARTH AND THE MOON: The Moon

Can you pick off Pickering's list?

1. Bright surroundings of Copernicus
2. Mare Nectaris
3. Mare Humorum
4. Bright surroundings of Kepler
5. Region of Gassendi
6. Notch in the Mare Tranquillitatis-Plinius region
7. Mare Vaporum
8. Light area around crater Lubiniezky
9. Sinus Medii
10. Shaded area near the walled plain Sacrobosco
11. Dark spot at the foot of the Apennines
12. Riphaeus Mountains

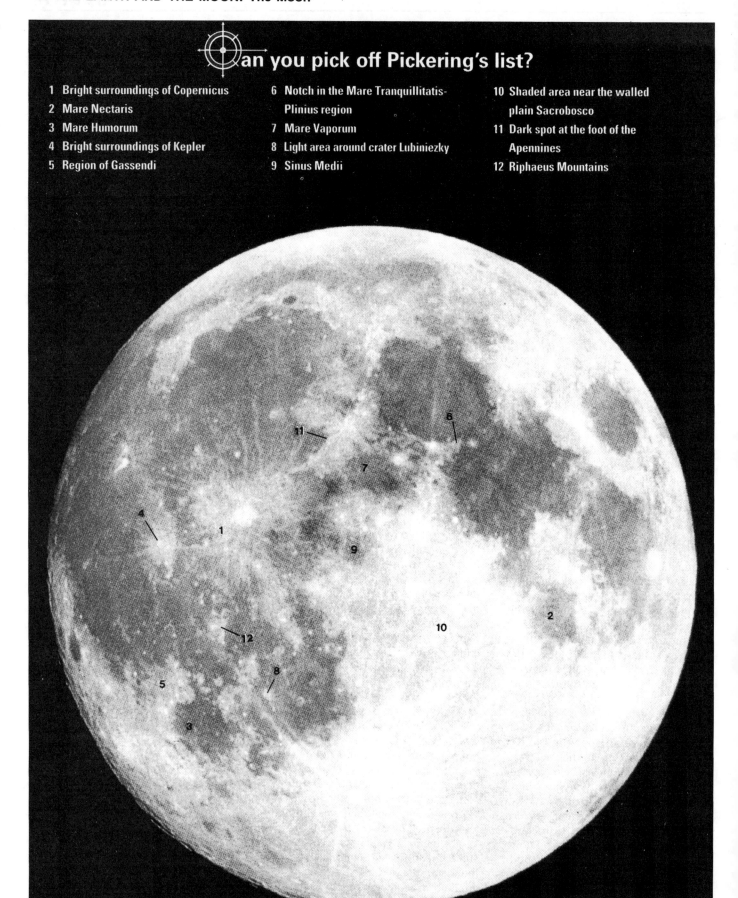

Ken Jents

list, known as Pickering's Dozen, was devised by William H. Pickering. He was brother and assistant to E. C. Pickering, longtime director of Harvard College Observatory, and used the list to test the eye's ability in discerning faint detail. While you might not spot them all, remember that many of these features can only be seen during evening and morning twilight, when glare from the moon is least intrusive.

First on Pickering's list and the easiest of the 12 to see is the crater Copernicus. Measuring 56 miles in diameter, it's located about midway between Mare Imbrium and Mare Nubium. Besides being the single finest example of a lunar crater, Copernicus also possesses a magnificent system of rays, surpassed only by the Tycho system. Look for it at full moon, when the rays are most prominent and the entire area appears as a bright spot against the gray expanse of eastern Oceanus Procellarum.

Next come Mare Nectaris and Mare Humorum. Mare Nectaris is roughly rectangular in shape and measures 180 miles across. You should be able to see it without too much trouble as a distinct spot below Mare Tranquillitatis. Mare Humorum is circular and somewhat larger (about 286 miles across), but it also lies close to the moon's western limb and due to foreshortening, appears oval. Its well-defined borders and dark floor make it only slightly less conspicuous than Mare Nectaris.

Pickering's fourth test feature is the crater Kepler, located just west of Copernicus. This 22-mile-wide object would be impossible to detect were it not for the brilliance of its ray system at full moon. Expect to see a smaller, dimmer version of Copernicus. Not far to the south, at the northern edge of Mare Humorum, lies the region of Gassendi. A walled plain some 70 miles in diameter, Gassendi is part of a larger area of cratered terrain that divides Mare Humorum from Oceanus Procellarum. Can you detect a brightening along the mare's northern shore?

Turn next to the Mare Tranquillitatis-Mare Serenitatis region. To the naked eye there is no clear division between these two seas, which seem to merge into a single oblong patch of gray. Look for a tiny notch about midway along the western edge. This indentation marks the southern rim of Serenitatis and is actually a blend of two features: the headland range of Promontories Acherusia and the 30-mile-wide crater Plinius. The peaks of the former rise almost 5,000 feet above Mare Serenitatis; Plinius, just to the east, lies on the northern shore of Tranquillitatis.

Once you've located the Plinius region, you're half-way home, but don't get too confident. As you might expect, the last six objects on Pickering's list are considerably more difficult — smaller, fainter, and less defined.

If your eyesight is good, and observing conditions cooperate, you should be able to glimpse the smallest of the naked-eye seas, Mare Vaporum, as a tiny smudge southeast of Imbrium. Try to spot the light area surrounding the crater Lubiniezky, on the northwestern shore of Mare Nubium. Lubiniezky is located in the region bright with streaks and rays—including one of Tycho's, which passes just to its southwest—this is best seen during a full moon.

You'll need excellent vision and possibly a good deal of practice to snare the next two objects. The first is Sinus Medii, a small marial bay a little below Mare Vaporum and on the eastern shore of Oceanus Procellarum. It will appear, if at all, as a very small, dark spot. If it seems too conspicuous, your eyes may be misleading you; it's easy to perceive the Mare Vaporum-Sinus Medii area as a single shadowy marking. Next on Pickering's scale is the faintly shaded area near the walled plain Sacrobosco. Situated in the lunar highlands southwest of Mare Nectaris, this feature lies just above one of Tycho's bright rays. You're a keen-eyed observer if you can detect this small variation in surface tone.

Pickering's eleventh test feature may be beyond your reach. Still, whenever observing conditions are ideal, make a point of scanning the Imbrium side of the Apennines for a tiny dip in brightness. That dip is a small portion of the Imbrium floor bound on either side by brighter surface material. If you don't find it at first, repeated attempts may bring success. Persistence, however, may not work for the last feature on Pickering's list, the Riphaeus Mountains. Pickering conceded that this small range in southern Oceanus Procellarum was probably below the threshold of human vision.

The moon is a challenging naked-eye object. To be sure, the larger maria are easy to spot, but searching out Luna's more elusive surface features can push your observing skills to the limit. With practice and persistence, your ability to detect these faint shadings and subtle tones will improve. Perhaps one day you'll even glimpse the Riphaeus Mountains.

Edmund Fortier's last article in ASTRONOMY *was* The Mars That Never Was *(December 1995).*

The Solar System

The Sun (Articles 19 and 20)
The Planets (Articles 21–26)
Comets (Articles 27–29)
Meteorites (Articles 30 and 31)
Asteroids (Articles 32 and 33)

Moving outward in scope with what we know, we cover the solar system next. Along with the Sun and planets, the "minor" players such as comets, meteorites, and asteroids are included.

The Sun, so central to our lives, often seems bypassed by students for more exotic objects such as black holes. Of course, the study of the Sun plays a central role in astronomy, if for no other reason than that it is the closest star to us. The unit begins with Kenneth Lang's two-part article, "Unsolved Mysteries of the Sun, Part 1 and Part 2." Lang focuses on several remaining unanswered questions about the Sun.

Study of other planets in our solar system depends heavily on the use of models. "Atmospheric Dynamics on the Outer Planets" by Peter Gierasch describes this mathematical model that is used to explain the behavior of the atmospheres on the outer planets. Then, Jeff Kanipe in "Planet in a Bottle" deals with another type of model—physical models. Physical models to explain the atmospheric dynamics of Jupiter and to simulate the winds on Mars in a wind tunnel are described.

The next three articles (Broad, Svitil, and Kluger) deal with a fascinating area of inquiry—the search for life on Jupiter's moon Europa (see "Magnetic Fields on Distant Moons Hint at Hidden Life," "Life in a Deep Freeze?" and "Water World"). The articles portray science as it happens. Data is gathered and analyzed, and the interpretations of the data offer intriguing clues that lead to further questions. We can certainly expect more developments in this area.

Comets have played a major role in the development of myths and legends. Bradley Schaefer's article "Comets That Changed the World," describes how human history has been influenced by these celestial visitors. In "The Comet's Gift: Hints of How Earth Came to Life," William Broad explains one currently debated topic: the role of comets in the origin of life on Earth. Were organic molecules "seeded" on Earth by comets?

"Bits of Mars and Pieces of the Moon" destroys the myth that all meteorites come from "space." According to this report, some meteorites come from the Moon and neighboring planets. The most famous meteorite of all, the Martian meteorite exhibiting apparent traces of primitive life, is addressed by J. Kelly Beatty, in "Life from Ancient Mars?" Evidence for an organic interpretation is presented, and alternatives are discussed.

Planetary impacts are not uncommon in Earth's geologic past. The one planetary impact studied most by geologists is the one that occurred approximately 66 million years ago at the end of the Cretaceous period. The impact is thought to have caused the extinction of the dinosaurs. In Ron Cowen's article, "The Day the Dinosaurs Died," a possible scenario for such an impact is described in detail. Then, "A Cosmic Collision" by John Lewis discusses the possibility of such an impact occurring during our lifetime. Lewis also provides alternatives to consider if such an impact proves imminent.

Looking Ahead: Challenge Questions

List five "mysteries" of the Sun that astronomers still need to solve.

Describe two physical models developed to explain or illustrate some aspect of the planets in our solar system.

What evidence suggests that there may be life on Europa?

How are organic molecules thought to form in comets?

What evidence supports the interpretation that the structures found in the Martian meteorite are organic in origin? What evidence indicates the structures are inorganic?

What possible procedures exist to prevent a comet or asteroid from striking Earth?

UNIT 3

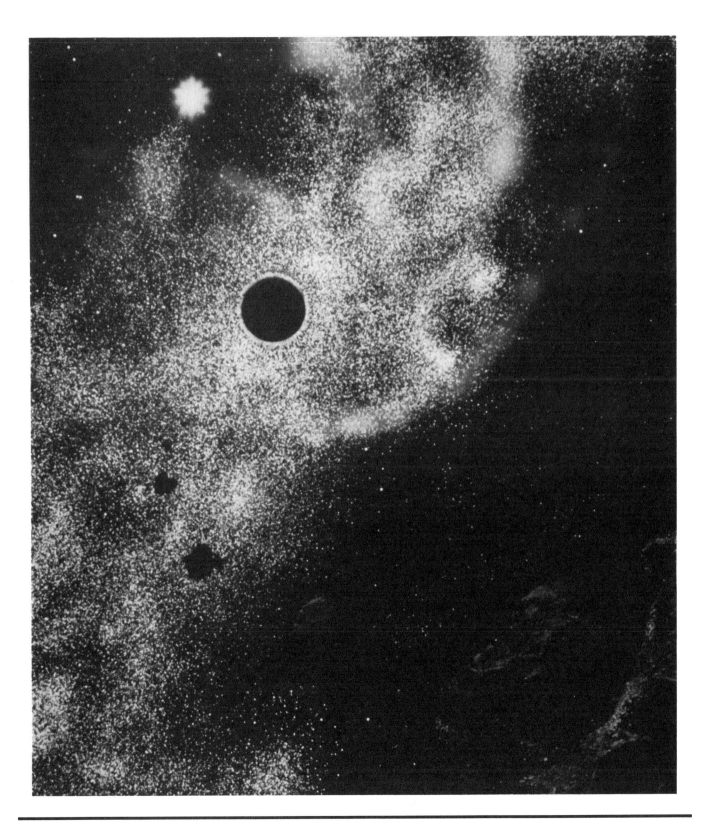

Unsolved Mysteries of the Sun—Part 1

Many problems confound our understanding of how the Sun works. But powerful new instruments observing the Sun night and day, from its unseen depths to its million-degree atmosphere, are providing new clues to long-standing mysteries.

Kenneth R. Lang

ASTRONOMERS once thought of the Sun and stars as simple objects, easily understood. They were smooth spheres of gas governed by simple laws of gravity, temperature, and pressure. The hydrogen fusion reactions powering most of them were worked out in 1939. Detailed models of stellar interiors followed in the 1950s and 1960s. The centers of stars, astronomy textbooks have long declared, are better known than the center of the Earth.

As we've examined the Sun more closely, however, it has turned out to be a bewildering turmoil of complex phenomena — a churning, quivering, ever-changing body that shows unexpected features and behavior at every scale. But now, after creating ever more problems for solar physicists, the Sun may soon yield up many of its secrets. Major new instruments in space, on the ground, and beneath the Earth's surface are poised to clarify several crucial solar mysteries.

Such instruments have extended our gaze from the visible solar disk deep into the Sun's interior and out through its tenuous atmosphere. The Sun is nearly transparent to neutrinos, particles that emerge directly from the nuclear-fusion reactor of the solar core. Surface oscillations, the signature of ultralow-frequency sounds trapped within the Sun, probe the rest of the solar interior. The Sun's outer atmosphere is routinely imaged by ultraviolet and X-ray telescopes lofted above the Earth's obscuring air.

Our new views of these domains have raised many questions. They include the enigmatic solar-neutrino problem, the Sun's strangely complex internal rotation, a crisis in the dynamo theory for generating the Sun's magnetic phenomena, unexplained explosions in the solar atmosphere, and the unknown mechanisms that heat the million-degree corona and accelerate the solar wind.

This is a decisive time for solar physics. A powerful new solar-neutrino detector, Superkamiokande in Japan, began taking data on April 1st, and the much-anticipated Sudbury, Ontario, neutrino observatory will begin operation in early 1997. The Solar and Heliospheric Observatory, or SOHO, a $1 billion spacecraft operated jointly by the European Space Agency and NASA, was launched on December 2, 1995, and is now observing the Sun nonstop. The 2-ton SOHO, the most ambitious solar observatory ever built, will provide at least a 2½-year span of coordinated studies extending from the center of the Sun nearly to the Earth. And ground-based telescopes in the Global Oscillation Network Group, or GONG, have just begun observing solar oscillations around the clock.

In this article I will describe what we are learning about the Sun's innermost and outermost regions. Next month's installment will cover helioseismology, the study of solar oscillations, and what they tell us about the regions in between.

THE SOLAR NEUTRINO PROBLEM

The Sun is immense; it has 109 times the diameter of Earth and 333,000 times the mass. At its center the solar gas is tremendously compressed by the weight of the material above. The temperature is 15.6 million degrees Kelvin and the gas density is 151 grams per cubic centimeter, more than 13 times the density of lead.

This is hot and dense enough for nuclear fusion reactions to occur — albeit slowly. The very fastest moving hydrogen nuclei, or protons, occasionally collide head on and fuse to form hydrogen-2 (deuterium). This starts a chain of events that ends up creating helium-4 and generating the Sun's heat. Outside the solar core, where the overlying weight and compression are less, the gas is cooler and thinner, and nuclear reactions cannot occur.

The reactions in the Sun's central furnace create prodigious quantities of neutrinos. These tiny particles travel at the speed of light almost unimpeded through the Sun, the Earth, and nearly any amount of ordinary matter as though it were not there at all. Massive neutrino detectors, tanks of fluid buried deep underground so that only neutrinos will

---

Kenneth R. Lang is professor of astronomy at Tufts University. His recent popular book, Sun, Earth and Sky *(Springer-Verlag, 1995), summarizes all aspects of the Sun and its interaction with Earth.*

19. Unsolved Mysteries of the Sun—Part 1

The Superkamiokande neutrino detector under construction deep below a mountain in Japan. When finished, the huge stainless-steel vessel (40 meters tall and 40 wide) was filled with 50,000 tons of ultrapure water. Lining its walls are 13,000 photomultiplier tubes, each 20 inches wide. They detect the telltale pulse of faint blue light made when a single neutrino from the heart of the Sun collides with an electron in a water molecule. Courtesy Yoji Totsuka, Institute for Cosmic Ray Research, University of Tokyo.

reach them, have managed to catch and count just a few of the estimated 60 billion solar neutrinos that pass through each cubic centimeter of the Earth each second. These captured neutrinos have opened a window directly onto the Sun's energy-generating core.

By finding solar neutrinos in roughly the predicted numbers, four pioneering neutrino detectors have now demonstrated that the Sun is indeed heated by hydrogen fusion. However, for almost 30 years these experiments have been finding only one-third to one-half the number of neutrinos that theory says they should, a discrepancy known as the solar-neutrino problem.

There are two possible explanations. Either we don't really know how the Sun and stars create their energy, or we don't understand neutrinos.

If the center of the Sun were just 1 million degrees cooler, nuclear reactions would slow down and produce fewer neutrinos, resolving the problem. However, this situation would have to be only temporary. Heat generated at the Sun's core takes about 170,000 years to work its way to the surface. If the nuclear reactions remained weak for anywhere near this time, the Sun would cool off. Yet it seems no such dimmings have occurred throughout geologic history.

In any case, after fine-tuning their models of the solar interior, astronomers have shown that the shortfall cannot be made to go away so easily. According to John Bahcall of the Institute of Advanced Study at Princeton, the relative numbers

The Sudbury Neutrino Observatory in Ontario, due to begin operations in early 1997. The central flask, made of clear acrylic plastic, will contain 1,000 tons of heavy water so that all three "flavors" of neutrinos can be detected — if they indeed arrive from the Sun. The flask will be surrounded by 10,000 photomultipliers. More than a mile of rock above the detector will block cosmic rays, and 7,000 tons of ordinary water surrounding the flask will block background radiation from the rock. Courtesy Los Alamos National Laboratory.

of solar neutrinos detected at various energies cannot be reconciled with *any* reasonable solar model. This suggests, he says, that "new physics is required to explain the observations."

Perhaps the neutrinos have an identity crisis on their way to us from the center of the Sun. Neutrinos come in three varieties, associated with the electron, the muon, and the tau particle. The electron neutrino is the type generated in the Sun's core and is the only type that solar-neutrino detectors can respond to as of now. Could some of the electron neutrinos be switching to another type during their 8½-minute journey from the center of the Sun to Earth, thereby escaping detection? One appealing idea is that neutrinos naturally "oscillate" between all three states. This would neatly account for why we find only about a third of the expected numbers. In one version, the switchover is modulated by the Sun's matter as the neutrinos pass through it.

Any such behavior would have profound implications for particle physics. In order to change states, a neutrino must have at least a tiny amount of substance, or rest mass. However, neutrinos are assumed to be completely massless in the present theory uniting the electromagnetic force with the "weak" force that governs neutrino interactions with other particles. The solar-neutrino results therefore may require a new physics of the sort that Bahcall implies, superseding the current

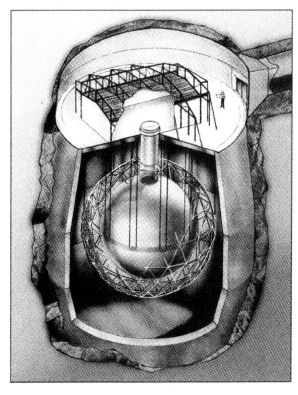

3. THE SOLAR SYSTEM: The Sun

electroweak model. If the oscillations tell us the actual neutrino mass, as they should if well observed, this in turn could specify the exact energy at which all the forces of nature become interchangeable in a Grand Unified Theory.

NEW NEUTRINO DETECTORS

Experiments currently or soon to be under way should settle the question of whether some neutrinos switch identities en route to Earth. These underground detectors include the Japanese Superkamiokande that began catching neutrinos last April, and the Sudbury Neutrino Observatory in Canada now nearing completion.

The Superkamiokande experiment, located in a mine deep under the Japan Alps to shield it from cosmic rays, uses 50,000 tons of ultrapure ordinary water. A previous 3,000-ton version verified that neutrinos are coming from the *direction* of the Sun. Occasionally a passing neutrino knocks a high-speed electron off of a water molecule. The electron moves through the water faster than light travels in water, generating an electromagnetic shock wave and a light cone of so-called Cherenkov radiation, somewhat in the manner a supersonic aircraft creates a sonic boom. Thousands of light detectors lining the water tank measure the axis of the light cone, which tells the direction of the incoming neutrino.

Superkamiokande will provide a 30-fold increase in the observed rate of neutrino-electron collisions. The neutrinos' energy spectrum may be measured via the energies of the scattered electrons. A distortion in the energy spectrum could reveal the way neutrinos transform from one type to another — if they do.

The Sudbury Neutrino Observatory, or SNO, is buried 2,070 meters (6,800 feet) underground in a working nickel mine near Sudbury, Ontario. The detector is a huge, spherical vat holding 1,000 tons of heavy water, a form of H_2O that contains deuterium, or heavy hydrogen (an atom whose nucleus consists of a proton and a neutron). The vat of heavy water is surrounded by a 7,000-ton jacket of ordinary water that shields it from weak natural radiation in the underground environment. The overlying rock blocks out cosmic rays.

The SNO should detect about 10 solar neutrinos per day, about 50 times the rate of existing experiments other than Superkamiokande. Unlike the ordinary-water detectors, which are sensitive only to electron neutrinos, Sudbury's heavy water will be sensitive to all three types of neutrinos, because they can knock the neutron from a deuterium nucleus to produce a different light signal. Thus we should soon know whether solar neutrinos change identity or not.

Other experiments have already put some limits on the neutrino's rest mass. The fact that high-energy neutrinos from Supernova 1987A arrived within just a few seconds of each other, after traveling 160,000 light-years from the Large Magellanic Cloud, sets the upper limit to the neutrino mass at about 16 electron volts (*S&T:* October 1988, page 348). By comparison the lightest known particle, the electron, has a rest mass of 511,000 eV.

An apparent breakthrough came in May, when scientists at the Los Alamos National Laboratory in New Mexico announced that they had observed muon antineutrinos turning into electron antineutrinos in a high-energy particle experiment. They claimed that at least one of these neutrino flavors seems to have a rest mass of at least 0.2 eV and possibly much more. This result firms up a claim made at Los Alamos last year, but it still needs further confirmation. The solar neutrino problem could be resolved with as little as 0.003 eV or less.

If neutrinos do have some mass, large numbers of them left over from the Big Bang might account for some of the invisible "missing mass" of the universe. They would be "nonbaryonic hot dark matter," just the kind that many cosmologists want (*S&T:* October 1994, page 28). Their collective gravity could influence the expanding universe's early history and ultimate fate — though cosmologists really wanted 25-eV neutrinos, which would have been heavy enough to balance the universe between eternal expansion and recollapse. In any case, a paradox that began at the heart of the Sun could have far-reaching consequences.

We now turn our attention from the Sun's innermost depths to its outermost atmosphere.

WHY IS THE CORONA HOT?

The Sun's rarefied, outermost atmosphere poses a mystery that has defied solar astronomers for even longer than the neutrino problem. How can the corona, so beautiful and familiar a sight at total eclipses, manage to exist?

The Sun has no surface; its gas just becomes more tenuous the farther out you go. The visible sharp edge of the photosphere is something of an illusion. It is merely the level at which the gas becomes thin enough to be transparent. By convention, the layers above this level are called the solar atmosphere. The corona is its most extensive part.

Somehow the corona is heated to a few million degrees Kelvin. This is one of the most fundamental, unsolved paradoxes of solar physics. The photosphere is hundreds of times cooler, even though it's closer to the Sun's center. Heat should not flow outward from a cooler to a hotter region. After all, when you sit far away from a fire, it warms you less.

Clearly some mechanism is carrying energy up into the corona and dumping it there. Despite more than a half century of investigation, this mechanism remains unknown.

Sunlight can't be it; sunlight passes right through the transparent corona without depositing substantial energy into it. In 1949 solar physicists proposed that sound waves from the turbulent photosphere would turn into shock waves in the thin corona and heat it, but observations in the 1970s from the orbiting Skylab showed that very little acoustic energy of any sort can get that high. Magnetic plasma waves (Alfvén waves) were then proposed, but these seem to pass right out of the corona without depositing much energy on their way through.

A step toward locating the heating process is to study the Sun at ultraviolet (UV), extreme ultraviolet (EUV), and X-ray wavelengths. These forms of radiation are emitted by very hot material and offer direct looks at the Sun's upper atmosphere.

Because this radiation is absorbed by air, it must be observed from telescopes in space. One very productive instrument has been the X-ray telescope aboard the Japanese Yohkoh satellite. Now we are getting a more comprehensive perspective with the ultraviolet and extreme ultraviolet instruments aboard SOHO, which is stationed 1.5 million kilometers Sunward from Earth and takes images continuously, 24 hours a day every day of the year. Such telescopes produce spectacular images of the Sun's hot upper atmosphere seen against the photosphere, which is relatively dark at these wavelengths.

At the high coronal temperatures, certain spectral lines are emitted by ions that are quite sensitive to their surroundings. Some spectral lines act as thermometers, yielding the temperature where they are formed. Others are sensitive to the local density. Velocities of moving material can

19. Unsolved Mysteries of the Sun—Part 1

be inferred from Doppler shifts and line broadening.

Observations in spectral lines that originate at different heights can be used to focus on different layers of the atmosphere. This way we can build up a three-dimensional understanding of the bulk motions, physical evolution, and wave motions in the different magnetic structures pervading the solar atmosphere. The suite of instruments aboard SOHO is performing such experiments right now, giving solar astronomers hope that the coronal-heating problem will be solved soon.

WHAT CAUSES THE SOLAR WIND?

The corona is forever expanding into interplanetary space, filling the solar system with a thin, perpetual outflow of ionized matter called the solar wind. At a certain distance from the Sun, where the solar gravity weakens, the gas pressure from the million-degree corona overcomes the gravitational attraction, producing a wind that accelerates away to supersonic speeds. As the outer corona disperses, it must be replaced by gases welling up from below to feed the eternal wind.

Spacecraft measurements show that the outflow has a fast component and a slow one. Some of the wind has a speed of about 400 kilometers per second; the rest travels about twice as fast. The slow-speed wind is an expected consequence of the corona's high temperature. But no one really knows what gives the high-speed wind its additional push.

The Ulysses spacecraft conclusively demonstrated that the high-speed component of the solar wind escapes from holes in the corona near the Sun's poles (*S&T:* March 1996, page 24). At least this was the situation near the current minimum of the Sun's 11-year magnetic-activity cycle. Magnetic lines of force in the coronal holes stretch straight outward, providing a fast lane for the high-speed wind.

The persistent coronal holes show up as large, irregular dark areas in EUV and X-ray images. The slower wind originates near the solar equator (at least around activity minimum), but its exact source

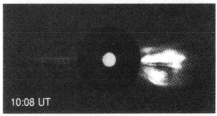
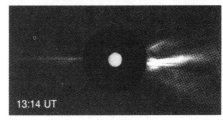
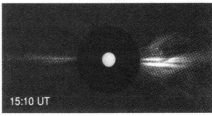

A coronal mass ejection, photographed in white light over the course of 8 hours by SOHO's Large Angle and Spectrometer Coronagraph. A black occulting disk blocks the glare of the Sun itself, whose size is represented by the circle at center. The roughly billion-ton bubble of hot gas at right grew larger than the Sun itself in just a few hours. At the same time streaks of gas appeared on the opposite side of the Sun; the two events may have been tied together magnetically. Courtesy Guenter Brueckner, Naval Research Lab.

remains a mystery to be solved by future observations.

Two instruments aboard SOHO use occulting disks to mask the Sun's glare the way ground-based coronagraphs do. They provide continuous side views of the corona from just above the photosphere out to 30 solar radii. These instruments are directly examining the regions where the corona is heated and the solar wind is accelerated. Both are providing new insight into the wind's origin and acceleration.

ENERGIZING SPACE NEAR EARTH

SOHO is stationed just outside the Earth's magnetic field, where another set of instruments is analyzing, *in situ*, the charged particles of the solar wind. When these data are combined with those from other instruments, the wind will be studied all the way from its source to the Earth.

The Earth's magnetic cocoon, or magnetosphere, is constantly buffeted and reshaped by the variable solar wind, and the gusty interplanetary "weather" can affect us significantly. By disturbing the Earth's magnetic field it can produce geomagnetic storms, create auroras, disrupt radio navigation and communication systems, endanger astronauts, destroy satellite electronics, and cause power blackouts on Earth. Already researchers are seeing signs in SOHO's data of the Sun preparing itself for coronal mass ejections days before they occur, offering the hope of improved forecasts. SOHO's investigations of the origin, acceleration, and propagation of the solar wind will have a direct impact on human activity.

In Part 2 (see next article), peeling apart the Sun's depths with SOHO and GONG will be presented.

Further Reading

Bahcall, John, et al. **"Progress and Prospects in Neutrino Astrophysics."** *Nature* 375, 29 (May 4, 1995).

Lang, Kenneth. **Sun, Earth and Sky.** Springer-Verlag, 1995 (Available from Sky Publishing Corp.)

Marsden, Richard G., and Edward J. Smith. **"Ulysses: Solar Sojourner."** *Sky & Telescope,* March 1996, 24–31.

SOHO information on the World Wide Web: http://sohowww.nascom.nasa.gov and http://umbra.nascom.nasa.gov/images/latest.html

Unsolved Mysteries of the Sun—Part 2

Kenneth R. Lang

THE STAR on which our lives depend, the enormous central body that contains more than 99.8 percent of the solar system's mass, is more complex and inscrutable than astronomers once thought. The Sun is full of mysteries — but some of them are nearing solution.

In [Part 1 (see Article 19)] I discussed the Sun's energy-generating core, which presents us with the solar-neutrino problem, and the Sun's outer atmosphere, where unknown processes heat the multimillion-degree corona and produce the solar wind. This second and final installment will discuss the regions in between, from the depths of the Sun to its visible surface (photosphere).

THE SUN'S PULSE

Solar astronomers have developed an ingenious and powerful way to see into the Sun's innards: by analyzing the weak, complex throbbing motions that cover its visible disk. These oscillations, discovered in the early 1960s, were shown by 1968 to result from pressure waves — sounds — that echo and resonate through the Sun's interior.

Any sound wave moving inside the Sun is refracted (bent) up toward the surface by the increasing speed of sound at greater depths. On striking the surface and rebounding back down, the wave causes the photosphere to move slightly up and down. These oscillations are extremely subtle. They typically cause the photosphere to rise and fall by less than 25 meters and to change temperature by just 0.005° Kelvin. Nevertheless they can be detected from tiny, periodic Doppler shifts in a well-defined spectral line — shifts that result from velocity changes of just centimeters per second — or from minuscule but regular variations in the Sun's total light output.

The loudest sounds in the Sun are extremely low-pitched, as befits an object so large and massive. The dominant frequencies cluster around one vibration per five minutes (0.003 hertz). This is 12½ octaves below the lowest note audible to humans, 20 vibrations per second (20 Hz). Many of the Sun's other notes are even lower.

If we could hear the Sun's infrabass ringing, the strong, pure tones would be superposed on a rumbling or roaring of about 10 million other, closely spaced frequencies. Every one of them has a unique path of propagation through the interior. Waves of different frequencies descend to different depths before being refracted back up to the surface (see the diagram next page). Observations of many different *oscillation modes*—resonances set up by the waves—can be combined to probe different depths of the Sun and calculate the temperatures, pressures, motions, and chemical compositions within each.

Because this technique is similar to analyzing the Earth's internal structure with seismic waves, this study is known as helioseismology (*S&T:* November 1987, page 470). When applied to other stars, notably white dwarfs, it is called astero-seismology (*S&T:* May 1995, page 14; February 1995, page 14; and April 1992, page 374).

Helioseismologists bring order to the cacophony by sorting out its harmonic components, or standing-wave modes. The distance between the oscillation nodes, for example, is specified by an integer, l, called the spherical harmonic degree. This indicates the number of times a sound wave bounces off the surface in one trip around the Sun. Sound waves of low degree (low values of l) are reflected only a few times during each circuit and travel deep within the Sun. Those of high degree (large l) hit the surface often and are confined to shallower depths, as illustrated at right.

Observations of the 5-minute oscillations have been compared with theoretical models to check the density, temperature, and composition at many levels inside the Sun. For example, early measurements of the sound speed were higher than specified by the models. This led to improvements in our understanding of the internal opacity that blocks the flow of radiation and governs the Sun's temperature structure.

A small but definite change in the observed sound speed has pinpointed the lower boundary of the convection zone, the turbulent outer region of the solar interior where circulating motions transport heat much like boiling water in a kettle. Convection happens where the Sun's material is too opaque for radiation to work its way outward most efficiently by diffusion. Just where this division occurs tells us a lot about the interaction of matter and radiation in stellar interiors. Helioseismology has located the base of the convection zone at 71.3 percent of the radius of the photosphere. Below this level lies the "radiative zone," where the Sun's gas is very calm and still and radiation "random walks" its way out from the energy-producing core.

Since sound waves move faster through higher-temperature gas, their speed can be used as a thermometer — and thus might bear directly on the solar-neutrino problem discussed last month. They might tell if something is cooling the

---

Kenneth R. Lang, a professor of astronomy at Tufts University, is the author of Sun, Earth and Sky *(Springer-Verlag, 1995), a popular-level book covering the Sun and its interactions with Earth.*

Sun's core below the expected 15.6 million degrees K and thus slowing neutrino production.

Since sound speed also depends on a gas's average molecular weight, measurements of the most deeply penetrating waves should also show just how sharply defined the Sun's helium-rich core really is. Already there are hints that the helium may be spread out more diffusely than expected, which would also affect nuclear reaction rates. If these results are confirmed, the core may not have a well defined outer edge.

DAYS WITHOUT NIGHT

A major obstacle to obtaining precise solar-oscillation data is the Earth's rotation, which keeps us from observing the Sun around the clock. The nightly gaps in the data create background noise that hides all but the strongest oscillations, especially at low frequencies. As a result, helioseismology has provided accurate results only about 80 percent of the way down to the Sun's center, corresponding to the outer 66 percent of the solar mass. It does not yet definitively probe the energy-generating core, which is usually considered to be the central 25 percent of the Sun's radius.

By watching the Sun continuously for days, weeks, and even months at a time, we can reduce the noisy confusion in the observed oscillations and probe the interior to greater depths and with unprecedented accuracy. This is now being done both from space and from the ground.

The Solar and Heliospheric Observatory (SOHO) spacecraft, launched last December, is the most advanced Sun observatory ever built. It has stationed itself about one percent of the way from Earth to the Sun, circling the Lagrangian L_1 point. From this vantage SOHO has a continuous, uninterrupted view of the Sun, 24 hours a day, 365 days a year.

Two of SOHO's helioseismology instruments, named GOLF and VIRGO, are beginning to garner long, undisturbed views of global oscillations. They will study sound waves with degrees l of 0 to 7, those able to penetrate the deep solar interior. They might also provide the first unequivocal detections of deep gravity waves. These are slow sloshings of material that are mediated by gravity, like waves on the ocean, rather than by pressure as in the case of sound waves. They are expected to arise from density variations in the Sun's core and should have longer periods than the oscillations measured so far.

The third SOHO helioseismology instrument, dubbed SOI/MDI, observes sound waves with very high values of l that dive just a short distance below the surface and return quickly. These cannot be measured effectively from the ground; turbulence in the Earth's atmosphere (poor seeing) limits the necessary image sharpness. With its superior view from space, the SOI/MDI can effectively measure small-scale subsurface flows.

Spacecraft have by no means rendered ground-based helioseismology obsolete. The Sun is now being observed around the clock by a worldwide network of observatories, known by the acronym GONG for the Global Oscillation Network Group. They form an unbroken chain that follows the Sun as the Earth rotates. As with the former British Empire, the Sun never sets on GONG. At each site, imaging spectrometers measure very precise radial velocities at more than 40,000 points on the Sun's surface. Two

20. Unsolved Mysteries of the Sun—Part 2

On the Scene at SOHO

INVISIBLE solar rays travel at the speed of light to a spacecraft parked just outside the Earth's magnetic domain. Instruments transform the radiation into high-resolution images as often as once per second. These are downlinked to terrestrial antennas and processed at the SOHO mission's Experiment Operations Facility. Here a dedicated team of international scientists watches the Sun, night and day, from a room without windows.

Each morning solar astronomers gather here to discuss scientific protocols and the day's events. Computers around a long table offer nearly instantaneous recall of all that SOHO has seen since its mission began in December 1995. Sometimes there is an electric sense of excitement and urgency as new images provide initial clues to unsolved mysteries on the Sun. Then, after the collective decisions have been made, subgroups retire to several small rooms, each dedicated to a particular instrument's perspective on our home star.

The facility is open to any interested astronomer. Here you can find solar physicists from throughout the world working together with striking collegiality. They have already overcome too many obstacles to bother with bureaucratic or political difficulties.

The images move nearly instantaneously from the SOHO Experiment Operations Facility to the information superhighway, so that anyone with access to the World Wide Web can watch the volatile Sun from home or office practically in real time. Point your Web browser to http://sohowww.nascom.nasa.gov or to http://umbra.nascom.nasa.gov/images/latest.html. Already there have been more than a million connections from high schools, corporations, universities, and government institutions.

My only concern is one of speed without control. Web-based astronomy is like accelerating a familiar car to tremendous speed in a possibly unexpected direction. Indeed, so much data is being collected so fast that NASA can barely transmit it all back to Earth, much less analyze it. So serious users need the scientists at SOHO's center of operations to help reduce and interpret the raw images and ensure quality control. These people are keeping the rest of us on the right track, heading with awesome speed toward resolution of the Sun's fundamental mysteries.

KENNETH R. LANG

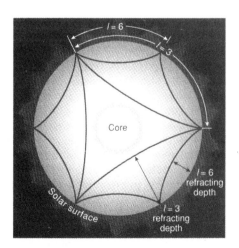

Sound waves in the Sun, like seismic waves in the Earth, do not travel in straight lines. The increasing speed of sound toward the Sun's center refracts inward-moving waves outward. At the same time, the Sun's surface reflects outward-moving waves back in. Shown here are ray paths for two fundamental oscillation modes (marked with their l values, which denote the number of times they hit the surface). The lower the l value, the deeper a wave penetrates.

75

3. THE SOLAR SYSTEM: The Sun

other helioseismology networks, which measure only the Sun's total brightness, have been operating for several years: the Birmingham Solar Oscillation Network (BiSON), a United Kingdom project started in 1981, and the French-led International Research on the Interior of the Sun (IRIS), which reached its full deployment of six instruments in 1994.

HOW FAST DOES THE SUN SPIN?

Sound waves reveal the motions of the Sun's material, such as inner convection currents and the Sun's deep internal rotation. Waves moving with the rotation will appear to move faster, like birds flying with the wind, so their measured periods will be shorter. Waves propagating against the rotation will be slowed and show longer periods. These opposite effects split an oscillation into a pair of close, but distinct, observed frequencies.

For over three centuries astronomers have known (from observations of sunspots) that the photosphere rotates faster at the Sun's equator than at higher latitudes, with a smooth variation in between. The Sun's sidereal rotation period ranges from 25.0 days at the equator to 27.8 days at ±40° latitude and even longer at higher latitudes. Helioseismology has recently shown, contrary to expectations, that this differential rotation persists right down to the base of the convection zone, 28.7 percent of the way to the Sun's center. At greater depths, the equatorial rotation slows down and the high-latitude rotation speeds up! The two rates become equal about halfway to the center. A little farther in, the rotation appears uniform and independent of latitude; that layer of the Sun rotates like a billiard ball or other solid body. There is no explanation for this bizarre rotational structure.

We know that young stars begin their lives spinning rapidly, since they conserve angular momentum while contracting from large interstellar clouds. An older star, like the Sun, has slowed with age. Astronomers have long assumed that the braking action is supplied by magnetic field lines carried outward by the stellar wind. They spin with the star's rotation like a giant pinwheel, flinging wind particles sideways from their outer parts. If this explanation is correct, the Sun ought to be braked from the outside in, and the interior should be rotating more rapidly than the outer shell. So far, this doesn't seem to be the case.

However, the picture is confused once you peer more than halfway down to the center. Some researchers report that the

SOHO'S SCIENTIFIC INSTRUMENTS

HELIOSEISMOLOGY

Instrument	Measurement	Team leader
GOLF	(Global Oscillations at Low Frequencies). Low-degree (l = 0 to 3) velocity oscillations.	Alan Gabriel (IAS, France)
VIRGO	(Variability of solar IRradiance and Gravity Oscillations). Low-degree (l = 0 to 7) oscillations in brightness, and the Sun's precise energy output.	Claus Fröhlich (PMOD/WRC, Switzerland)
SOI/MDI	(Solar Oscillations Investigation/ Michelson Doppler Imager). High-degree (up to l = 4,500) velocity oscillations	Philip Scherrer (Stanford Univ.)

SOLAR ATMOSPHERE

SUMER	(Solar Ultraviolet Measurements of Emitted Radiation). Temperatures, densities, and velocities in the chromosphere and corona.	Klaus Wilhelm (MPAe, Germany)
CDS	(Coronal Diagnostic Spectrometer). Temperatures and densities in the transition region and corona.	Richard Harrison (RAL, United Kingdom)
EIT	(Extreme-ultraviolet Imaging Telescope). Activity in the chromosphere and corona; full-disk EUV imaging.	Jean-Pierre Delaboudinière (IAS, France)
UVCS	(UltraViolet Coronagraph Spectrometer). Coronal electron and ion temperatures, densities, and velocities (1.3–10 solar radii).	John Kohl (SAO)
LASCO	(Large Angle Spectroscopic COronagraph). Coronal activity, mass, momentum and energy transport (1.1–30 solar radii).	Guenter Brueckner (NRL)
SWAN	(Solar Wind ANisotropies). Variations and inhomogeneities in the solar wind.	Jean-Loup Bertaux (SA, France)

SOLAR WIND IN SITU

CELIAS	(Charge, ELement and Isotope Analysis System). Mass, charge, composition, and energy distribution of particles (0.1–1,000 keV/e).	Peter Bochsler (Univ. of Bern, Switzerland)
COSTEP	(COmprehensive SupraThermal and Energetic Particle analyzer). Energy distribution of protons and helium ions (0.04–53 MeV/nucleon) and electrons (0.04–5 MeV).	Horst Kunow (Univ. of Kiel, Germany)
ERNE	(Energetic and Relativistic Nuclei and Electron experiment). Energy distribution and isotopic composition of ions (H–Ni, 1.4–540 MeV/nucleon) and electrons (5–60 MeV).	Jarmo Torsti (Univ. of Turku, Finland)

deep interior is spinning faster than the overlying layers; others find that the core rotates more slowly. These divergent opinions stem from inadequate measurements of low-degree modes. Only modes with l = 1 to 3, propagating nearly vertically, reach the central parts of the Sun. The rotational splitting of these oscillation modes is slight, comparable to their natural frequency width. This means the two components of a split frequency overlap, and the controversy may not be settled until SOHO and/or GONG obtain very long, uninterrupted stretches of low-noise data.

In the meantime, helioseismologists have their hands full studying rotation, flows, and turbulence nearer the surface, in the regions where the Sun's magnetism may be generated and sustained.

THE ORIGINS OF SOLAR ACTIVITY

There's no doubt about the new solar paradigm. The Sun is a churning caldron of activity, with ever-changing structures molded by turbulent, concentrated magnetic fields.

The Sun's gas is hot enough to be a *plasma* — an ionized gas, in which electrons are stripped from atoms and fly loose. Any plasma conducts electricity. When a conductor moves across a magnetic field, electric current is generated within it. This current creates more magnetism, which adds to the original magnetic field, reshaping it. Conversely, a moving magnetic field will tend to drag a conductor along with it. The upshot is that gas and magnetic lines of force are intimately locked together in the Sun, as they are in plasmas throughout the universe.

A plasma slides easily along magnetic lines of force, like beads on wires. But it resists moving sideways across the field. The study of how conducting gas be-

20. Unsolved Mysteries of the Sun—Part 2

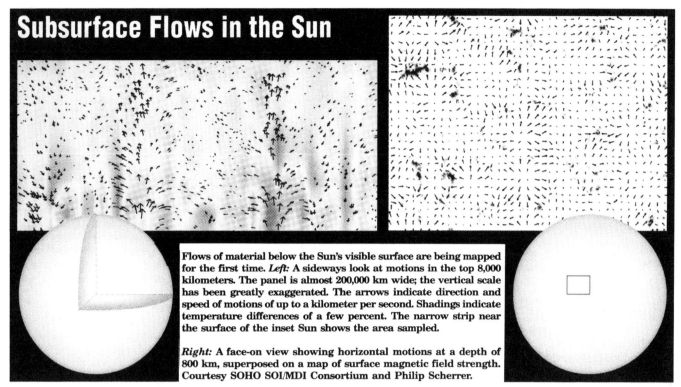

Subsurface Flows in the Sun

Flows of material below the Sun's visible surface are being mapped for the first time. *Left:* A sideways look at motions in the top 8,000 kilometers. The panel is almost 200,000 km wide; the vertical scale has been greatly exaggerated. The arrows indicate direction and speed of motions of up to a kilometer per second. Shadings indicate temperature differences of a few percent. The narrow strip near the surface of the inset Sun shows the area sampled.

Right: A face-on view showing horizontal motions at a depth of 800 km, superposed on a map of surface magnetic field strength. Courtesy SOHO SOI/MDI Consortium and Philip Scherrer.

haves in magnetic fields is called magnetohydrodynamics. It is a daunting, highly mathematical science that dominates almost everything that happens in the Sun's outer layers, on its visible surface, and in its atmosphere.

The Sun's ever-changing magnetism produces gaseous unrest on an awesome scale. Among many other effects, flares release stored magnetic energy equivalent to billions of nuclear explosions, raising the temperature of Earth-size regions tens of millions of degrees. Magnetically energized coronal mass ejections, or CMEs, hurl billions of tons of high-speed material into space. And yet astronomers do not know exactly how or why these two types of eruptions occur and cannot reliably predict them.

Solar flares and CMEs ebb and flow in step with the Sun's 11-year cycle of magnetic activity. This period is often called the "sunspot cycle" because it was discovered (in the early 1840s) from a periodic variation in the number of sunspots — blemishes that we now know to be caused by intense magnetic fields. During solar maximum the Sun is heavily peppered with big, complex, often bipolar spot groups concentrated into *active regions* — disturbed areas that emit intense X-rays and are prone to flares. At sunspot minimum, much smaller magnetized regions continuously well up all over the solar surface. We are in the minimum of a sunspot cycle in 1996.

The Sun's tangled surface magnetism originates from an unseen generator in the interior. Hot, circulating gases moving through the Sun's magnetic field generate electric currents that in turn amplify the magnetism, just as in a power-plant dynamo. (The same thing happens in the fluid interiors of many planets.) If we could understand how this *dynamo effect* works in the Sun, we might explain the cyclic winding up and relaxation of solar activity.

Unfortunately, today's dynamo models may be inconsistent with observed motions inside the Sun. The electrified gas seems to be moving up and down in deep, large-scale convection currents rather than circulating horizontally near the solar surface. For the time being, theorists have avoided this problem by placing the dynamo down closer to the Sun's center, where the differential rotation seems to change into a uniform spin.

High-resolution helioseismology should help show exactly what sunspots are and how solar activity originates. Sunspots absorb as much as half the power of the sound waves that propagate through them — why? The frequencies of the Sun's oscillations vary slightly with the 11-year cycle of sunspot activity. Does magnetism alternately stiffen and loosen the internal material?

The SOI/MDI package aboard SOHO measures subsurface motions, as well as surface magnetism, at up to a million points on the Sun's disk. Driven by heat from below, the electrified gas seems to be moving up and down in great, sweeping motions thousands of kilometers in extent. In some localized regions the solar gases seem to spiral down like water in an unplugged bathtub. In others they apparently compress the magnetic fields, pushing them together near the surface to confine sunspots.

The results accumulating from SOHO and GONG will provide a new picture of interior magnetism that is intimately linked to surface activity. This may supply the crucial missing link in our understanding of the Sun's multifaceted, ever-changing behavior.

Further Reading

Fleck, Bernhard, Vicente Domingo, and Arthur I. Poland, eds. **"The SOHO Mission."** *Solar Physics* 162: (1995): Nos. 1–2.

Gough, Douglas O., *et al.* **"Perspectives on Helioseismology"** and many related articles. *Science,* May 31, 1996, beginning on 1281–1283.

Harvey, John. **"Helioseismology."** *Physics Today,* October 1995, 32–38.

Lang, Kenneth. **"Sun, Earth and Sky"** Springer-Verlag, 1995. (Available from Sky Publishing Corp.)

Atmospheric Dynamics on the Outer Planets

Peter J. Gierasch

Jupiter and the other outer planets exhibit vigorous and complicated atmospheric flows (1). Averaged over time, the mean flows are alternating east-west jets with characteristic velocities of about 100 m s^{-1} on Jupiter and Uranus, and about 400 m s^{-1} on Saturn and Neptune. On Jupiter and Saturn there are several alternating jets in each hemisphere (see figure). On Uranus and Neptune there are westward equatorial jets and one eastward jet at high latitudes in each hemisphere. It is not known what determines the speed, the symmetry, and the magnitude of these flows. The situation can be contrasted with that on Earth, where to a first approximation the atmospheric temperature is governed by radiative balance. An equator-to-pole gradient is established in response to insolation. A dynamical regime becomes established, with one major tropospheric jet in each hemisphere. The speed and geometry of the jets are related in a straightforward way to the externally imposed heating. The jets turn out to be unstable, and quantitative details of the turbulent flow are very difficult to predict with accuracy, but the general nature of the velocity and temperature regime is well understood.

There are two possible forcings for flows on the outer planets, but it is not known which is dominant. One is the latitudinal gradient of insolation and the other is heating from below. Except possibly for Uranus, where the internal heat flow appears to be small, internal heat sources generate about the same amount of energy as insolation. For either energy source, the thermal drive for atmospheric motions is of global scale. The emergence of multiple alternating jets on Jupiter and Saturn means that an internally determined length scale arises. The general circulations on these planets are therefore responding to the external forcing in a more indirect manner than on Earth, and the fluid mechanics correspondingly is more subtle.

Our knowledge of the vertical structures and depth of outer planetary flows is very limited. Observations by remote sensing are limited to the stratospheres and the upper tropospheres where atmospheric pressures are approximately 1 bar or less. The major cloud systems, which make the atmospheric jets visible on Jupiter and Saturn, are near the 1-bar level. The deepest available information comes from the Galileo probe, which penetrated to about the 24-bar level on Jupiter (2), but this represents only a single profile of atmospheric properties. In the terrestrial case, experience shows that knowledge of the surface boundary condition and of the depth of the troposphere are absolutely essential to understanding the flow regime. The latitudinal temperature gradient only constrains the vertical wind shear, not the speed of the flow. As is discussed in meteorology texts (3), the depth of the flow regime must also be known in order to determine characteristic velocities.

Thermodynamic aspects of the outer planetary flows are also not well known. The major constituent of the atmospheres is H_2. Within the outer few hundred kilometers it behaves as an ideal gas, but at greater depths it becomes more liquidlike. In the conventional view, heat transfer from the interiors is by convection, and the thermodynamic structure of the deep atmospheres and the interiors is close to isentropic, but recent calculations of radiative opacities have led to the suggestion that there may be a substantial subadiabatic (stably stratified) layer at a depth of a few thousand kilometers (4), at least on Jupiter and Saturn.

In addition to the stratification, another important thermodynamic issue is the energy storage mechanism involved in heat transfers. The ultimate drive for dynamics must be buoyancy, which arises because of density contrasts that are associated with heat transfer. A wide range of temperatures exists within the envelopes of the outer planets, approximately from 100 K to a few thousand kelvin, and several constituents are candidates for phase changes. Examples include water, methane, and even silicon compounds at deeper levels where temperatures are highest. Each phase change gives a possible buoyancy effect, either through latent heat release or by precipitation and molecular weight alteration. Another possible thermodynamic

The author is in the Department of Astronomy, Cornell University, Ithaca, NY 14853, USA. E-mail: gierasch@astrosun.tn.cornell.edu

21. Atmospheric Dynamics on the Outer Planets

effect is the conversion of para hydrogen to ortho hydrogen, which are known to be out of equilibrium on Jupiter and Neptune (5). Either mechanism, phase change or hydrogen conversion, could generate buoyancy contrasts of a few percent or more, which is ample to affect dynamics.

In view of all the complexities, how does one approach the question of why the outer planets have jet systems, long-lived ovals, and all the other observed richness of meteorological behavior? Cho and Polvani, in this issue (6), describe the behavior of an extremely idealized mathematical model and show that the model reproduces several features of the observed planetary circulations. The model represents a thin homogeneous "ocean" of depth H, on a planet of radius a, surface acceleration of gravity g, and rotation period P. It has no thermodynamic forcing and is initialized with a random velocity field. As Cho and Polvani discuss, other workers have studied similar models (7), but this is the first time that a series of experiments have been carried out for an unforced flow in full spherical geometry and for a range of parameter values spanning all the outer planets. The idea is to discover the key processes at work by isolating them in a very simple calculation.

The calculations represent extremely interesting fluid dynamical results. But what does one learn about the planets from qualitative agreement with observation in this model? The model has not been demonstrated to be unique in showing agreement, and therefore any conclusions must be tentative. One point of importance is the width of the jets that emerge. The model contains three scales: the planetary radius, the "deformation radius," and the Rhines scale. The deformation radius, from meteorology, is $L_D = \sqrt{gH}/\Omega$, where the rotation rate is $\Omega = 2\pi/P$. The Rhines scale (L_β) is given by $L_\beta = \sqrt{Ua}/\Omega$, where U is the magnitude of the flow speed. Rhines (8) has shown that in two-dimensional flow on a rotating sphere, an inverse turbulent cascade of energy to large scales is interrupted at scale L_β, and alternating jets can arise. The spacing of jets in the Cho and Polvani experiments, after initial transients, turns out to be on the order of L_β. But then, what sets the flow amplitude U on which the Rhines scale is based? This may depend on thermodynamics and remains an unanswered question. It is also possible that the new simulations are not based on the relevant deformation radius, and that the wrong regime, in terms of the ratio of L_β to L_D, is being explored. As Cho and Polvani point out, it is not at all clear what value of L_D (if any) is appropriate to simulate the correct planetary dynamics in a two-dimensional model.

But if the Cho and Polvani calculations have indeed captured the essential physics of

Images of Jupiter, Saturn, Neptune, and Uranus (clockwise from top left) obtained by the NASA Voyager spacecraft. The contrasts associated with cloud features are largest on Jupiter because the condensing ammonia clouds are relatively high in the atmosphere and are not obscured by overlying haze or Rayleigh scattering gas. With image enhancement, features can be identified and mean flow drifts measured even on Uranus. The resulting zonal velocity profiles are exhibited by Cho and Polvani [see figure 2 of (6)]. The structure of small-scale turbulent motions is clear only on Jupiter.

jets and eddies on the outer planets, then the thermodynamic complexities described above for deep atmospheres are incidental, and fluid dynamics controls the gross structure and the visual appearance of the outer planets. If true, this would be a striking conclusion, simultaneously simplifying and complicating. The fluid dynamics is turbulent and nonlinear, yet leads to highly organized and persistent mean flows.

The simulations do not produce eastward currents at low latitudes on Jupiter and Saturn. Observations show strong eastward equatorial jets, which are particularly puzzling because they represent concentrations of angular momentum (more rapid rotation than the average). An angular momentum pumping process is needed to maintain them. Because these jets are on the equator, they cannot be produced by poleward drift of gas that conserves angular momentum, the way eastward mid-latitude jets on Earth can be produced. As Cho and Polvani remark, the fact that none of their numerical experiments produces these jets suggests that another mechanism, beyond the scope of the simple model, may be necessary. Stratification and the third dimension might be the missing ingredients.

Future progress will depend on new information from the planets. Numerical modeling has become very powerful, but the physical system is so ill-defined that modeling is not well constrained. It would be useful to have detailed maps of velocity fields within Jupiter's clouds, so that statistical properties could be compared with numerical simulations. The NASA Galileo orbiter may obtain such data during the next 2 years. It would also be useful to have more probes beneath the clouds of the outer planets, to better define the depth and the stability properties of the flows.

References and Notes

1. A. P. Ingersoll, *Science* **248**, 308 (1990).
2. The first set of reports on the NASA Galileo Jupiter entry probe measurements appeared in the 10 May 1996 issue of *Science*.
3. J. R. Holton, *An Introduction to Dynamic Meteorology* (Academic Press, New York, 1992).
4. T. Guillot, D. Gautier, G. Chabrier, B. Mosser, *Icarus* **112**, 337 (1994).
5. B. J. Conrath, P. J. Gierasch, E. A. Ustinov, paper presented at the XXI General Assembly of the European Geophysical Society, The Hague, Netherlands, May 1996.
6. J. Y.-K. Cho and L. M. Polvani, *Science* **273**, 335 (1996).
7. For example, T. E. Dowling and A. P. Ingersoll, *J. Atmos. Sci.* **46**, 3256 (1989); P. S. Marcus, *Nature* **331**, 693 (1988); G. P. Williams, *J. Atmos. Sci.* **35**, 1399 (1978).
8. P. B. Rhines, *J. Fluid Mech.* **69**, 417 (1975).

Planet in a bottle

There's only so much you can discover by peering through a telescope or running a computer simulation, says *Jeff Kanipe*. But build a scaled-down planet, and you can put your theories to a tougher test

WHEN it comes to being part of the action, astronomers certainly drew the short straw. Meteorologists get to chase hurricanes, biologists can watch mayflies live and die in the palms of their hands, but astronomers can't visit newborn stars to watch fledgling planets form. Nor can they walk across the dusty plains of Mars to feel the strength of storms there. Apart from the occasional spacecraft mission to a lonely outpost of the Solar System, they have to watch the drama of the Universe unfold on a distant stage.

But many astronomers are now trying to bring the complex worlds beyond Earth back to the laboratory. They are recreating dust storms on Mars and watching the seeds of planets grow from replicas of interstellar ice and dust. They are even building miniature models of Jupiter's atmosphere to try to answer questions that have nagged astronomers for years—why, for instance, does the giant planet have stripes?

In a laboratory at Johns Hopkins University in Baltimore, Peter Olson and his colleague Jean-Baptiste Manneville have shed some light on that mystery. Over the past two years they have been trying to recreate the striking banded structure that appears in the outer layer of Jupiter's clouds. Jupiter has between 10 and 12 light and dark bands that circle the globe in both hemispheres. Observations from Earth show that these bands are produced by strong winds circulating in alternating directions, and that they have not changed their latitudes for at least 100 years.

But how do these winds arise? According to Olson, two models have been in the running to explain the alternating jet system. One theory is that the winds are almost entirely driven by heating from the Sun, much like the Earth's trade winds. These occur because heat builds up at the surface in the equatorial region, warming the air and making it rise. The warm air flows at high altitudes towards the poles. On its way it cools, sinks and flows back to the equator. Because the Earth spins towards the east, the trade winds flow westwards.

Big bands

A similar process could create the bands on Jupiter. Olson says that numerical models show that if the thin outer layer of clouds originally contained a lot of turbulence, then the rotation of the planet could indeed turn these patterns into the familiar bands. But the Sun's heat doesn't penetrate the outer layer of clouds on Jupiter, so this would be possible only if the winds were confined to a very thin upper layer of its atmosphere.

However, Olson suspected that an alternative model, first suggested in the 1970s by Fritz Busse of the University of Bayreuth in Germany, might be correct. Busse's idea was that the jets of wind may arise from convection currents whipped up by the intense heat deep within the planet. Jupiter radiates about twice as much energy as it receives from the Sun. Some of this heat is left over from the planet's creation and some is generated by drops of helium raining down into the interior.

"We wanted to know whether or not the deep convection model could produce a banded structure," says Olson. So the researchers decided to construct an experiment that would replicate at least some of the conditions in Jupiter's atmosphere. The average density of Jupiter is only about 1.3 times that of water, so they chose water as a good approximation. They had to create a temperature difference across the water to generate convection patterns, and then simulate the force of Jupiter's powerful gravity, which would exert a pull on the atmosphere towards the planet's centre.

To do this Olson and Manneville put together a copper sphere, 25 centimetres wide, nested inside a 30-centimetre Plexiglas sphere. They filled the gap between the copper and the Plexiglas with water at room temperature, then circulated chilled antifreeze within the inner copper sphere to create a temperature difference between the inner and outer layers of several degrees centigrade.

By spinning the whole apparatus at a steady 13 revolutions per second to create a centrifugal force, the researchers simulated gravity. The direction of the force in the laboratory Jupiter is outward, whereas in the real planet, obviously, it is inward. But with the direction of the heat flow reversed—from the outer surface to the centre—the "gravitational" force and the heat flow are in opposite directions, just as they are on Jupiter.

Once everything was in motion, Olson and Manneville injected a fluorescent dye near the outer boundary of the Plexiglas sphere. When it was bathed in ultraviolet light, several alternating dark and bright bands emerged in the flow patterns. Bright zones appeared where the dye was flung to the outer surface of the Plexiglas by the motion of the sphere. Where convection was strong, however, the dye was drawn away from the surface toward the copper sphere, creating a dark belt.

Olson reported in August last year that the experiments show that an inner heat source could easily be driving the Jovian winds (*Icarus*, vol 122, p 242). What's more, they also found that the number of bands in the model Jupiter was proportional to the so-called Rayleigh number, a measure of the temperature difference across the convecting regions. The Rayleigh number for Jupiter is far too large to recreate in the lab, but when the researchers projected their findings to Jupiter's value, they worked out that there should be 10 or 12 bands in each hemisphere, just as observed.

In a spin

Olson says that the results support Busse's idea that convection takes place in narrow columns, hundreds or even thousands of kilometres long, that are aligned from north to south. Individually, these evolve in a complex, chaotic way. But on a large scale, each creates a bulk convection pattern that moves at a unique rate in several giant cylinders, concentric with Jupiter's rotation axis. Where the edges of each of these cylinders emerges at Jupiter's surface, it creates one of the distinctive bands (see Diagram). In other words, the winds are not just a surface effect—Olson's model suggests that they rage right through the planet.

Compelling as the results seemed, they were far from conclusive. After all, just how "real" is this model? Jupiter is over 430 million times larger than the laboratory replica. Olson and Manneville did not include the effects of the different amounts of solar heating at different latitudes. Nor did they take account of Jupiter's magnetic field, which is likely to exert a pull on the metallic hydrogen that makes up much of the inner three-quarters of the planet. And unlike the real Jupiter, their model had a rigid outer boundary.

"Lab modelling never quite covers all the conditions," says Olson. "But it does play a provocative role in that it can either complement or contradict numerical models." And whether his model is realistic or not, results from the probe carried by the Galileo spacecraft that plunged into the atmosphere of Jupiter in December 1995 are suggesting that it may well be correct. If the bands were produced by superficial solar heating, they would be expected to run only a few tens of kilometres deep. But during its descent, the Galileo probe recorded constant wind speeds of around 650 kilometres per hour from the outer cloud layer to at least 130 kilometres inside.

Harry Swinney of the University of Texas at Austin agrees that simulation is an important "reality check" on the traditional numerical models. "They have their place, but sometimes they try to do too much," he says. "They'll put everything into it including the kitchen sink."

'By rights, Jupiter's Red Spot shouldn't exist. It should have lost its coherency and been pulled apart long ago'

A numerical weather model might try to make predictions for every factor influencing global circulation, such as precipitation, temperature variations, wind patterns and cloud reflectivity, then calculate how circulation should evolve over hundreds of hours. Swinney says that, sure enough, you can see new patterns evolve, but exactly why a given pattern arises is far from clear because the model depends on so many things.

Swinney and his colleagues Joel Sommeria and Steven Meyers have turned to laboratory simulation to try to work out another Jovian mystery—why its Great Red Spot, a titanic cyclone rotating anticlockwise at speeds of about 400 kilometres an hour, has managed to rage for so long. "There are other persistent long-lived vortices on Jupiter and the other planets," says Swinney. "But nothing like the Great Red Spot, which has been observed for more than 300 years." By rights, the Red Spot shouldn't exist. Sandwiched between fast-moving easterly and westerly jets in Jupiter's southern hemisphere, it should have been pulled apart long ago.

Drifting spot

Three decades ago, scientists suggested that the Red Spot could be an atmospheric disturbance positioned over an isolated mountain. Yet this idea bit the dust when it became clear that the spot sometimes drifts both east and west. Later theories drew analogies with terrestrial hurricanes, which are driven by heat rising from the Earth's surface or solar heat absorbed near the equator. But no numerical model could explain why the Jovian storm is so stable.

So in 1988, Swinney and his colleagues began trying to create vortices like the Red Spot in the laboratory. They use a rotating doughnut-shaped tank, 86 centimetres in diameter, filled with water pumped through a series of inlets and outlets in its floor. Again, the water mimics the atmosphere, with dye used to track currents, and the rotation of the tank simulating that of the planet at mid-latitudes. The researchers have found that they can create large vortices that appear spontaneously and, like the Red Spot, are oval shaped and about twice as long as they are wide. They can survive indefinitely, even if the rotating tank slows down or speeds up a little.

Swinney's simulations cannot clear up all the mysteries of Jupiter's atmosphere. But they do confirm that you can simulate the turbulent atmosphere realistically on a small scale. For instance, the model managed to replicate a sequence of images from the Voyager spacecraft, which visited Jupiter in 1979. The images showed the striking patterns that emerge over several days when large vortices merge with smaller

3. THE SOLAR SYSTEM: The Planets

ones. "It's very similar to ours," says Swinney.

Jupiter's atmosphere is not the only one to have come under the spotlight. For many years, Ronald Greeley of Arizona State University has been reconstructing the Martian winds in a wind tunnel at the NASA Ames Research Center in Moffett Field, California. Once used to test the structures of rockets at low pressure, it now houses the Martian Surface Wind Tunnel (MARSWIT).

"There isn't any other facility that can simulate the surface of Mars like ours," says Greeley. The 14-metre-long wind tunnel looks like a chimney lying on its side. The walls are made of plywood, except in the middle, where two Plexiglas panels allow scientists to peek inside the tunnel. The end at which air flows into the tunnel is conical, like the horn of a trumpet, to focus the wind. To create the same kind of surface wind turbulence that exists on Mars, the wind passes over small pebbles fixed to the tunnel floor.

Features on the surface of Mars constantly evolve as sand and dust storms sculpt long corduroy-like grooves and leave sand dunes on its surface. They also leave bright and dark streaks up to 200 kilometres long. Sometimes dust storms can engulf the entire planet, and Greeley and his colleagues have been using the tunnel to investigate one of the biggest mysteries about Mars—how the dust storms can happen at all. The atmospheric pressure on the planet is only about 6 or 7 millibars, less than a hundredth of the average pressure at the Earth's surface.

Jovian bands: columns of convection within several cylinders round Jupiter's rotation axis add up to give jets of wind in alternating directions

Visits by the Viking landers in 1976 revealed that the winds kick up fine dust made up of particles a few micrometres thick. To simulate windblown sand in the MARSWIT, Greeley and his team use silica microspheres mixed with natural silt. To mimic the Martian dust, they use finely ground walnut shells, which have a similar shape. They are less dense than their Martian counterparts, and this compensates for the fact that Mars has just two-thirds the gravity of the Earth.

'Researchers in Germany are simulating the conditions in the Solar System long before Mars and Jupiter came into existence'

To find out the minimum windspeed needed to whip up the dust storms, Greeley evacuated the wind tunnel to pressures similar to that on Mars. They then increased the windspeed until the dust started to rise from the surface. Their results, which they have submitted to *Icarus,* suggest that the minimum windspeed for kicking up dust is 450 kilometres per hour over a smooth flat surface. When the surface was strewn with boulders, however, the dust could be stirred up by a wind of only 155 kilometres per hour. "The minimum velocity to move things around was one of our best results," says Greeley. "We were able to nail these down more."

No one yet knows whether these windspeeds actually match those on Mars. The fastest winds measured by the Viking lander had speeds of only about 130 kilometres per hour. But NASA's Pathfinder spacecraft, which is now on its way to Mars and due to land in July, carries a series of windsocks that will give the best measurements to date of windspeeds. Greeley, who is involved in the Pathfinder mission, is eagerly anticipating the spacecraft's data. "We can't wait to apply our lab results to [Pathfinder's] results," he says. The MARSWIT was used to check that Pathfinder's windsocks would not be aerodynamically unstable and "flutter" in Martian conditions.

Greeley has also used the wind tunnel to show that the light streaks on Mars must be formed when dust settles out of the atmosphere, while the dark streaks form in the wake of obstructions to the wind, such as large craters. Like Olson, he feels that joining forces with computer modellers is crucial for getting to the bottom of planetary geology. "It takes this multi-prong approach to let you gain on the problems in a systematic fashion," he says. "Nature is always more complicated than any model. So it really takes both approaches to come up with good answers."

Researchers in Germany are simulating the conditions in the early Solar System long before Mars and Jupiter came into existence. Five billion years ago, the Solar System grew out of a disc of hydrogen and helium gas and countless grains of silicate dust, typically a micrometre or two across. Astronomers believe that the planets began to form as the dust particles collided as a result of their random, Brownian motion. Some stuck together. As the particles grew bigger than 10 micrometres across, friction and turbulence came into play, increasing the number of collisions and making the particles bigger still. Once the clumps were about a kilometre in diameter, they would have exerted a large enough gravitational pull to suck in more particles.

Model spanners

But numerical models throw a spanner in the works, say Jürgen Blum and Torsten Poppe of the University of Jena, Germany. These suggest that dust grains in the early Solar System would have moved too fast to adhere to each other after impact. Instead of sticking together, small particles would ricochet off one another. This rebound effect would prevent clumps from growing larger than about 10 micrometres across *(New Scientist,* Science, 27 July, p 15).

So why then are fully grown planets flying round the Sun today? In 1994, Blum and Poppe began constructing a device that would allow them to fire single grains of silicon dioxide, between 0.5 and 2 micrometres across, at a flat, polished silica target. This allowed them to measure the upper and lower velocity limits for particle sticking.

'We may see faint planets forming round distant stars, thanks to a little box of dust and gas'

They found that particles in the middle size range of their test sample stuck together at velocities of about 1.2 metres per second—four times higher than predicted by numerical models. Moreover, they found that these particles often became electrostatically charged when they collided. This electrostatic effect could help glue them together, and may explain why planets exist.

Blum and Poppe now hope to take artificial planet-building a step further. To simulate more accurately the low-gravity conditions that existed in the primordial Solar System, Poppe and Blum have designed the Cosmic Dust Aggregation Experiment (CODAG), to fly aboard the space shuttle in 1997. A stream of gas and 2-micrometre particles will be injected into a 1.5-litre vacuum chamber on the shuttle, forming a weightless, diffuse cloud. The experiment should reveal the shapes of the particles and how they clump together.

The CODAG experiment could also point the way to the birthplaces of planets far beyond the Sun. Blum and Poppe will measure how light is scattered by the fluffy aggregates in the canister, and compare this to the way light is scattered by the gas discs around young stars.

"Our results should enable astronomers to conclude whether the discs of dust around young stellar objects contain the seeds of planets," says Poppe. In a few years, we may be able to focus on faint planets forming round distant stars, thanks to nothing more than a little box of dust and gas.

Jeff Kanipe *is a science journalist in Dallas, Texas.*

Magnetic Fields on Distant Moons Hint at Hidden Life

WILLIAM J. BROAD

When scientists wonder about the hidden depths of a moon or planet, they often look for signs of a magnetic field.

On Earth, the lines of magnetic force that envelop the planet indicate the hot churning of a molten core. At the surface, geologic effects of this deep commotion help make Earth a habitable place, constantly shaping the land and sea, powering earthquakes and volcanoes, recycling the crust and keeping planetary ingredients in a stir beneficial to life.

In orbits around Jupiter, pulses of geologic activity.

But what of other worlds? Scientists have searched for magnetic fields around rocky extraterrestrial bodies partly as a way to gauge the presence and prevalence of deep geologic activity, but until now the hunts have mostly been unsuccessful. Nothing substantial has been found around the Moon, or Mars, or Venus—all of which have been written off as geologically dead or, at the very least, lost in deep slumber. And Earth, with its churning core, has seemed more and more unique.

But now, the National Aeronautic and Space Administration's Galileo probe of Jupiter has found that three of the giant planet's large moons—Io, Europa and Ganymede—are pulsating with signs of magnetic fields, suggesting that layers of thick ice on those distant worlds hide interiors that are potentially alive geologically, and possibly alive biologically as well.

The lines and ridges, furrows and fissures that deeply wrinkle the surfaces of these icy moons (and in the case of Io, the volcanoes) had already hinted at a kind of interior drama. But the magnetic clues now suggest that activity is widespread and going on now, rather than ages ago.

Indications of magnetic fields around Io and Ganymede are analyzed in the current issue of the journal Science, and the Europa findings are to be announced in the journal on Friday. (Experts writing last week in the journal Nature declared that the fourth and most distant of the large Jovian moons, Callisto, was magnetically barren.)

The positive findings about the three Jovian moons, scientists say, raise the odds that their depths may be habitable and may harbor alien life, perhaps swimming through dark seas. Such speculation is an increasingly popular topic among planetary scientists.

Magnetic fields could be beneficial to life not only because they imply inner heat and geologic vitality, scientists add, but because they would help shield the moons and any alien creatures from cosmic rays, the streams of highly penetrating particles that speed through space. Earth's magnetic field deflects many of these deadly radiations.

"It's very exciting," Dr. Margaret G. Kivelson, an astrophysicist at the University of California at Los Angeles who is the lead scientist interpreting the magnetometer readings of the Galileo probe, said in an interview. "The images we receive show only surfaces, and sometimes you can infer a bit about what's going on below, like with Europa and all the floes. But when you measure a magnetic field, it gives you important clues. They're not always unambiguous, but they give you a

23. Magnetic Fields on Distant Moons

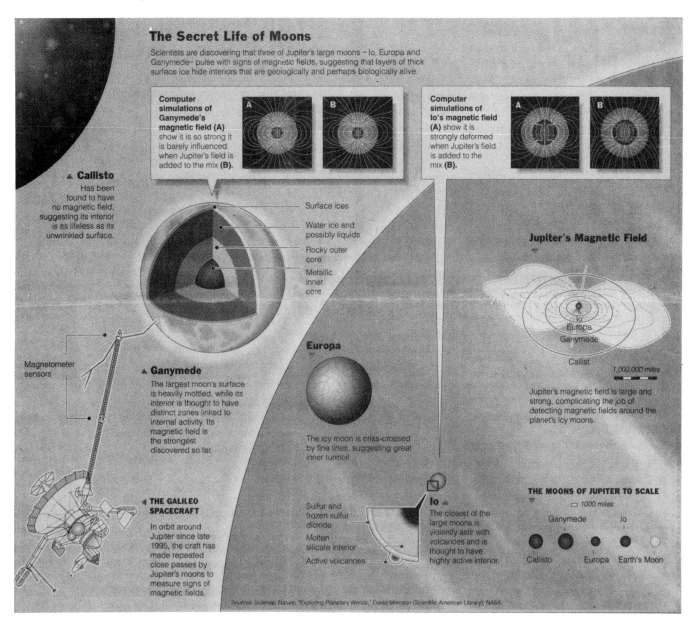

The New York Times; Illustration by Juan Velasco

sense of something going on deep inside."

Jupiter is the solar system's biggest planet, with 16 known moons, four of them quite large. (Ganymede, the biggest moon in the solar system at 3,269 miles in diameter, is larger than Mercury.)

Increasingly, scientists say, the large Jovian moons are proving to be more surprising and interesting than Earth's planetary neighbors.

"This stream of data is strengthening our thinking of moons as worlds in their own right," said Dr. Louis Dr. Friedman, executive director of the Planetary Society, a private group in Pasadena that backs space exploration. "That change started to happen with Voyager, as we saw volcanoes on Io." The Voyager spacecraft flew past Jupiter in 1979.

"Now," Dr. Friedman added, "it's really happening as we see hints of all these processes going on."

People have long known of magnetic fields. The ancient Chinese, Arabs and Italians are variously credited with inventing the magnetic compass, which aligns itself with magnetic lines of force. By the 12th century, seamen used the instrument to navigate the Mediterranean.

A magnetometer is a modern tool that measures the strength of magnetic fields, simple ones being akin to compasses. Magnetometers hurled into space in the last few decades failed to find signs of substantial fields around the Moon, Mars and Venus. But in 1974 and 1975, the Mariner 10 spacecraft while studying Mercury found signs of a field whose strength was less than 1 percent of Earth's,

3. THE SOLAR SYSTEM: The Planets

suggesting the planet has some degree of inner activity.

All the gas giants of the outer solar system—Jupiter, Saturn, Uranus and Neptune—were found to have magnetic fields, though the means of generation is thought to be the inner movement of materials much lighter than the molten iron on Earth and Mercury. And that rules out the kind of geology that interests life scientists. Of all the planets, Jupiter was found to have the most powerful magnetic field, which stretches through space for millions of miles, and is the solar system's largest planetary feature.

This potent force complicates the job Galileo's magnetometer is now facing in trying to find and measure the magnetic fields of the Jovain moons. The spacecraft, launched in 1989, reached Jupiter in December 1995 and ever since has been engaged in detailed studies.

Last December, the space agency announced that Galileo had found a strong magnetic field around Ganymede—the first for any moon in the solar system. It appears to be about 10 percent as strong as Earth's, or fairly substantial. The implication is that Ganymede has a molten metallic core.

Galileo also found signs of a magnetic field around Io, which lies much closer to Jupiter's own powerful magnetism, making the clues more ambiguous. Moreover, the data were scant. Galileo flew by Ganymede four times taking magnetometer readings, but did so only once for Io.

In the current issue of Science, experts from the University of California at Los Angeles report computer simulations of these moons' interiors based on Galileo's readings. They concluded that Io's magnetic field was possibly derivative of the main Jovian field, and might collapse in its absence.

But Ganymede, the team found, was "almost certainly operating as a dynamo in its own right." The dynamo theory holds that planetary magnetism arises as inner flows of metals produce electromagnetic fields.

"It's all very intriguing," Dr. Gerald Schubert, an author of the paper who is a planetary scientist at the University of California at Los Angeles, said in an interview. "I wouldn't have predicted this prior to Galileo."

A riddle is what powers Ganymede's strong field. Io, much closer to Jupiter and moving in a slightly elliptical orbit, is alternately squeezed and stretched. These so-called tidal forces heat its interior, powering its intense volcanism.

But Ganymede experiences much less tidal heating. Dr. David J. Stevenson of the California Institute of Technology has proposed that Io, Europa and Ganymede were once in orbits that exerted a much greater gravitational pull on Ganymede, heating it up many millions of years ago and generating a strong magnetic field that lingers today.

"This is a very important and exciting discovery," he said of Galileo's findings about Ganymede. "It's very exciting because we have so few examples of internal fields" in rocky planets and moons.

And now Europa, one of the solar system's most enigmatic bodies, is also reportedly showing signs of magnetism. Its bright surface of water ice is broken into a chaos of lines and ridges, looking something like shattered glass. The moon is almost entirely free of impact craters, suggesting they were wiped clean by ice movement and turnover.

Friday's issue of Science is to carry a report on Europa's magnetism by a team of researchers based at the University of California at Los Angeles and led by Dr. Kivelson, scientists said.

Europa, closer to Jupiter's potent magnetic fields than Ganymede, is a more difficult target to study. Moreover, Galileo had its magnetometer in working order only once in a Europa flyby, last December. A cosmic ray apparently knocked the magnetometer out of action temporarily just before a February pass, which was meant to bolster the readings.

Even so, Galileo found sketchy signs of a field.

"It's much weaker" than Ganymede's, said a scientist who spoke on condition of anonymity because researchers are barred by the journal from talking about their findings before publication. "But it's intriguing."

The possibility is that Europa is geologically alive beneath its craggy surface ice, its core hot and astir, creating a deep warm ocean perhaps teeming with alien life.

Dr. Kivelson, who refused to comment on the coming Science report, said the Galileo team was eagerly looking forward to future data that might clarify the situation.

The space agency recently decided to extend the Galileo mission so the hardy spacecraft could zero in on Europa in its final days. Originally scheduled to end in December 1997, the mission is now to go through 1999 so the probe can repeatedly swing past the icy moon for a series of close-ups.

"We believe that will settle the question," Dr. Kivelson said, of whether Europa has a strong magnetic field of its own, and perhaps a deep extraterrestrial sea as well.

Article 24

LIFE IN A DEEP FREEZE?

The discovery of possible seas on Jupiter's moon Europa heats up the search for alien organisms—and there are many moons still to explore

JEFFREY KLUGER

THE SOLAR SYSTEM, TO BE BRUTALLY honest, has turned out to be something of a bust. There was a time when the planets seemed to have a lot of potential, but only if scientists didn't look too closely. Once they did, things got ugly fast. The planets were either flash-frozen or deep-fried, uninhabitable gas giants or uninhabitable rocky pellets, smothered by a toxic atmosphere or almost totally airless—altogether poor company for a glamour world like Earth.

But planets aren't all there is to the sun's family. The solar system is also packed with moons—more than five dozen of them. Increasingly, astronomers are appreciating that these cosmic offspring may be far more remarkable than the parent worlds they orbit. Unlike most of the planets, the moons have oceans, the moons have continents, the moons even have active volcanoes.

And now, it seems, a moon may be the best place yet to look for the most remarkable thing of all: extraterrestrial life. Last week the sturdy Galileo space probe finished beaming back the sharpest images ever taken of Jupiter's ice-covered satellite Europa. The pictures revealed more clearly than ever before that the moon's frosty rind is nothing more than a planet-wide ice cap floating atop a globe-girdling ocean of ordinary water.

What's more, the spacecraft spotted brown stains on the ice that could conceivably be a mix of hydrogen cyanide and other life-related chemicals. "If this is indeed hydrogen cyanide," says Richard Terrile, a planetary scientist at NASA's Jet Propulsion Laboratory in Pasadena, California, "we have organic chemicals mixed into a bath of water. That's a recipe for life."

Whether there's anything so dramatic waiting on any of the other moons in the

COSMIC VESUVIUS Constant gravitational tugging from three nearby moons churns up Io's innards, causing volcanoes to erupt all over its surface

3. THE SOLAR SYSTEM: The Planets

solar system is unclear, but NASA wants to find out. The agency is contemplating five or more missions to the planetary satellites in about the next 10 years. Says Terrile: "We're beginning to appreciate that within our solar system, there are all these mini-systems worth exploring."

It was in 1979, during the Voyager 1 spacecraft's first encounter with the Jovian moon Io, that astronomers began to suspect there might be more to the moons than met the telescopic eye. While 43 of the 61 satellites measure less than 300 miles or so in diameter, most of the others are more than 1,200 miles across. Bodies with this kind of bulk are capable of supporting an atmosphere—a big plus when you're trying to incubate life. What just about all the moons lacked was the heat needed to get biological chemistry going.

Or so it seemed. In 1979, however, when the first pictures of Io were beamed to Earth, NASA got a shock. Rising from Io's face were the unmistakable plumes of up to 10 erupting volcanoes. "Suddenly," says Torrence Johnson, project scientist for the Galileo mission, "it was clear that other bodies in the solar system could be geologically active."

Ordinarily, a body the size of Io should have cooled off long ago, making volcanoes impossible. But for every pass the moon makes around Jupiter, it makes several passes by its large, slower-orbiting sister moons: Europa, Ganymede and Calisto. Every time Io does that, the gravitational tug of these nearby satellites gives it a twang. On Earth, the gravity of just one moon is sufficient to cause the oceans to rise and fall in great crashing tides. On Io, the gravitational influence of three nearby moons is enough to distort the shape of the world itself, causing it to pulse with a heartbeat-like *lub-dub*. This rhythmic motion churns up internal heat, which in turn stirs up moonwide volcanoes.

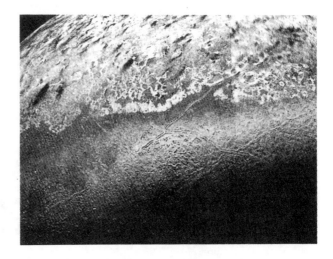

BIG CHILL Triton is the coldest known place in the solar system, with volcanic plains formed by icy slush, not lava

Though all such otherworldly erupting is dramatic, it amounts to little more than geological pyrotechnics. On Europa, however, tidal heating may have produced something truly remarkable. The formations Galileo spotted last week are definitely icebergs, though less jagged-looking than those found on Earth. Astronomers don't know why Europan ice and terrestrial ice would not fracture the same way, but they admit they have no experience with the kinds of cracks that are produced when an entire world is frozen over. More to the point, the bergs are small, rising just 300 to 600 ft. above the surrounding ice. Since only 10% of an iceberg shows above the water, that means these measure a mile or so from top to bottom—and so, therefore, does the planet-wide ice crust from which they came. On the scale of a 2,000-mile-wide moon, that's not much of a crust at all.

No matter how thick the ice is, the waters beneath it must still be liquid, thanks to tidal heating. This is good news for biology. Scientists don't pretend to know how warm a Europan ocean might be, but even waters that are just a degree above freezing would feel downright balmy to organisms that evolved in it.

While Europa may be the solar system's most promising Petri dish, it is by no means the only one. Saturn's Titan, larger than both Mercury and Pluto, has an atmosphere fully 60% denser than Earth's, forming a sort of photochemical haze that appears to be full of the stuff of prebiology. The problem is that Titan is cold. With temperatures hovering near –290°F and no signs yet of significant heat to drive chemical reactions, the moon could be awash in organics that are nevertheless unable to combine in biologically useful ways.

"I expect fantastic chemistry on Titan," says astronomer Steven Squyres of Cornell University. "I don't expect a trace of life." Others aren't so sure; if there's lightning in the Titanian atmosphere, it could energize organic molecules in a hurry. "I would be surprised if there is life on Titan," says astronomer Toby Owen of the University of Hawaii, "but we've been surprised by the solar system before."

Jupiter's Io and Neptune's Triton could also prove surprising. Though Io appears largely dehydrated, planetologists don't

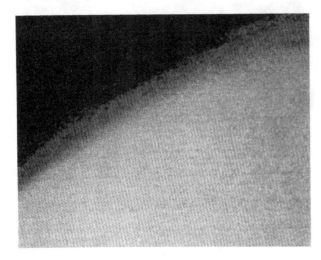

MISTY WEATHER Titan's dense photochemical haze makes picture taking hard but complex biology possible

24. Life in a Deep Freeze?

rule out the possibility of sub-surface water, particularly since they think that ordinary steam might provide some of the propulsive muscle behind the moon's volcanoes. Triton presents a greater organic hurdle. At –391°F, the moon is the coldest known object in the solar system. Nevertheless, it appears heavy with sub-surface ice, which seems to have got warm enough, in the past at least, to flow over the landscape in a lava-like slurry. More tantalizing, dark streaks near the poles suggest that occasional geysering on the frozen moon may have spouted carbon or some other organic material. "We don't fully understand what's going on inside Triton," Terrile says, "but something is pumping a lot of energy."

Whether any of the moons will ever be understood fully, of course, is open to question. Before long, however, they will certainly be understood better. Galileo could be functioning until late 1999, with more than 20 passes through the Jovian system still to come. Next fall NASA plans to launch the new Cassini-Huygens spacecraft on a seven-year odyssey to Saturn. In addition to making at least 36 orbital slalom runs through five of Saturn's inner moons, the ship will fire off a probe that will puncture Titan's cloud cover, parachute to its surface and send environmental readings back to Earth.

Even before Cassini's work begins and Galileo's ends, other ships could be on the way to join them in the outer solar system. NASA is tentatively planning several new Europa probes, including one that will photograph its surface and take radar soundings beneath its crust. If the radar picks up the telltale echoes of liquid water, another spacecraft would be sent to land on Europa and release a heated probe designed to melt through the ice layer and look for signs of life in the seas below.

None of these proposed missions will come cheap. Even with NASA's new commitment to building smaller, less expensive spacecraft, interplanetary ships still cost at least $200 million each. Planetologists, however, insist that the potential discoveries could be well worth the money.

"These moons make up one of the most eccentric cosmic families imaginable," says Terrile. "As with any other family, some individuals are under-achievers, some are overachievers, and a few may be up to something truly fantastic." It's these last that NASA wants to get to know better.

Water World

Beneath the six-mile-thick shell of ice that encases the moon Europa may lie a vast liquid ocean. And in its dark, alien depths, we may—if we look—find something swimmingly alive.

KATHY A. SVITIL

WHEN GALILEO GALILEI AIMED a telescope at Jupiter one dark night in 1610, he spied four large, bright satellites, lost to the naked eye in the glare of the gassy giant planet. He could hardly have guessed that one of those moons—Europa, second closest to Jupiter—might one day shine light on the origin of life on Earth.

Each of the four Galilean satellites is different. Ganymede, the largest, has its own magnetic field and even an atmosphere, albeit a very thin one. Io, the closest to Jupiter, is so deformed by the mother planet's gravity and that of the other Galilean moons that it is heated into the most volcanically active object in the solar system. Callisto, the farthest out, is a dead moon pockmarked with craters.

Europa's surface displays pronounced bands where the crust has cracked and filled with darker material. Some suggest that the dark stuff may hold clues to a living ocean below.

25. Water World

Compared with the other three, Europa is almost boring. It is a little smaller than our own moon. It sports no volcanoes spewing molten rock, has no atmosphere to speak of, few large craters, and little topography. Its density, about three times that of water, indicates that it is made mostly of rock. But the surface is clean ice, with a fluffy topping of frost. From a distance, Europa looks as white and smooth as a giant cue ball, but up close, the cue ball shows cracks—dark linear features, some a thousand miles long, criss-crossing the crust. Researchers don't know what the dark material is or how the cracks formed, nor are they sure what lies beneath the icy shell. But what they suspect is an ocean.

An ocean some 500 million miles from the sun sounds far-fetched, but the story gets stranger still. If Europa does indeed hold an ocean, those waters might harbor life. In the search for evidence of past life on other worlds, many researchers have put even odds on Europa and Mars. And Europa may in fact boast better odds when it comes to claiming the grand prize of exobiology—organisms that are still alive today.

At least one proposal for a mission to explore that tantalizing possibility has already been submitted to NASA. The plan, devised by engineer Henry Harris of the Jet Propulsion Laboratory in Pasadena, California, involves sending an orbiter to Europa to fling a 22-pound metal sphere at the mysterious dark streaks in its ice. Those streaks may be the result of "contaminants" in ocean water that has flowed up through cracks in the ice. The orbiter would fly through the resulting plume of debris and capture samples to bring back to Earth, where they would be examined for organic material. Other researchers are suggesting that the possibility of Europan life deserves not just a single probe but an entire series of missions. In that case, the first order of business becomes proving that an

[P]icture emphasizes Europa's different textures of ice. The dot in the lower right is Pwyll, a young impact crater about 31 miles wide.

ocean really exists beneath the ice shell.

The circumstantial evidence is compelling. "It's extremely likely that Europa had liquid water near the surface at some point," says planetary scientist Steven Squyres of Cornell, who has been speculating on the likelihood of a Europan ocean for more than a decade, ever since the *Voyager* spacecraft beamed back the first images of the moon's fractured ice. "You are not going to start out with a moon that is pure dry rock and suddenly at the end of its evolution slap a lot of water on the outside. Instead it's going to begin as rocky material with some water dispersed throughout—maybe as ice, maybe as water captured within minerals." As time passed, radioactive compounds generated heat, thus melting and dehydrating the rock. Eventually the denser rock became concentrated in the center of the satellite, and the less dense stuff—water—moved toward the outer part of the moon.

For Europa to have an ocean now, some of that water had to stay liquid. The water at the surface, where temperatures are estimated at –230 degrees, obviously froze. Once in place, however, the ice shell could have protected water beneath it from the cold and the vacuum of space, and calculations performed by Squyres and others suggest there might have been enough heat to keep that water liquid. The key is tidal heating, the same force that deforms Io into a volcanic frenzy. The gravitational forces of Jupiter and the nearby companion moons tug on Europa like dogs worrying a rubber bone, causing it to bend back and forth. That strain is released as heat, and combined with radioactive heat from the core, Squyres says, it could be enough to maintain an ocean beneath the ice.

The surface of the moon also shows signs of a deep ocean. Europa has very few large craters, though Ganymede and Callisto are littered with them (craters that form on Io are rapidly paved over by molten rock). One good way to erase craters—and all the other topography missing from Europa—involved a process called viscous relaxation: if there is a warm, mobile, deformable layer under the frost (either water or warm ice), the surface features gradually fade away, just as a ball of Silly Putty at room temperature will eventually flatten. According to calculations by astronomer Gene Shoemaker of the Lowell Observatory in Flagstaff, Arizona, viscous relaxation almost certainly erased Europa's large craters—those more than about six miles across. The ice, Shoemaker also suggests, is probably no more than six miles thick. If there's an ocean beneath it, it's probably ten times as deep; and the moon's rocky interior probably has a diameter of about 1,800 miles.

The most prominent features on Europa, the dark bands that form a mesh across the surface, also support the notion of an ocean. "If you sort of rotate them back together, they close up very nicely," Squyres says. "It looks as if they have spread apart and dark stuff has welled up from beneath. That suggests that while you have an upper layer that is cold and brittle, you really don't have to go down very far before there is something much more mobile." And then there's Europa's fluffy frost. "It looks like what you'd get if you cracked the ice to expose liquid water to a vacuum, which would cause the water to vaporize and condense on the surface," says Squyres.

None of this, of course, proves anything. Scientists had hoped that the *Galileo* spacecraft, when it zoomed to within 436 miles of Europa last December, would yield some evidence of an ocean, perhaps a geyser gushing through a crack of the ice. That didn't happen, and it's unlikely that any of *Galileo*'s scheduled visits to Europa over the next year will produce the smoking gun. "*Galileo* doesn't carry the right tools to do the job," Squyres says.

Some researchers advocate sending another *Galileo*-type remote-sensing orbiter with new tools. It could carry better imaging equipment (that could spot water in a crack, for example) and perhaps long-wavelength radar, which is very good at penetrating ice. Since water reflects radar better than rock does, the signal bounced back would have a distinct look if an ocean lay beneath the ice. Alternatively, the orbiter could bounce laser pulses off the moon to measure the effect of tidal stresses—how much Europa flexes back and forth as its orbit takes it closer to Jupiter, then farther away. A moon with a shell of ice over a layer of water will flex more than one with solid ice on top of rock.

Another option is to drop sensors directly into the moon. For example, a magnetometer might be able to detect variations from tides, and a seismometer could pick up vibrations from ice quakes (a likely effect if an ocean is underneath the frozen surface). Or, taking a page from polar research on Earth, the orbiter could drop penetrators into the ice. "A penetrator is basically a high-tech dart," says JPL engineer Joan Horvath. "You eject them from orbit and they wham into the surface, and then you watch to see where they go. They could tell you how much the ice is moving."

Should one or more of these techniques prove that Europa has an ocean, Horvath and her colleagues at JPL, in collaboration with several teams of polar researchers, propose sending a "cryobot" to melt through the ice. The design, patterned after probes used in Greenland, is simple: a thin metal cylinder, about five feet long and six inches in diameter, with a plutonium-powered thermoelectric generator inside to melt a path for the probe. The cryobot would be connected to the surface by a communications cable so that engineers on Earth could receive data and perhaps even direct its actions. Once the probe finally reaches water—if the ice is only six or so miles thick, that would take about ten months—it would release its pay-

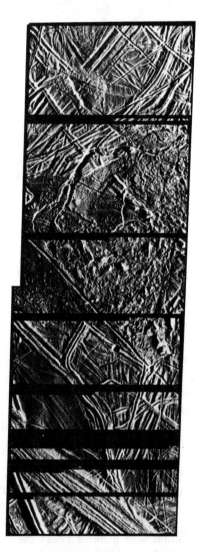

Europa's surface sports many more ridges and cracks than craters, as this composite of two photos taken by the *Galileo* spacecraft shows.

All images courtesy JPL

load, a five-inch-long mini-submersible, or hydrobot, to explore the foreign sea.

The hydrobot would contain instruments—precisely what sort hasn't been decided—to capture images and detect hints of life. If all went well, the results could be in by 2015. Nobody, however, expects a Europan whale to swim up and swallow the probe. In fact, even if Europa does have an ocean, it might not be compatible with life. "The three things necessary for life," Squyres says, "are liquid water, the right biogenic elements, and a biologically useful source of energy. The first two can be satisfied if Europa has an ocean: besides water, you would have salts, soluble organics—a broth of all the ingredients you'd need to create living material—that leached out of the rock along with the water. The big question is the energy."

On Earth, sunlight provides energy for most living things. On Europa, far from the sun, the most likely source is tidal heating, but only if it's energetic enough to produce submarine volcanism as well. A little warmth won't do. "For example, if you go to the Earth's seafloor," Squyres says, "there is heat leaking out everywhere. But only at the hydrothermal vents, where you have very high local temperatures, is there enough energy for life to run its metabolisms."

If submarine volcanism is supporting life on Europa, the organisms may resemble the high-temperature-loving microbes that thrive on the effluent spewed out of Earth's hydrothermal vents. From just such life, it is now believed, all organisms on Earth actually began. "The fact that we may all be descended from these guys who lived at hydrothermal vents doesn't necessarily mean life arose there and not at the surface," Squyres says. "If somebody took some seawater and hot basalt in the lab and made living organisms from stuff that wasn't living before, that would make the case for me." Or, perhaps, if living critters are someday found swimming in a dark Europan sea.

The Stars of Mars

After the rover's success, will man be next?

SHARON BEGLEY

AS THE MARTIAN ROVER PAID COURtesy calls up and down Ares Vallis last week, she showed the exemplary manners expected of any ambassador. Before reaching out with her Alpha Proton X-Ray Spectrometer for a 10-hour handshake with "Barnacle Bill," the rover first rolled toward this blue-gray stone at a stately half inch per second. Tipping a bit as she went over a small rock, she recovered gracefully, backed up about 12 inches and stuck her spectrometer onto Bill. She collected data on what he was made of. Then, commanded by controllers at NASA's Jet Propulsion laboratory in Pasadena, Calif., 119 million miles away, she set off on her next visit: to Yogi. Despite a slight fender bender en route that left her hung up on Yogi's face for two days, she was expected to relay more data to JPL. Those results, and the stark photos of the rock-strewn Ares Vallis region, made the Pathfinder mission seem like the best thing to ever hit NASA's Mars program. Except that some NASA officials believe that honor belongs to a very unlikely candidate. In August 1993, just three days short of its orbital rendezvous with the Red Planet, the Mars Observer spacecraft mysteriously disappeared. It had been, as NASA Administrator Daniel Goldin puts it, "the last ship out of port." Stuffed with scientific instruments, the $980 million mission exemplified the old NASA: loading up a spacecraft with every instrument scientists could think of, but then not sending another probe for a decade or more. A few months after the loss, Goldin visited a lab at Arizona State University. The scientists and graduate students there had just seen eight years of hard work on Mars Observer vanish. As the students stood around looking glum, Goldin announced the dawn of a new era, his listeners recalled. The loss of Observer, he declared, was the best thing ever to happen to America's Mars program. Now NASA had an open field to explore the solar system in a new way.

That new way is encapsulated by three words that Goldin has made the agency's mantra under his five-year reign: better,

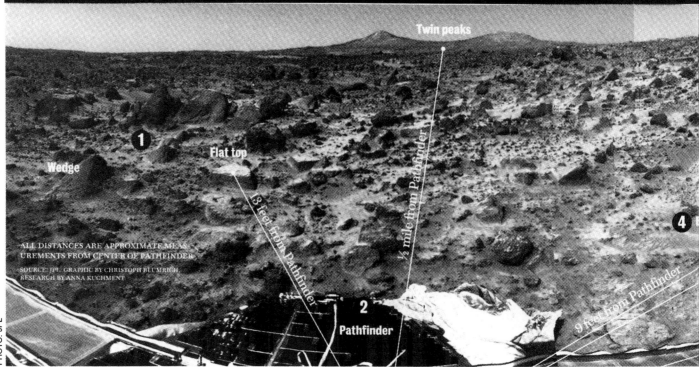

Making Tracks on Mars

Pathfinder landed in Ares Vallis, where a huge flood washed down rocks from the highlands 1 billion to 3 billion years ago. The rover examined two of the rocks last week for clues to Mars's past.

1 Flood plain: The tilt of the rocks shows that the flood, hundreds of miles across, washed down the valley from the southwest

26. Stars of Mars

faster, cheaper. Pathfinder's success, from its never-before-attempted landing on 17-foot airbags through the first-ever mobile exploration of another world, shows that the right stuff can be bought for a supersaver fare. It can also pump up public, media and presidential interest in the space program. Pathfinder Web sites recorded 220 million hits the first five days. CNN seemed to be gunning for the position of official Martian network. Bill Clinton, channel-surfing in Warsaw, "could not turn it off. It's just thrilling." And suddenly, there was serious talk about sending astronauts to Mars early in the 21st century.

At first many NASA vets were skeptical of the cut-rate approach. "When we got close to launch and were still under the [budget] cap," said Pathfinder manager Tony Spear, "other [NASA] managers told me they hadn't thought we'd make it." But they did, even returning hundreds of thousands of dollars to the NASA kitty. It helped the bottom line to let well-paid, long-serving space jocks retire and make way for kids happy to work long hours, to put their personal lives on hold, and even to plan things so their first baby isn't born for a good two months after Pathfinder landed. "The idea was to grab young scientists who didn't know the job was 'impossible'," says flight-software engineer Steve Stolper of JPL. Can this "better, faster, cheaper" be replicated on future missions?

Like Pathfinder, tomorrow's missions will be designed in NASA's new risk-taking culture. "We can tolerate failures," says Goldin. "We want to encourage people to take risks in spacecraft design." Risk means that NASA's days of quintuple redundancy are gone. There is only one radio receiver on the Pathfinder lander, for instance. If it fails, the rover cannot execute commands radioed up from JPL to the lander and then relayed. A tolerance for risk also means forgoing some certainty. Every previous lunar or planetary lander had been carried up to 120,000 feet at Mach 2 and then dropped to see how the parachutes worked. Viking, which touched down on Mars in 1976, underwent such tests, at $10 million a pop. Partly because Pathfinder was using the same parachute design as Viking, and partly to save money, "we didn't do these tests," says JPL's Brian Muirhead.

But accepting risk didn't mean that the rover team would take a chance that Sojourner would get stuck in some Martian cul-de-sac, or flip over and become as immobile as a supine turtle. Every Sunday night David (The Gremlin) Gruel locked himself inside Pathfinder's test bed—a sandbox—and pulled the blinds. There, alone, he took out his shovel and bag of rocks and did his worst. He built sand piles, rocky hills and boulder clusters, practically taunting the rover engineers to navigate them safely the next week. "The one that stumped the guys the most," says Gruel, 27, "actually happened." That was when he inflated the airbags under the lander's "petals" so they blocked the rover's exit down the ramp. The Sojourner team solved the puzzle by repeatedly lifting and lowering the petals and deflating the bag—exactly as they had to do the day after landing.

Another cost-saving move is to buy more parts off the shelf (diagram). The telecommunications team for the Sojourner rover realized that they couldn't design and make modems any better than the Motorola RNET 9600 and still meet the cost cap and launch date. In 1995 they bought 30. The engineers subjected them to frigid Martian temperatures, replaced plastic connectors with wires and substituted fiberglass and aluminum tape for the original metallic outer box. The $695 Spacetech IMC "spaceball" used to control the rover from JPL like a character in a videogame didn't even have to be modified.

Upcoming missions to Mars—one orbiter and one lander every 26 months—follow the Pathfinder script. Take the Global Surveyor, which begins mapping Mars next March 15. Its infrared camera captures

2 Lander: Renamed Sagan Station, Pathfinder opened its 'petals' after landing and extended its ramps. The rover rolled out and went to work.

3 On a roll: The spectrometer on the rover determined the elemental composition of rocks and soil by bombarding them with alpha particles

4 Close encounter: On July 6, the rover planted its spectrometer on Bill. It contains abundant quartz, formed by melting and remelting.

5 Moving violation: After dislodging a wheel that became stuck July 9, the rover prepared to chemically analyze Yogi

3. THE SOLAR SYSTEM: The Planets

unique radiation signatures emitted by different minerals, says Arizona State's Philip Christensen. Among the targets: carbonates, cherts and opals—all characteristic of thermal hot springs. Thermal springs may be the likeliest places to find signs of life on Mars. But not even a goal as grand as *finding where life lived on Mars* meant Christensen could spend like a teenager with his father's credit card. He had to design, build and deliver the camera for $8 million, compared with $22 million for an identical one on the lost Mars Observer. Why the big differential? First, says Christensen, "this is just a copy. So that argues for flying the same [kind of] instruments on several missions." Second, the price of the first spectrometer shot up when NASA, at the last minute, ordered the Arizona team working on Observer to use better quality parts. "We were told, 'This is a very expensive mission. It has to work'," recalls Christensen. "So I bought thousands of spare parts and tested them all. That raised [the camera's] odds of success from 95 percent to 98 percent—at a cost of several million dollars."

David Paige has also felt NASA's belt-tightening. The UCLA geologist and colleagues are providing the scientific payload for the next Martian lander, due to touch down on the Red Planet's equivalent of Antarctica in December 1999. The Mars Volatiles and Climate Surveyor would be the first Martian prospector cum chef. A six-foot robot arm will dig as deep as 18 inches, scoop up soil and feed it into an oven. The guiding principle here is that you can tell what's in something by smelling what wafts out when you cook it. (Paige's instrument will determine the chemical composition of emissions from the heated sample.) "We had $20 million to build and deliver it," says Paige. "We managed to do it with off-the-shelf technology and our own inventions." They had to invent the oven, but are adapting the rover's electronic controls and motors for the robot arm. And in a tectonic cultural shift, the 1999 mission will be run from several sites—JPL, but also UCLA. That way NASA doesn't have to spring for so many airfares and hotel rooms.

How low can a space budget go? Theoretical physicist Carl Kukkonen, who heads up JPL's "microdevices" lab, is readying $10 million probes that can be launched from the space shuttle on a rocket the size of a five-gallon gasoline can, or from the ground with a 10-foot-tall rocket "that three of us could carry," says Kukkonen. "We could be launching in the next decade." Each craft, hardly bigger than a birthday cake, would have its own navigation system. Each would be able to set its course by the stars and automatically correct it with tiny thrusters—all with no input from ground controllers. "They're fire-and-forget spacecraft," says Kukkonen. Equipped with a camera to shoot photos of a planet or asteroid and a spectrometer to determine chemical composition, they could be launched at a rate of one a month.

More Than the Sum of Its Parts

Sojourner was fitted with many items that were purchased off the shelf and extensively adapted for use on Mars. Here are some of them:

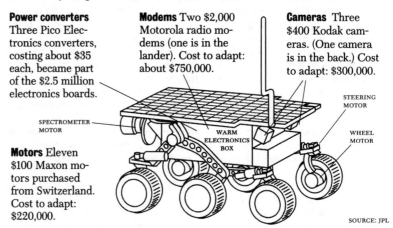

Power converters Three Pico Electronics converters, costing about $35 each, became part of the $2.5 million electronics boards.

Modems Two $2,000 Motorola radio modems (one is in the lander). Cost to adapt: about $750,000.

Cameras Three $400 Kodak cameras. (One camera is in the back.) Cost to adapt: $300,000.

Motors Eleven $100 Maxon motors purchased from Switzerland. Cost to adapt: $220,000.

SOURCE: JPL

O F COURSE, IF PATHFINDER had failed, then "better, faster, cheaper" would have looked like the dumbest idea since allowing the O-rings on the Challenger to become brittle. But it succeeded. Now NASA is applying that philosophy even to missions carrying astronauts. When President Bush asked NASA to cost out a crewed Mars mission, the agency came back with the idea-killing figure of $300 billion. Goldin has directed that the Johnson Space Center do it for $20 billion, tops.

A key to hitting this price point is to forget about carrying along fuel for a return flight. Extra weight means a bigger, costlier spacecraft and launcher. Instead, a lander arriving before the astronauts would make rocket fuel out of the carbon dioxide in Mars's atmosphere. It would also make oxygen, to supply the astronauts with breathable gas. The first oxygen factory will fly in 2001. Oxygen and fuel produced by a factory on the 2003 mission might be used to power a small rocket that would lift off from the surface of Mars, shoot some pictures, then land. If that works, the 2005 mission may use made-on-Mars fuel to power a spacecraft that would bring rock and soil samples back to Earth.

Perhaps as early as 2011, if all of this works, Earthlings might staff a permanent Martian station. "Eventual Mars settlement," said Elric McHenry of NASA's Advanced Projects Office, "is a guiding principle of our current planning." But the rent will have to be cheap, too. Goldin has so embraced the "living off the land" philosophy that he says astronauts will grow their own food on Mars, too. Other schemes: zapping the Martian permafrost with microwaves to extract water, mixing the water with Mars's claylike red dust to make bricks, turning the calcium and sulfur in the soil into gypsum.

If Sojourner's limited exploration is any indication, such a self-reliant settlement could be nestled in a landscape of quiet majesty. Scientists knew that Ares Vallis had been swept by a flood of Biblical proportions more than 1 billion years ago, so the "discovery" of such a flood, as trumpeted by headlines last week, wasn't really one. But the rover did turn up some surprises. Her analysis of Barnacle Bill found a rock as rich in silicon dioxide—quartz—as any in Earth's continental crust, said geologist Harry McSween of the University of Tennessee. Quartz can form from a complex melting and remelting of rock—"or of melting in the presence of a lot of water," says McSween. It's too early to know whether the arid, barren surface of Mars hides a watery interior— and hence more chances to create Martian life as well as to support, one day, Earthlings. But the real payoff of Sojourner and Pathfinder has been to drive home something that no space mission ever has. "It's being shoved in our face that there is a whole world up there, with mountains and hills and craters," says engineer Robert Zubrin, who helped conceive live-off-the-land. "What Pathfinder is doing is making Mars sensuous." And changing the face of space exploration. Within 15 years, astronauts on Mars may stumble upon the still and silent Pathfinder and Sojourner, their batteries long dead and their mission long over, and recognize them as the pioneers that put Earthlings on course for the Red Planet.

The Kuiper Belt

Rather than ending abruptly at the orbit of Pluto, the outer solar system contains an extended belt of small bodies

Jane X. Luu and
David C. Jewitt

After the discovery of Pluto in 1930, many astronomers became intrigued by the possibility of finding a 10th planet circling the sun. Cloaked by the vast distances of interplanetary space, the mysterious "Planet X" might have remained hidden from even the best telescopic sight, or so these scientists reasoned. Yet decades passed without detection, and most researchers began to accept that the solar system was restricted to the familiar set of nine planets.

But many scientists began seriously rethinking their notions of the solar system in 1992, when we identified a small celestial body—just a few hundred kilometers across—sited farther from the sun than any of the known planets. Since that time, we have identified nearly three dozen such objects circling through the outer solar system. A host of similar objects is likely to be traveling with them, making up the so-called Kuiper belt, a region named for Dutch-American astronomer Gerard P. Kuiper, who, in 1951, championed the idea that the solar system contains this distant family.

What led Kuiper, nearly half a century ago, to believe the disk of the solar system was populated with numerous small bodies orbiting at great distances from the sun? His conviction grew from a fundamental knowledge of the behavior of certain comets—masses of ice and rock that on a regular schedule plunge from the outer reaches of the solar system inward toward the sun. Many of these comparatively small objects periodically provide spectacular appearances when the sun's rays warm them enough to drive dust and gas off their surfaces into luminous halos (creating large "comae") and elongate tails.

Astronomers have long realized that such active comets must be relatively

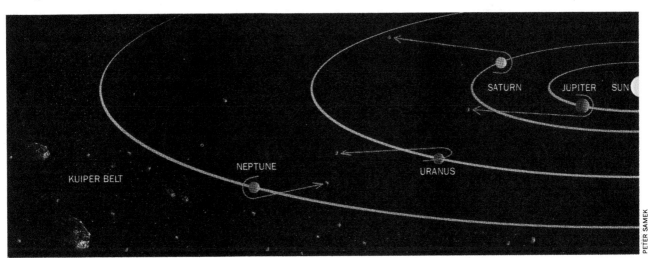

GRAVITY OF THE PLANETS acted during the early stages of the solar system to sweep away small bodies within the orbit of Neptune. Some of these objects plummeted toward the sun; others sped outward toward the distant Oort cloud (*not shown*).

3. THE SOLAR SYSTEM: Comets

new members of the inner solar system. A body such as Halley's comet, which swings into view every 76 years, loses about one ten-thousandth of its mass on each visit near the sun. That comet will survive for only about 10,000 orbits, lasting perhaps half a million years in all. Such comets were created during the formation of the solar system 4.5 billion years ago and should have completely lost their volatile constituents by now, leaving behind either inactive, rocky nuclei or diffuse streams of dust. Why then are so many comets still around to dazzle onlookers with their displays?

Guiding Lights

The comets that are currently active formed in the earliest days of the solar system, but they have since been stored in an inactive state—most of them preserved within a celestial deep freeze called the Oort cloud. The Dutch astronomer Jan H. Oort proposed the existence of this sphere of cometary material in 1950. He believed that this cloud had a diameter of about 100,000 astronomical units (AU—a distance defined as the average separation between Earth and the sun, about 150 million kilometers) and that it contained several hundred billion individual comets. In Oort's conception, the random gravitational jostling of stars passing nearby knocks some of the outer comets in the cloud from their stable orbits and gradually deflects their paths to dip toward the sun.

For most of the past half a century, Oort's hypothesis neatly explained the size and orientation of the trajectories that the so-called long-period comets (those that take more than 200 years to circle the sun) follow. Astronomers find that those bodies fall into the planetary region from random directions—as would be expected for comets originating in a spherical repository like the Oort cloud. In contrast, Oort's hypothesis could not explain short-period comets that normally occupy smaller orbits tilted only slightly from the orbital plane of Earth—a plane that astronomers call the ecliptic.

Most astronomers believed that the short-period comets originally traveled in immense, randomly oriented orbits (as the long-period comets do today) but that they were diverted by the gravity of the planets—primarily Jupiter—into their current orbital configuration. Yet not all scientists subscribed to this idea. As early as 1949, Kenneth Essex Edgeworth, an Irish gentleman-scientist (who was not affiliated with any research institution) wrote a scholarly article suggesting that there could be a flat ring of comets in the outer solar system. In his 1951 paper, Kuiper also discussed such a belt of comets, but he did not refer to Edgeworth's previous work.

Kuiper and others reasoned that the disk of the solar system should not end abruptly at Neptune or Pluto (which vie with each other for the distinction of being the planet most distant from the sun). He envisioned instead a belt beyond Neptune and Pluto consisting of residual material left over from the formation of the planets. The density of matter in this outer region would be so low that large planets could not have accreted there, but smaller objects, perhaps of asteroidal dimensions, might exist. Because these scattered remnants of primordial material were so far from the sun, they would maintain low surface temperatures. It thus seemed likely that these distant objects would be composed of water ice and various frozen gases—making them quite similar (if not identical) to the nuclei of comets.

Kuiper's hypothesis languished until the 1970s, when Paul C. Joss of the Massachusetts Institute of Technology began to question whether Jupiter's gravity could in fact efficiently transform long-period comets into short-period ones. He noted that the probability of gravitational capture was so small that the large number of short-period comets that now exists simply did not make sense. Other researchers were, however, unable to confirm this result, and the Oort cloud remained the accepted source of the comets, long and short period alike.

But Joss had sown a seed of doubt, and eventually other astronomers started to question the accepted view. In 1980 Julio A. Fernández (then at the Max Planck Institute for Aeronomy in Katlenburg-Lindau) had, for example, done calculations that suggested that short-period comets could come from

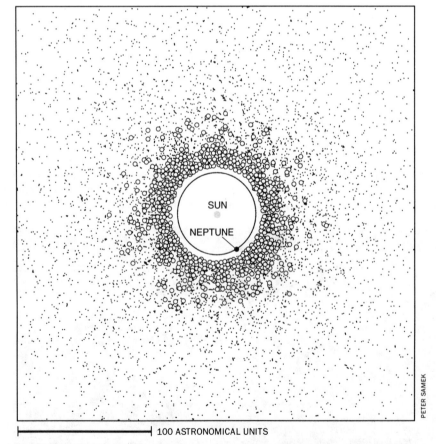

COUNTLESS OBJECTS in the Kuiper belt may orbit far from the sun, but not all of those bodies can be seen from Earth. Objects (*circles*) that could reasonably be detected with the telescope on Mauna Kea in Hawaii typically lie near the inner border of the belt, as seen in this computer simulation of the distribution of distant matter.

Kuiper's proposed trans-Neptunian source. In 1988 Martin J. Duncan of the University of Toronto, Thomas Quinn and Scott D. Tremaine (both at the Canadian Institute for Theoretical Astrophysics) used computer simulations to investigate how the giant gaseous planets could capture comets. Like Joss, they found that the process worked rather poorly, raising doubts about the veracity of this well-established concept for the origin of short-period comets. Indeed, their studies sounded a new alarm because they noted that the few comets that could be drawn from the Oort cloud by the gravitational tug of the major planets should be traveling in a spherical swarm, whereas the orbits of the short-period comets tend to lie in planes close to the ecliptic.

Duncan, Quinn and Tremaine reasoned that short-period comets must have been captured from original orbits that were canted only slightly from the ecliptic, perhaps from a flattened belt of comets in the outer solar system. But their so-called Kuiper belt hypothesis was not beyond question. In order to make their calculations tractable, they had exaggerated the masses of the outer planets as much as 40 times (thereby increasing the amount of gravitational attraction and speeding up the orbital evolution they desired to examine). Other astrophysicists wondered whether this computational sleight of hand might have led to an incorrect conclusion.

Why Not Just Look?

Even before Duncan, Quinn and Tremaine published their work, we wondered whether the outer solar system was truly empty or instead full of small, unseen bodies. In 1987 we began a telescopic survey intended to address exactly that question. Our plan was to look for any objects that might be present in the outer solar system using the meager amount of sunlight that would be reflected back from such great distances. Although our initial efforts employed photographic plates, we soon decided that a more promising approach was to use an electronic detector (a charge-coupled device, or CCD) attached to one of the larger telescopes.

We conducted the bulk of our survey using the University of Hawaii's 2.2-meter telescope on Mauna Kea. Our strategy was to use a CCD array with this instrument to take four sequential, 15-minute exposures of a particular segment of the sky. We then enlisted a computer to display the images in the sequence in quick succession—a process astronomers call "blinking." An object that shifts slightly in the image against the background of stars (which appear fixed) will reveal itself as a member of the solar system.

For five years, we continued the search with only negative results. But the technology available to us was improving so rapidly that it was easy to maintain enthusiasm (if not funds) in the continuing hunt for our elusive quarry. On August 30, 1992, we were taking the third of a four-exposure sequence while blinking the first two images on a computer. We noticed that the position of one faint "star" appeared to move slightly between the successive frames. We both fell silent. The motion was quite subtle, but it seemed definite. When we compared the first two images with the third, we realized that we had indeed found something out of the ordinary. Its

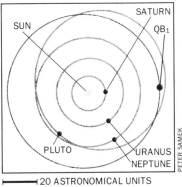

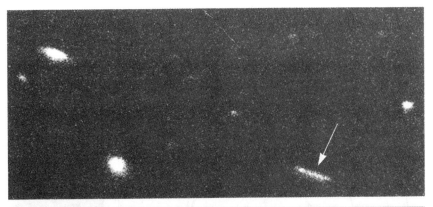
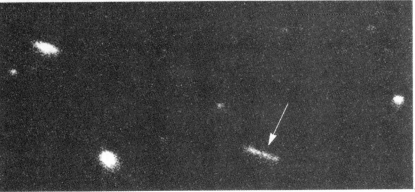

SEQUENTIAL CCD EXPOSURES from 1992 revealed Kuiper belt object QB$_1$ clearly against the background of fixed stars (*middle* and *bottom*). This pair of images covers only a small part of the complete CCD frame (*top right*) that had to be analyzed before the authors could identify QB$_1$ (*arrows*) and determine its orbit (*top left*).

3. THE SOLAR SYSTEM: Comets

slow motion across the sky indicated that the newly discovered object could be traveling beyond even the outer reaches of Pluto's distant orbit. Still, we were suspicious that the mysterious object might be a near-Earth asteroid moving in parallel with Earth (which might also cause a slow apparent motion). But further measurements ruled out that possibility.

We observed the curious body again on the next two nights and obtained accurate measurements of its position, brightness and color. We then communicated these data to Brian G. Marsden, director of the International Astronomical Union's Central Bureau of Astronomical Telegrams at the Smithsonian Astrophysical Observatory in Cambridge, Mass. His calculations indicated that the object we had discovered was indeed orbiting the sun at a vast distance (40 AU)—only slightly less remote than we had first supposed. He assigned the newly discovered body a formal, if somewhat drab, name based on the date of discovery: he christened it "1992 QB_1." (We preferred to call it "Smiley," after John Le Carré's fictional spy, but that name did not take hold within the conservative astronomical community.)

Our observations showed that QB_1 reflects light that is quite rich in red hues compared with the sunlight that illuminates it. This odd coloring matched only one other object in the solar system—a peculiar asteroid or comet called 5145 Pholus. Planetary astronomers attribute the red color of 5145 Pholus to the presence of dark, carbon-rich material on its surface. The similarity between QB_1 and 5145 Pholus thus heightened our excitement during the first days after the discovery. Perhaps the object we had just located was coated by some kind of red material abundant in organic compounds. How big was this ruddy new world? From our first series of measurements, we estimated that QB_1 was between 200 and 250 kilometers across—about 15 times the size of the nucleus of Halley's comet.

Some astronomers initially doubted whether our discovery of QB_1 truly signified the existence of a population of objects in the outer solar system, as Kuiper and others had hypothesized. But such questioning began to fade when we found a second body in March 1993. This object is as far from the sun as QB_1 but is located on the opposite side of the solar system. During the past three years, several other research groups have joined the effort, and a steady stream of discoveries has ensued. The current count of trans-Neptunian, Kuiper belt objects is 32.

The known members of the Kuiper belt share a number of characteristics. They are, for example, all located beyond the orbit of Neptune, suggesting that the inner edge of the belt may be defined by this planet. All these newly found celestial bodies travel in orbits that are only slightly tilted from the ecliptic—an observation consistent with the existence of a flat belt of comets. Each of the Kuiper belt objects is millions of times fainter than can be seen with the naked eye. The 32 objects range in diameter from 100 to 400 kilometers, making them considerably smaller than both Pluto (which is about 2,300 kilometers wide) and its satellite, Charon, (which measures about 1,100 kilometers across).

The current sampling is still quite modest, but the number of new solar system bodies found so far is sufficient to establish beyond doubt the existence of the Kuiper belt. It is also clear that the belt's total population must be substantial. We estimate that the Kuiper belt contains at least 35,000 objects larger than 100 kilometers in diameter. Hence, the Kuiper belt probably has a total mass that is hundreds of times larger than the well-known asteroid belt between the orbits of Mars and Jupiter.

Cold Storage for Comets

The Kuiper belt may be rich in material, but can it in fact serve as the supply source for the rapidly consumed short-period comets? Matthew J. Holman and Jack L. Wisdom, both then at M.I.T., addressed this problem using computer simulations. They showed that within a span of 100,000 years the gravitational influence of the giant gas-

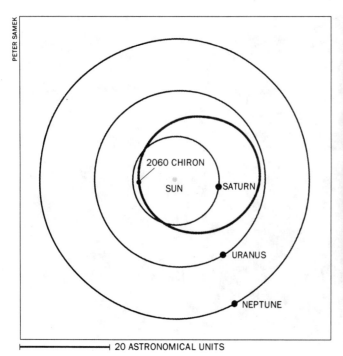

2060 CHIRON may have escaped from the Kuiper belt into its current planet-crossing orbit (*left*). Although quite faint, the subtle glow surrounding 2060 Chiron (*far right*) marks this object as a celestial cousin to other "active" bodies, such as Comet Peltier (*above*).

eous planets (Jupiter, Saturn, Uranus and Neptune) ejects comets orbiting in their vicinity, sending them out to the farthest reaches of the solar system. But a substantial percentage of trans-Neptunian comets can escape this fate and remain in the belt even after 4.5 billion years. Hence, Kuiper belt objects located more than 40 AU from the sun are likely to have held in stable orbits since the formation of the solar system.

Astronomers also believe there has been sufficient mass in the Kuiper belt to supply all the short-period comets that have ever been formed. So the Kuiper belt seems to be a good candidate for a cometary storehouse. And the mechanics of the transfer out of storage is now well understood. Computer simulations have shown that Neptune's gravity slowly erodes the inner edge of the Kuiper belt (the region within 40 AU of the sun), launching objects from that zone into the inner solar system. Ultimately, many of these small bodies slowly burn up as comets. Some—such as Comet Shoemaker-Levy 9, which collided with Jupiter in July 1994—may end their lives suddenly by striking a planet (or perhaps the sun). Others will be caught in a gravitational slingshot that ejects them into the far reaches of interstellar space.

If the Kuiper belt is the source of short-period comets, another obvious question emerges: Are any comets now on their way from the Kuiper belt into the inner solar system? The answer may lie in the Centaurs, a group of objects that includes the extremely red 5145 Pholus. Centaurs travel in huge planet-crossing orbits that are fundamentally unstable. They can remain among the giant planets for only a few million years before gravitational interactions either send them out of the solar system or transfer them into tighter orbits.

With orbital lifetimes that are far shorter than the age of the solar system, the Centaurs could not have formed where they currently are found. Yet the nature of their orbits makes it practically impossible to deduce their place of origin with certainty. Nevertheless, the nearest (and most likely) reservoir is the Kuiper belt. The Centaurs may thus be "transition comets," former Kuiper belt objects heading toward short but showy lives within the inner solar system. The strongest evidence supporting this hypothesis comes from one particular Centaur—2060 Chiron. Although its discoverers first thought it was just an unusual asteroid, 2060 Chiron is now firmly established as an active comet with a weak but persistent coma.

As astronomers continue to study the Kuiper belt, some have started to wonder whether this reservoir might have yielded more than just comets. Is it coincidence that Pluto, its satellite, Charon, and the Neptunian satellite Triton lie in the vicinity of the Kuiper belt? This question stems from the realization that Pluto, Charon and Triton share similarities in their own basic properties but differ drastically from their neighbors.

A Peculiar Trio

The densities of both Pluto and Triton, for instance, are much higher than any of the giant gaseous planets of the outer solar system. The orbital motions of these bodies are also quite strange. Triton revolves around Neptune in the "retrograde" direction—opposite to the orbital direction of all planets and most satellites. Pluto's orbit slants highly from the ecliptic, and it is so far from circular that it actually crosses the orbit of Neptune. Pluto is, however, protected from possible collision with the larger planet by a special orbital relationship known as a 3:2 mean-motion resonance. Simply put, for every three orbits of Neptune around the sun, Pluto completes two.

The pieces of the celestial puzzle may fit together if one postulates that Pluto, Charon and Triton are the last survivors of a once much larger set of similarly sized objects. S. Alan Stern of the Southwest Research Institute in Boulder first suggested this idea in 1991. These three bodies may have been swept up by Neptune, which captured Triton and locked Pluto—perhaps with Charon in tow—into its present orbital resonance.

Interestingly, orbital resonances appear to influence the position of many Kuiper belt objects as well. Up to one half of the newly discovered bodies have the same 3:2 mean-motion resonance

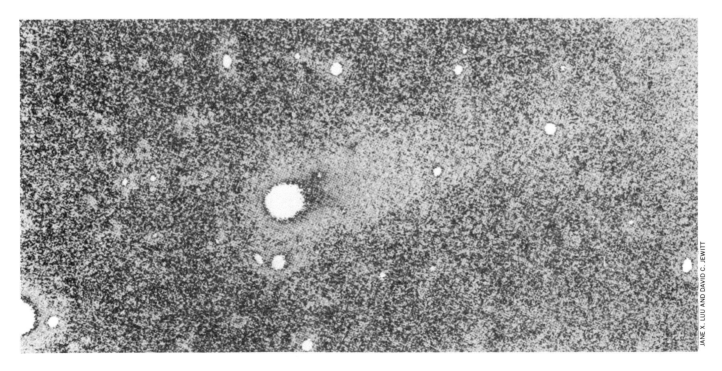

3. THE SOLAR SYSTEM: Comets

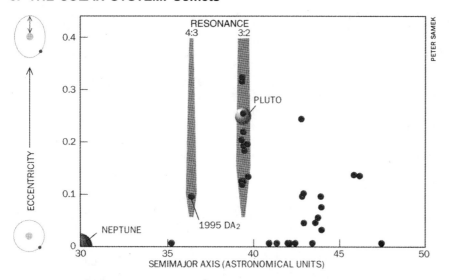

MEAN-MOTION RESONANCE governs the size and shape of the orbits of many Kuiper belt objects. Orbits are described by eccentricity (deviation from circularity) and semimajor axis (*top arrow*). Like Pluto, about half the known Kuiper belt bodies (*grey points*) circle the sun twice while Neptune completes three orbits—a 3:2 resonance. The object 1995 DA$_2$ orbits in one of the other resonances. Renu Malhotra of the Lunar and Planetary Institute in Houston suggests that this pattern reflects the early evolution of the solar system, when many small bodies were ejected and the major planets migrated away from the sun. During these outward movements, Neptune could have drawn Pluto and a variety of smaller bodies into the resonant orbits that are now observed.

as Pluto and, like that planet, may orbit serenely for billions of years. (The resonance prevents Neptune from approaching too closely and disturbing the orbit of the smaller body.) We have dubbed such Kuiper belt objects Plutinos—"little Plutos." Judging from the small part of the sky we have examined, we estimate that there must be several thousand Plutinos larger than 100 kilometers across.

The recent discoveries of objects in the Kuiper belt provide a new perspective on the outer solar system. Pluto now appears special only because it is larger than any other member of the Kuiper belt. One might even question whether Pluto deserves the status of a full-fledged planet. Strangely, a line of research that began with attempts to find a 10th planet may, in a sense, have succeeded in reducing the final count to eight. This irony, along with the many intriguing observations we have made of Kuiper belt objects, reminds us that our solar system contains countless surprises.

The Authors

JANE X. LUU and DAVID C. JEWITT came to study astronomy in different ways. For Jewitt, astronomy was a passion he developed as a youngster in England. Luu's childhood years were filled with more practical concerns: as a refugee from Vietnam, she had to learn to speak English and adjust to life in southern California. She became enamored of astronomy almost by accident, during a summer spent at the Jet Propulsion Laboratory in Pasadena. Luu and Jewitt began their collaborative work in 1986 at the Massachusetts Institute of Technology. Jewitt was a professor there when Luu became a graduate student. Jewitt moved to the University of Hawaii in 1988. It was during Luu's postdoctoral fellowship at the Harvard-Smithsonian Center for Astrophysics that Luu and Jewitt discovered the first Kuiper belt object. In 1994 Luu joined the faculty of Harvard University.

Further Reading

THE ORIGIN OF SHORT PERIOD COMETS. Martin Duncan, Thomas Quinn and Scott Tremaine in *Astrophysical Journal*, Vol. 328, pages L69–L73; May 15, 1988.

THE KUIPER BELT OBJECTS. J. X. Luu in *Asteroids, Comets, Meteors 1993*. Edited by A. Milani, M. Di Martino and A. Cellino. Kluwer Academic Publishers, 1993.

THE SOLAR SYSTEM BEYOND NEPTUNE. D. C. Jewitt and J. X. Luu in *Astronomical Journal*, Vol. 109, No. 4, pages 1867–1876; April 1995.

THE ORIGIN OF PLUTO'S ORBIT: IMPLICATIONS FOR THE SOLAR SYSTEM BEYOND NEPTUNE. Renu Malhotra in *Astronomical Journal*, Vol. 110, pages 420–429; July 1995.

COMETS THAT CHANGED THE WORLD

BY BRADLEY E. SCHAEFER

GO OUT TONIGHT AND WATCH Comet Hale-Bopp flaming in the sky. Do you feel fear? Historically, humanity has believed that comets signaled catastrophes such as droughts, earthquakes, epidemics, death, and the destruction of cities.

The most famous comet is Halley's, and its best-known premodern apparition is that of A.D. 1066, which was associated with William the Conqueror. The comet was first sighted in April of that year, and it remained visible until June. Both the English and the Normans were preparing for the invasion, and both saw the spectacle as a dire omen, though no one knew whom the celestial sword was pointing at. It was only after the Battle of Hastings in October that the chroniclers decided the comet must have been bad for the English.

But that is a story of how a comet was noted in a historical context, whereas numerous apparitions substantially affected many people as they occurred. Comets have the power to impel people to action because they are taken to be signs in the sky placed there by gods, and any such messages must therefore be acted upon. So here are five case studies of comets that changed history.

THE *SIDUS JULIUM*

Julius Caesar was assassinated by about two dozen aristocratic knifemen in the Roman senate on the Ides of March, 44 B.C. He had been the greatest Roman conqueror and statesman, though his murderers claimed moral superiority because Caesar had ended political freedoms by declaring himself lifetime dictator a month earlier. His death started volatile factional rivalries, with much of the action centering on Caesar's will. His testament included a substantial sum to every Roman citizen and most of the rest of his huge estate to his nephew Octavian, whom he also adopted as his son. However, Mark Antony had seized control of Rome and had taken Caesar's wealth and papers.

Octavian was an untried youth of 18 in whom his uncle must have seen some spark of greatness. He heard the news of the assassination while studying in Illyria. After consulting with friends, he decided to return cautiously to Rome. Upon crossing the Adriatic Sea, he learned of his adoption and immediately changed his surname to Caesar, fully aware of the danger that placed him in.

But his political genius shone as he worked with all sides while helping none. Before reaching Rome, he had respectful and wary meetings with various factional leaders, leaving an impression of restraint and wisdom. Next, in Rome, he formally accepted his adoption, thus becoming a leader of the Caesarian party. Then he ingratiated himself to the Roman people by paying Caesar's bequest with money scraped together from relatives and property sales. Finally, he held public games previously promised by Julius Caesar to celebrate the latter's victories.

These games were dedicated to Venus Genetrix and ran from July 20th to 30th. On the first day a bright comet appeared in the northern sky; it was visible in the evening for seven days. Octavian immediately declared the comet to be the soul of Julius Caesar (*sidus Julium*) being taken up to the heavens. This inspired an effective campaign aimed at Caesar's deification. The youth ordered a memorial statue erected with the sidus Julium over Caesar's head as a symbol of divinity. The comet was repeatedly invoked on coins, finger rings, Octavian's helmet, and all of Julius's statues, as well as in songs and poems. Within a year and a half, the Senate declared Julius Caesar to be a god.

This comet had various far-reaching

3. THE SOLAR SYSTEM: Comets

implications. First, it stood as a rallying symbol for the Caesarian forces. Symbols make effective propaganda, and their competitors had no such potent icon. Second, the apotheosis of Julius made Octavian the son of a god. As such, his authority became accepted as inevitable. Third, the deification based on the comet was the first for any Roman leader. This soon set the style and politics for upcoming emperors, eventually leading to the "divine "rights" of medieval monarchs. Octavian had started out with only a famous name and a comet, but he levered these into control over the entire Roman Empire as Augustus Caesar.

NERO'S PARANOIA

With the sidus Julium as an example, the Romans regarded comets as omens for the deaths of emperors. The comet of A.D. 54 was thus associated with the death of Emperor Claudius, though it was not a natural death but one of poisoning by his wife, Agrippina. She placed her son, Nero, on the imperial throne at the age of 16. The government was effectively run by the philosopher Seneca and the praetorian prefect Burrus, who kept the empire functioning smoothly. Seneca was Nero's tutor and an important Latin writer. He also amassed a fortune through corruption, including usury in Britain that led directly to the revolt of its queen, Boudicca. Within a few years of his coronation, Nero gradually asserted his independence, most prominently with public performances on the lyre, charioteering exhibitions, and readings of his poetry.

In the year 60 a bright comet blazed in the sky from August to December. The historian Tacitus (c. 56–117) tells us: "The general belief is that a comet means a change of emperor. So people speculated on Nero's successor as though Nero were already dethroned." The most likely candidate was sent into exile, where he was soon assassinated by the emperor's soldiers. To placate his ex-pupil, Seneca wrote a scholarly comet study with a surprisingly modern outlook. It centered on physical theories to explain comets with no mention of divine implications. He tried to distinguish the comets of Claudius and Nero by the directions they traveled in the sky, with the subtext being that Nero need not be frightened. But the emperor became progressively more depraved and bloodthirsty, ultimately murdering his own mother, brother, and two wives, as well as virtually all other relatives and friends.

Then, toward the end of 64, another bright comet appeared. Tacitus says that the comet was "atoned for by Nero by aristocratic blood." The historian Suetonius (c. 70–130) says Nero's "astrologer Balbillus observed that monarchs usually avoided portents of this kind by executing their most prominent subjects and thus directing the wrath of the heavens elsewhere; so Nero resolved on a wholesale massacre of the nobility." Many dozen victims from the highest level of Roman society were executed, and even Seneca committed suicide on the orders of his former student.

Nero's comets were interpreted as fatal omens, and the emperor's reaction proved their accuracy with the decimation of the Roman upper class.

MONTEZUMA'S COMET

The Aztec Empire covered much of Central America after it started expanding in 1427. There was perpetual warfare to supply slaves and human sacrifices. An estimated 20,000 victims were ritually slaughtered every year, often with hideous tortures. The capital was Tenochtitlán, on the current site of Mexico City, which was among the largest cities in the world in the early 1500s. The metropolis bustled with activity and commerce, had immense well-ordered markets, boasted botanical and zoological gardens, and was highlighted by monumental architecture and great temple pyramids. Montezuma II became emperor in 1502 and continued the endless wars, in part to supply the blood lusts of his priests.

The Aztec culture borrowed heavily from the earlier Toltec civilization, including many of its gods and myths. Chief among these are legends about the great Quetzalcoatl, a priest/hero of the Toltecs. Quetzalcoatl was transformed by the Aztecs into a god of wisdom in the form of a plumed serpent. Centuries earlier the legendary hero had departed eastward, setting sail on a raft of snakes. Before leaving, he promised to someday return out of the eastern seas to reclaim his land, and the Aztec tradition was that this would occur in the year called 1 Reed.

In 1517 many chroniclers reported that a "flaming ear of corn" appeared in the sky every midnight and stayed visible until the Sun's glare hid it in the morning. The apparition was variously described as "wide at the base and narrow at the peak," "a brilliant white cloud," and "a great column of flame." Montezuma saw the comet before any of his soothsayers or sorcerers, and after questioning them he had them tortured to death for their lack of attentiveness (their houses were razed and their families enslaved, too).

Montezuma was sorely frightened, and his people wept and shouted in terror. In consultation with the prophetic king of Tezcoco, he determined that the ill omen foretold death on a massive scale as well as the fall of the empire. Montezuma confronted his terror with yet more gruesome rituals, capital punishment of more retainers, and construction of even larger sacrificial altars. The continued menace of the comet depressed the emperor, who even attempted to flee the throne and hide in a cave.

Into this setting the conquistadors landed from the eastern sea in the Aztec year 1 Reed. This was 1519, and the 508 Spanish soldiers were led by Hernán Cortés and a greed for gold. Over the next half year the Spaniards fought battles and recruited allies from among the Aztecs' vassals. Their guns, horses, and hunting dogs made an irresistible force when combined with the allied native armies. Gathering strength, the force marched on Tenochtitlán.

With the first reports of a strange people on the Caribbean shore, Montezuma immediately realized that this must be the prophesied return of Quetzalcoatl. The coincidence of the year and the omen of the comet was strong evidence. The emperor sent envoys bearing fabulous gifts befitting a returning divinity. Further ambassadors and magicians could not turn back the Spaniards who had seen the Aztec gold. When Cortés reached Tenochtitlán on November 8, 1519, Montezuma gave splendid gifts and handed over control of the capital. Essentially, a comet had spooked the emperor into letting his enemies march in and take control with no opposition. After Montezuma's kidnapping, his massive ransom in gold, his death, treacherous massacres, and a smallpox epidemic, the Aztec Empire fell.

The doom foretold by the comet was a self-fulfilling prophecy, as Montezuma allowed Cortés's army to occupy and control Tenochtitlán without resistance.

THE MILLERITE COMET

The northeastern United States was a hotbed of religious fervor and revival

28. Comets That Changed the World

VIRGIN-SACRIFICE HOAX

The following story appeared on page 2 of the May 19, 1910, issue of the *Los Angeles Times* with a dateline of Aline, Oklahoma:

> Miss Jane Warfield, 16 years old, was today rescued from a band of religious fanatics twenty-five miles southwest of this place. She was about to be offered up as a sacrifice to make a blood atonement that the sins of the world might be forgiven.
>
> The Sheriff of Dewey county had been informed that the band of religious enthusiasts known as the Select Followers had given out that their leader, Henry Heinman, had received a revelation from God that the world was to end today and that the heavens would be rolled up like a scroll following contact with the tail of the comet; that the only thing that would avert the disaster was a blood sacrifice; that in order to save the world a sacrifice had been planned and the lot had fallen to Jane Warfield.
>
> Sheriff Hughes with a posse of six men reached the dell in the Glass Mountains just as the sacrifice was being prepared, and only in time to rescue the girl. The Warfield girl was clad in spotless white, with a wreath of white roses about her head. Her hands were bound and Heinman was standing in front of her with a long, keen hunting knife in his hand. About him were grouped about forty of his followers.
>
> Sheriff Hughes placed the girl in the hands of safe parties today and Heinman will be held to await the action of investigating officers.
>
> The sect came here two years ago, led by Heinman, who is said to be from Leesburg, O., and a graduate of some Ohio university. At one time he was a disciple of Harmon, the Free Thinker.

Wow! I would love to see a painting by someone in the style of Rembrandt or Albert Bierstadt of the evocative scene in the dell.

As part of my research for this article, I spent much time trying to corroborate this story. I called up sheriff's offices, I scanned through many regional and state newspapers on microfilm, I called local historical societies, I checked census records, and I contacted world experts on American cults. Nowhere could I find support for the article. Then I got a lead to a woman in Aline (born in the year 1910) who had previously tracked down the story with local material. She found that the virgin-sacrifice story was a hoax. It was perpetrated by a lawyer named Edgar Marchant, the former owner of the *Aline Chronoscope*, who had a history of sending out tall tales.

Even though the story is not true, it is nonetheless a valid indicator of the mood produced by Comet Halley. That is, people were disposed to believe such stories, and comet scares stem from people's beliefs. The concocted story appeared in various newspapers even as the *Los Angeles Times* printed a long editorial deriding the morals and ignorance of the age.

B.E.S.

---

during the early 1800s. Many sects and cults were founded, which flourished then floundered. A prominent sect was the Millerites, named after William Miller (1782–1849). His study of the Bible convinced him that the second advent of Christ was due about the year 1843. He started preaching in 1831 and rapidly picked up followers as the appointed end of the world approached. By 1843, an estimated 50,000 to 100,000 people were Millerites.

The Millerites relied on more than just Biblical prophecy applied to ancient history; they also had current signs in the sky. They were frequently pointing out halo complexes, sundogs, and auroras. The great Leonid meteor storm of 1833 was an extraordinary and strong omen of the end of the world. Then, as the fated year came, a great comet appeared in the sky. This Sun-grazer was visible in daylight, and it ranks in a tie as the brightest comet of all time. The sight was astounding to all. One person saw cryptic runes in the tail and asked Miller to decipher their meaning. The fearsome omen persuasively supported the coming doomsday.

An example of this dread took place in Skaneateles, New York. A Millerite preacher on the streets received a hostile reception from villagers, who even threw snowballs and tossed pails of water on him. However, when the evangelist pointed at the fire blazing across the sky, their ridicule changed to apprehension. The preaching that followed was then closely and respectfully considered and had a great effect on the townsfolk.

This comet appeared without warning and had an unknown physical nature — both essentials for it to be perceived as a divine omen — and thus could serve as a warning to the faithful up until the day of the Great Disappointment.

COMET PANICS

Fear of comets can spread widely, even reaching the point where a large fraction of the population acts hysterically. Famous panics occurred in 1773, 1798, 1843, and 1857. I suspect that the existence of an active press was required for fears to propagate and intensify, though perhaps it is simply that the press allows us to learn about individual scares. The hysteria need not be related to anything in the sky, since the panics in 1798 and 1857 had no comets visible.

The last great comet scare was in 1910, when Comet Halley made its much-heralded return. This apparition was a bright one, because the comet passed close to the Earth. The resulting large and bright fire in the sky could frighten people, while the proximity suggested an impact with our planet. For example, the *Chicago Daily Tribune* headlined "Comet to Bounce off Earth Today" at the top of its front page.

The Earth even passed through the comet's tail on May 18th. To make mat-

COMET TO BOUNCE OFF EARTH TODAY

From *The Chicago Daily Tribune*, May 18, 1910

ters worse, astronomers had just discovered the poisonous gas cyanogen in the tail's spectrum. Newspaper headlines warned: "Comet's Tail Poisonous: French Astronomer Thinks Gases Might Affect the Earth's Atmosphere." Such reports were often rejected but nonetheless spotlighted. This led to widespread construction of gas-proof rooms around the world, sometimes with instructions from scientists. A typical headline was "Chicago is

3. THE SOLAR SYSTEM: Comets

Terrified: Women Are Stopping Up Doors and Windows to Keep Out Cyanogen." I have found other reports of extensive construction in South Africa, Haiti, and France. The sale of comet pills, inhalers, and conjure bags at exorbitant prices was brisk in Haiti, Texas, and Georgia.

Generalized terror was found everywhere across the United States. Front-page headlines screamed: "See Comet Then Die," "Several Driven Insane: Others Commit Suicide," and "Killed Watching For Comet." From the New York, Los Angeles, and Chicago newspapers alone, I find reports of eight suicide attempts (seven successful), five people committed to insane asylums, two murders, five accidental deaths, one murder confession, and one virgin-sacrifice hoax (see above) that were all directly attributable to the comet. The toll throughout America and the world must have been much higher. Work stoppages were widespread, for example by most farm workers over the entire South, miners across the nation, immigrants in major cities, and chorus girls in Chicago. Religious revivals did their biggest business since the doomsday of the Millerites, and even professional astronomers attracted crowds of thousands. All evidence suggests that some small fraction of Americans were terrorized to hysteria, while a larger percentage were at least deeply frightened.

Nevertheless, perhaps the majority of people were not really worried. This is best reflected in the newspapers, which devoted at most one page a day to the return of Comet Halley. Also, I found no notices of cancellations, so sporting and theatrical events kept their schedules as if nothing were going on. Cartoons of the day showed a zeal for poking fun at both astronomers and the public's fears. Indeed, every major and minor city had numerous "comet parties" for an evening of gaiety, while a thousand Dartmouth College students performed a snake dance in pajamas at midnight and then went to serenade the astronomers.

The proximity in time and place to our own culture suggests that we too are susceptible to such panics.

ARE COMETS EVIL?

The question is occasionally raised whether comets are generally viewed as good or bad signs. To answer this for Western civilization, I have compiled a catalog of 35 comets viewed as omens from Roman (including Byzantine) times across the central and eastern Mediterranean region. Of these, only two were of definitely positive character. In a study of Swedish records, 90 out of 90 accounts saw comets as negative. Elsewhere in the world, comets are universally regarded as bad omens as reported by ethnographers for Tibetans, Aztecs, Chinese, Incas, and North Americans, as well as all western-European-based societies. All seven comets discussed in the case studies on the previous pages related the omen to either the death of emperors or the end of the world. This universality demonstrates that comet phobia is based deep in the psyche of humanity.

The rationale for comet phobia has changed over the millenniums. The original basis for such dread is likely that unpredictable events (comets, eclipses, and meteors) are transients that break the cyclic harmony of nature. Thus, when hairy stars are sinister omens of whimsical gods, they are predictors of doom. Seneca failed to alter this formula by rejecting the connection with the divine. Even after Edmund Halley and Isaac Newton demonstrated that comets are predictable events that obey the laws of nature, the essential character of comets was still totally unknown, thus allowing irrational fright. Their physical structure began to be understood in the early 1900s, but the cyanogen poison was latched onto as an alternative threat. Soon after this, comets were realized to be dirty snowballs emitting near-vacuum tails with no divine connections, and so the "modern scientific view" of comets evolved with no menace. But this belief has changed again in the last decade, with the realization that the rare collisions of comets with our planet can have megacatastrophic results. It seems as though humans are fated to be scared of comets.

So the Hale-Bopp question of "Do you feel fear?" depends on your background. Rural Third World workers might yet have panics, while *Sky & Telescope* readers are likely to enjoy a comet's beauty and majesty — but only after confirming that the high-velocity snowball will give the Earth a wide berth.

Bradley E. Schaefer, an astrophysicist at Yale University, has been studying the peak brightnesses of historical supernovae to constrain their luminosities and hence the Hubble parameter.

The Comet's Gift: Hints of How Earth Came to Life

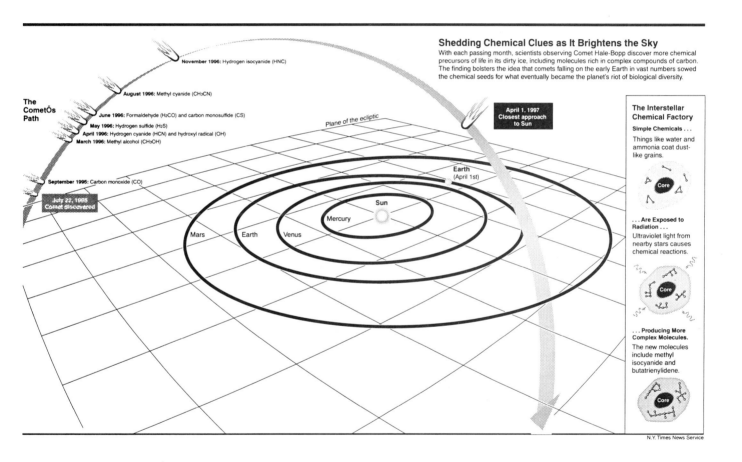

WILLIAM J. BROAD

A TRILLION or so comets are thought to lurk beyond the planets on the dark fringes of the solar system. Lost to a realm where the Sun's faint rays do nothing to lessen the interstellar chill, most of these whirling chunks of dirty ice orbit endlessly.

But over recorded time, a few comets have left this deep freeze and sped into the inner solar system, at times lighting Earth's skies. Past civilizations often saw them as harbingers of death and doom, and irrational reactions to comets still occur, as evidenced by the grim suicides in Rancho Santa Fe, Calif., where 39 people recently died in the

3. THE SOLAR SYSTEM: Comets

apparent belief that Hale-Bopp was their ticket to aliens and extraterrestrial bliss.

Modern astronomers are fascinated by Hale-Bopp for other reasons. Never before have they witnessed anything quite so spectacular. Its icy core is estimated at 25 miles wide, more than 10 times the size of the average comet and big enough to swallow many Manhattans. Its great size makes it unusually bright and easy to study. Since it was first discovered 20 months ago beyond the orbit of Jupiter, astronomers have scrutinized it unceasingly as the Sun warmed its outer layers, causing the gargantuan ice ball to shed many tons of clues every second about the nature of its chemical makeup.

Now, the first comprehensive findings are in and give support to a remarkable theory. It suggests that cometary ices bear the chemical precursors of life and that comets fell on the aboriginal Earth in vast numbers and sowed these precursors for what eventually became the planet's riot of biological diversity. The same mechanism is thought to be at work throughout the cosmos, sowing the seeds of life on untold worlds.

Portents of doom may really be sowing chemical seeds of life.

This view is now getting major support as telescopes around the globe find Hale-Bopp spewing not just tons of water but methanol, formaldehyde, carbon monoxide, hydrogen cyanide, hydrogen sulfide and many compounds rich in carbon—in other words, the basic ingredients thought to be necessary for the origin of life.

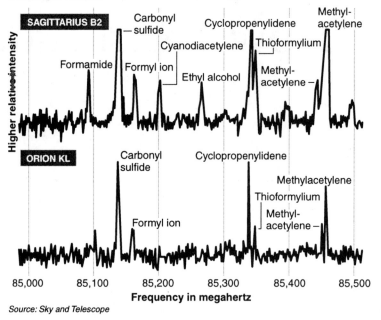

N.Y. Times News Service

"This is the ironclad link to the new paradigm," Dr. Dale P. Cruikshank, an astronomer at the National Aeronautics and Space Administration's Ames Research Center in California, said in an interview. "We've never had such a panorama of important molecules."

Dr. Cruikshank is the first author in a series of eight articles on the Hale-Bopp findings in the current issue of the journal Science. The new discoveries are seen as a milestone in the developing field of bioastronomy, which looks to the heavens for the chemical forerunners of life. And the tidings of cosmic fertility are strikingly at odds with the old view of comets as portents of doom.

Hurtling through space at more than 27 miles a second, the comet today makes its closest approach to the Sun. Scientists say its return trip to the outer limits of the solar system will probably produce even more insights into its nature, as well as striking nighttime shows in the Northern Hemisphere throughout April and early May.

"It's not over yet," said Dr. Brian G. Marsden of the Harvard-Smithsonian Center for Astrophysics in Cambridge, Mass., who helps track comets and other heavenly bodies for the International Astronomical Union. "It's going to be improving its visibility in the evening sky."

Such a comet appears "once every 200 years or so," Dr. Marsden added. "What makes it great is its persistence."

Dr. Harold A. Weaver, an astrophysicist at John Hopkins University and lead author on one of the Science papers, said Hale-Bopp might actually outperform itself while heading back toward its icy abode.

"The nucleus sometimes stores heat and might remain at elevated temperatures for longer outbound than inbound," he said in an interview. "We already have this beautiful set of data and we want to watch it go back out again. There's clearly a lot of excitement."

Some telescopes must stop tracking Hale-Bopp as it approaches the Sun or be damaged by bright sunlight. This is particularly true for the Hubble Space Telescope, orbiting Earth. Later, as the comet races toward the dim hinterlands, astronomers will redouble their efforts to gather clues to Hale-Bopp's chemical and physical makeup in hope of strengthening the uncanny link between comets and the first stirrings of life on Earth.

Such thinking is a radical departure from the traditional view developed over the decades by scientists, mainly geologists and geophysicists.

In the old picture, Earth was thought to have coalesced out of primordial dust as a bare sphere with no atmosphere, basically a rocky desert. The gases and water vapors and carbonaceous brews that formed the atmosphere and filled the seas were seen as having come from within Earth in an early period of intense volcanism. Lightning storms then stirred the primordial soup and created carbon-rich molecules that organized themselves into self-replicating units, or crude forms of life.

Not so, says a newer theory, which astronomers, astrophysicists and planetary scientists tend to advance.

More than four billion years ago in its early days, Earth was hot enough to expel into space most of the water and lighter materials and chemicals, this theory holds. So the planet remained a barren rock.

For the makings of life, the new theory looks to the wastelands out among the stars, especially to the dark clouds that loom against starry backdrops. Such dim zones are common among the countless stars that make up the Milky Way, Earth's galaxy. On a clear dark night, the naked eye can easily see the voids amid the milky brightness.

These dark interstellar clouds turn out to be peppered with grains of matter about the size of talcum particles that are virtual factories for the production of complex chemicals. To date, scientists have identified nearly 100 molecular species.

A light show in space supports a remarkable theory.

Interstellar water was found in 1968, formaldehyde in 1969, methyl alcohol in 1970, hydrogen sulfide in 1972, methylamine in 1974, ketene in 1976, methane in 1978, ethylene in 1980, methyl diacetylene in 1984, methyl isocyanide in 1987 and butatrienylidene in 1990. Such molecules were discovered to make up about half of the interstellar matter, the rest being mainly atomic hydrogen.

These chemical finds were made as astronomers turned radio, infrared, ultraviolet and visible-light telescopes on the icy wastelands and studied signals that showed molecules giving off their own precise signatures amid a chaos of electromagnetic waves.

One pioneer of such analysis is Dr. Yvonne J. Pendleton of the Ames Research Center. She works there with Dr. Cruikshank, her husband and occasional co-author. In the March 1994 issue of Sky and Telescope magazine, the couple wrote about finding evidence of chemical precursors to life in space.

No signs of amino acids—the constituents of DNA, life's master heredity molecule—have yet been found amid the interstellar wastes. But lots of forerunners have, including ammonia and hydrogen cyanide. The leap to amino acids is said to take only the addition of liquid water.

Today, scientists see the interstellar factory as working in several steps. First, dying stars lace the void with an array of relatively simple molecules such as methane (CH_4), water (H_2O) and carbon monoxide (CO), as well as tiny grains of silicon, the element found in rocks, sand and semiconductors. Over eons, the molecules coat the silicon grains with icy mantles.

Dense clouds of such materials condense in places to form new stars that irradiate nearby grains with bursts of untraviolet light, transforming the simple molecules into more complex ones like formaldehyde (H_2CO) and methyl alcohol (CH_3OH).

In theory, such complex interstellar dusts then become the raw material for new generations of stars, perhaps accompanied by planets as well as trillions of icy comets.

Astronomers on Earth observe comets rushing through the inner solar system only fleetingly. In contrast, the interstellar wilds are always visible. Despite the difficulty, astronomers, aided by recent advances in telescopes and instrumentation, have begun to learn some of the chemical secrets of the icy visitors.

The first glimmers came in 1985 and 1986 with Halley's comet, which yielded signs of water, carbon dioxide and formaldehyde (used on Earth as a disinfectant and preservative). In 1989, another comet showed evidence of methyl alcohol (used as antifreeze), while one in 1990 gave signs of hydrogen sulfide (the poison that smells like rotten eggs).

The big advance came last year with Comet Hyakutake, which was extremely bright and easy to analyze. Its signals gave evidence of many different molecules akin to those of the interstellar wastes.

Now astronomers studying Hale-Bopp, discovered on July 22, 1995, have topped even that. The findings are significant both for the sheer number of molecules and, as important, for a detailed description of when the chemicals evaporated from the giant ice ball as it raced inward from the deep freeze toward the Sun.

In Science, a 12-member team headed by Dr. Nicolas Biver at the Observatoire de Paris-Meudon, in Meudon, France, describes nine chemicals sighted by radiotelescopes: Carbon monoxide (CO) in September 1995, methyl alcohol in March 1996, hydrogen cyanide (HCN) in April 1996, hydroxyl radical (OH) in April 1996, hydrogen sulfide (H_2S) in May 1996, formaldehyde in June 1996, carbon monosulfide (CS) in June 1996, methyl cyanide (CH_3CN) in August 1996 and hydrogen isocyanide (HNC) in November 1996.

"It's a landmark," Dr. Cruikshank said of the list. "It shows when and how these molecules turn on, and shows the evaporation sequences. They're also found in space. This is the link that really clinches the connection to the interstellar medium and proves that this material is essentially unaltered."

Put differently, it shows that comets are heavenly vans carrying tons of carbon-rich materials from interstellar space to addresses throughout the solar system.

Dr. Cruikshank said bombardments of cometary ice undoubtedly hit not only the early Earth (heavily for a period of perhaps a billion years or so) but also Mercury and the Moon. Both those bodies have recently yielded signs of hidden ice.

Even today, he added, microscopic dust particles from comets and rocky asteroids rain down on Earth at the rate of about 300 tons a year.

The prevailing wisdom, bolstered by Hale-Bopp, is now that comets played a pivotal role in begetting life on Earth.

"Thirty years ago we thought you had to have lightning storms," Dr. Michael J. Mumma, a longtime comet expert at NASA's Goddard Space Flight Center in Maryland, said in an interview. "Now it's recognized that the materials are delivered automatically."

"That doesn't mean you get life out of that," Dr. Mumma added. "But it means you deliver large amounts of simple chemicals, and maybe some complex ones, that can lead to life directly."

Bits of Mars and pieces of the moon

A recent study by Cornell University scientists helps to confirm that some meteorites found on Earth originated on other planets in our solar system as well as the moon. Planets and their satellites in the inner solar system have been sharing bits and pieces with each other for billions of years, the scientists claim. Even today, rocks and particles shorn off from ongoing collisions continue their interplanetary voyages.

"We found that planets and satellites are transferring material all the time," said Cornell's Joseph A. Burns, professor of engineering and astronomy and director of the recent studies. The work of Burns and his colleagues accurately describes the history of planetary ejecta for the first time and moves scientists a step closer to understanding why an equal number of meteorites on Earth come from the moon and Mars.

Material is thrown out of craters whenever there is a collision — such as the July 1994 crash of Comet Shoemaker-Levy 9 into Jupiter or the event near Chicxulub, Mexico, that may have triggered the dinosaurs' demise on Earth. Such collisions are forceful enough to launch a fraction of the rocks and particles into space at speeds so fast they escape the gravitational grasp of their parent planet. Since the moon is so much closer to us than Mars and has a gravitational pull much easier to escape, one might expect much more lunar debris on Earth. Yet a dozen meteorites from the moon and a dozen from Mars have been identified — mostly since 1980 and mostly on the Antarctic ice sheet, where they are relatively easy to spot.

During the past several years, Burns and his colleagues used high-speed desktop computers to simulate throwing thousands of rocks off the moon, Mars, Venus, and Mercury. The scientists followed the rocks' paths through space over millions of years, assuming a variety of ejection speeds from their home planets and the gravitational tugs of the various other planets.

Meteorites in orbit

Results of their simulations showed that half of the lunar material gets to Earth very fast — within the first 50,000 years — while material from Mars tends to take much longer, up to 15 million years. The earliest-known Martian meteorite on Earth took 700,000 years to arrive.

Like billions of celestial pinballs, these extraterrestrial rocks dance about the solar system, buoyed by gravitational kicks from the planets. Some of the meteorites may be destroyed by catastrophic collisions in the asteroid belt. Over millions of years, their orbits become far different from what they were when they began.

"After a few million years [of] wandering, they've forgotten where they are from," says Burns' colleague Brett J. Gladman, whose doctoral dissertation comprises this study. The research team also included Cornell doctoral candidate Pascal Lee, Martin Duncan of Queen's University, Ontario, and Harold F. Levison of the Southwest Research Institute in Boulder, Colo. They reported their research in the March 8, 1996, issue of *Science*. The results appear to confirm the properties and ages of lunar and Martian meteorites found on Earth.

"Long-range gravitational effects strongly influence the orbits of many meteoroids, increasing their collision rates with other planets and the sun," the research team wrote in *Science*. "These effects and collisional destruction in the asteroid belt result in shortened time scales and higher fluxes than previously believed, especially for Martian meteorites." The scientists also believe it's possible that some meteorites on Earth may, in fact, have originated on Mercury or elsewhere.

The study, funded by the National Aeronautics and Space Administration, shows that about 40 percent of the material launched from the moon eventually lands on Earth. Yet only 4 percent of material ejected from Mars makes it to Earth. Less than 1 percent of particles launched from Mercury find their way to this planet. "The details remain to be worked out, but we are hot on the trail," says Burns.

Implications

The study of these celestial dynamics and of meteorites in general could shed light on the formation of the solar system. While most meteorites on Earth come from asteroids, scientists now are aware that a few others have originated on nearby planets or moons, said Burns. Furthermore, material ejected into space from Earth during collisions millions of years ago could — and likely did — land on other planets. The study, therefore, has practical implications for unmanned space travel to other planets. "One must question whether it makes sense to spend billions of dollars to sterilize a spacecraft going to another planet," said Burns, "when that planet already may have been 'contaminated' by Earth through this natural ejecta process."

Life from ancient MARS?

A UNIQUE METEORITE RAISES THE STAKES IN THE EXPLORATION OF THE RED PLANET

J. KELLY BEATTY

ON JULY 20th, current and former NASA employees gathered in Washington, D.C., to celebrate the 20th anniversary of the Viking landings on Mars. The mission had tackled a centuries-old question: Does life exist on the red planet? Twin landers toiled for months before scientists concluded that the Martian surface was completely sterile, at least in the few square meters within reach of the landers' arms.

Even as the Washington reunion carried on, space scientists elsewhere were putting the final touches on a profound finding about Mars, one that would reverberate worldwide like a thunderclap. On August 7th, after two years of painstaking analysis, a team of nine researchers revealed evidence that primitive life may have once inhabited a chunk of the Martian crust. Moreover, they had not ventured to Mars to retrieve this special stone—it had come to us, free, as a meteorite.

In the Vikings' day, few scientists would have believed that pieces of Mars (or the Moon) could fall from the sky. But now they readily accept that such stones can and do reach Earth after being launched from our neighboring worlds by large impacts. The current tally includes 11 recognized falls from the Moon and 12 from Mars.

The rock hailed by scientists in August has several stories to tell. One is the way it got here in the first place. Based on how many cosmic rays it absorbed while in space, which alter the isotopic balance of its minerals, this Martian expatriate apparently spent 16 million years wandering through the inner solar system before falling onto the Allan Hills ice field in Antarctica. Then, 13,000 uneventful years later, an American search team chanced upon it on December 27, 1984, and later designated it ALH 84001. Roberta Score, one of the searchers, recalls that it was the greenest meteorite she'd ever seen among the team's hundreds of Antarctic finds. "I've always thought that rock was weird," she says.

A second tale concerns initial confusion over ALH 84001's lineage. For nearly a decade the stone was considered a diogenite, a class of igneous meteorites believed to be chips of the asteroid Vesta. But a reanalysis three years ago revealed its Martian identity (*S&T:* June 1994, page 14). A feeding frenzy ensued, as laboratory teams worldwide clamored for samples to analyze. They soon concluded that the stone was truly ancient—4½ billion years old—and thus probably part of the planet's initial crust. By contrast, all the other Martian meteorites then known were no older than 1.3 billion years.

Here, then, is a stone that experienced virtually the whole of Martian history. Analysts quibble over some of the details, but they agree that ALH 84001 must have crystallized slowly from magma. It was then fractured throughout by the shock from a nearby impact 3.8 to 4.0 billion years ago—eons before being blasted toward Earth. The sample later spent some time under water abundantly charged with carbon dioxide, perhaps more than once. This immersion allowed small beads of carbonate to form along internal cracks.

THE EVIDENCE, PLEASE

From the outset, cosmic chemists were fascinated by the strange-looking carbonate "globules." Orange-brown in their centers, many have alternating dark and bright outer rims that have been likened to Oreo cookies. These layers betray changes in composition and thus the character of the fluid from which they precipitated. The closer researchers looked, the more amazing the carbonates became, and it didn't take long to realize that ALH 84001 might be telling more than a purely geologic story.

In *Science* for August 16th, the team led by David S. McKay and Everett K. Gibson Jr. (NASA/Johnson Space Center) present the evidence they've amassed that this meteorite carries a biological calling card from Mars—and not from Earth.

First, images obtained with a scanning electron microprobe show clusters of elongated shapes no more than 100 nanometers long (4 millionths of an inch) in and near the carbonates. Looking like tiny sausage links, these shapes could just be flecks of the mineral. But they bear a striking resemblance to the earliest microfossils on Earth, which formed 3.45 billion years ago.

Second, the team found that the dark rims on the carbonate nuggets are due to tiny embedded grains of magnetite (Fe_3O_4) and iron sulfide (FeS). The crystals are all remarkably free of structural defects or elemental impurities, and they are incredibly small. "You can fit about a billion of these on the head of a pin," notes Kathie L. Thomas-Keprta (Lockheed Martin). Under most conditions these iron compounds would not coexist, especially in concert with carbonate. But certain bacteria on Earth, particularly anaerobic ("oxygen-hating") strains, synthesize them simultaneously with relative ease.

Third, and considered most compelling, was the discovery that the carbonates are infused with organic molecules known as polycyclic aromatic hydrocarbons, or PAHs (*S&T*: July 1995, page 12). Ordinarily their presence would not automatically be seen as a marker of biologic activity—PAHs are observed often in such diverse bodies as meteorites and interstellar clouds, presumably as a consequence of star formation. But as Richard N. Zare (Stanford University) explains, it's not a matter of quantity but of quality. "We've looked at other meteorites that have PAHs, and this distribution [of PAH types] is much simpler—it very much resembles what you'd expect when simple organic matter decays."

Team leader McKay admits that none of these findings, by itself, offers definitive proof that primitive life once permeated ALH 84001. All of them can be mimicked by purely inorganic mechanisms. However, he says, "The relationship of all these things in terms of location, found within a few hundred-thousandths of an inch of one another, is the most compelling evidence."

Understandably, everyone involved—from the researchers, to the broader scientific community, to NASA administrator Daniel Goldin—underscores the need to bolster this tentative result, however persuasive, with something more ironclad. Even President Clinton has weighed in, cautioning the space agency to "ensure that this finding is subject to a methodical process of further peer review and validation."

"This is not easy science," says microbiologist J. William Schopf (University of California, Los Angeles). While the NASA-sponsored team has made a convincing case that no terrestrial contamination has occurred, Schopf is concerned that PAHs stem ubiquitously from many completely lifeless processes. And he's cast a skeptical eye on the carbonate "microfossils," which are 100 times smaller than their ancient terrestrial counterparts.

There's also debate on when and how the carbonates formed. An initial study pegged their age at 3.6 billion years, and that has become the NASA party line. But a more recent dating my Meenakshi Wadwha (Field Museum, Chicago) and Günter W. Lugmair (Scripps Institute of Oceanography) argues for the much younger and more provocative age of 1.3 to 1.4 billion years. The right answer could prove critically important, since water has apparently not flowed across the Martian landscape in quantity over the last 3 billion years (*S&T*: December 1995, page 18). Wadwha and Lugmair point out that some other Martian meteorites crystallized 1.3 billion years ago—a coincidence that may flag a period of heightened volcanic and hydrothermal activity on Mars.

A final issue is the temperature at which the carbonates formed. Keeping the bacteria happy would have required conditions no hotter than about 150° Celsius, above which life cannot survive. Perhaps fortuitously, some studies suggest that the carbonates materialized from a fluid heated at most to 80° C. But a minority of geochemists think the temperature was more likely at least 650°. And others argue that if the carbonates formed under pleasantly warm and wet conditions, some of the surrounding rock should have turned to clay minerals—which are almost completely absent.

Whatever the outcome, ALH 84001 has reshaped our perspective of Mars and the objectives of future spacecraft missions. Officially, managers at NASA have adopted an attitude of "skeptical fascination" with the result, and that the scientific community alone should assess its believability. But in quiet moments they must surely be rubbing their hands together in glee, for should all this enlightened speculation prove true the prospects of full-scale Martian exploration will be very bright indeed.

The Day the Dinosaurs Died

For more than 4 billion years, it orbited the Sun—a pockmarked, misshapen body roughly the size of Manhattan. Over time, it wandered into the inner solar system, occasionally passing near our planet as it journeyed around the Sun.

Then, one day about 65 million years ago, Earth and the rough-hewn body reached the same place at exactly the same time.

Ron Cowen

The solar system is a hazardous place. Countless fragments of rock and ice tumble about in orbits that occasionally produce impacts on planets. Most of the time these are minor events. Once in a great while, they are catastrophic events, affecting planets as a whole and leaving long-lasting effects. The most famous large impact on planet Earth, the one th[at] killed the dinosaurs, is now coming into clear focus.

After 15 years of fractious debate, most scientists now agree that an asteroid or comet collided with Earth 65 million years ago. Most of the nonbelievers threw in the towel in 1991 when scientists discovered the smoking gun for the impact—a vast, buried crater in the Yucatan Peninsula. Known as Chicxulub (Cheek-shoe-lube), a Mayan word that ironically means "devil's tail," this huge hole was formed during the interval of geologic time known as the K-T boundary, which marks the end of the Cretaceous (K) period and the beginning of the Tertiary (T). (See "Tracking Down Chicxulub.")

But if there's consensus that a huge object smacked into Earth, critics have found another battleground. A catastrophic solar system impact has leaked into other fields as geologists, paleontologists, astronomers, and atmospheric scientists debate exactly how lethal the long-ago collision was, and which of its many aftereffects proved most deadly.

Before the impact, dinosaurs still roamed Earth and numerous species of phytoplankton—which serve as food for tiny sea animals—filled the oceans. A geologically short time later, the last dinosaurs had gone belly up, along with two-thirds of all living species.

Was the K-T impact entirely responsible? And if so, exactly how did it exert its lethal influence?

To understand the K-T impact, scientists are drawing on their knowledge of all solar system impacts. "The ramifications of a large impact are still poorly understood," note planetary scientists Owen Toon and Kevin Zahnle of NASA's Ames Research Center in Mountain View, Calif.

Says Buck Sharpton of the Lunar and Planetary Institute in Houston: "The focus here is to try to understand how you can have an impact event [that was so catastrophic]. Bear in mind that we always thought that a [comet or asteroid] would deposit most of its energy right at ground zero. So how can it, as energetic as it was, cause a global extinction? It had to wipe out just about everything."

"It's not so much that the impact itself was so lethal," agrees paleontologist Peter Ward of the University of Washington in Seattle. "It squashed a few moths, a few this and a few that. That's not what killed everything."

"My own philosophy about the K-T impact is that everybody wants to discover the one thing that killed

32. Day the Dinosaurs Died

the dinosaurs," says Toon. "But the fact is it's a very complicated situation, in which multiple things happened. It's not like everything died from one cause."

Chronology of a Catastrophe

A grand planetary impact is an amazingly destructive thing. Imagine the energy unleashed if the entire world's arsenal of nuclear weapons suddenly exploded. Now multiply that by 10,000. The blast, equivalent to 10 trillion megatons of TNT, is only a conservative estimate for the destructive force generated by the smash. Here's the chronology of the catastrophe, according to Toon and several other scientists.

Time zero: As the object cannonballed through Earth's atmosphere, it generated a shock wave that blew a hole in the air.

Seconds to first hour: Reaching the ground about three seconds later, the bolide plowed through a 1 kilometer-deep ocean and into Earth's crust, generating earthquakes at every exposed fault line over the next few hours. Within the first hour of impact, 100-foot-high walls of water — ocean waves from what is now the Gulf of Mexico — flooded coastal plains. But these were only the local effects, devastating just a few percent of Earth's surface. Minutes after the impact, billions of tons of debris blasted into space, three to four times the mass of the bolide. Rocketing upward and outward at speeds nearly great enough to escape Earth's gravity, the dusty debris reentered Earth's atmosphere far from the site of the impact with a vengeance, reheating and glowing red-hot.

For 30 minutes to an hour after the impact, "the entire sky was like a big, glowing sheet of rock," says Toon.

"Imagine the effects of a thousand shooting stars suddenly entering the atmosphere and ablating at an altitude above 60 km. The sky would turn from its normal transparent blue to a brilliant red sheet of glowing lava," note Toon and Zahnle.

Toon adds, "The shooting stars wouldn't have hit you; they stop at 60 km or so above the ground. But there's so many of them that you're standing there in this radiation bath, hot enough to set paper on fire.

The collision of Comet Shoemaker-Levy 9 with Jupiter drove this point home, several scientists assert. Fragments of the comet no bigger than half a kilometer in diameter — far smaller than the size of the impactor that created Chicxulub — produced a plume of debris that reached 5,000 kelvins.

More to the point, the hot plume splashed down over an area on Jupiter that exceeded that of Earth.

"Comet Shoemaker-Levy 9 provided proof that the K-T impact ignited global fires and scorched Earth," says Jay Melosh of the University of Arizona in Tucson.

"The bigger animals would have been broiled in their tracks," he notes. But, says Toon, some organisms that had natural shelter might still survive. "If you're a crocodile underwater — no problem. If you're a mouse in your hole — no problem."

From the first hour to six months: After an hour or so, according to Toon and Zahnle, the red sky would

Tracking Down Chicxulub

It all began in 1978 when Walter Alvarez, a geologist at the University of California, Berkeley, returned from a field trip to northern Italy's Apennine Mountains. Near the town of Gubbio, he had collected some 400 pounds of rock from a curious, pencil-thin layer of gray clay. Like a bookmark, this layer seemed to indicate the boundary between the end of the Cretaceous period and the beginning of the Tertiary.

Alvarez's father, Nobel Laureate physicist Luis Alvarez, suggested he measure the amount of iridium, a member of the platinum group of elements, in the clay layer. Iridium is common in comets, asteroids, and meteorites, but exceedingly rare in Earth's crust. Thus, a substantial amount of iridium might mean that many small meteorites had the chance to bombard the boundary layer, an indication that the layer lasted for an extended time.

To their astonishment, the scientists found that the clay contained 30 times more iridium than the gradual deposition of micrometeorite dust could account for. Berkeley researchers Luis Alvarez, Walter Alvarez, Frank Asaro, and Helen V. Michel proposed an alternative explanation. They suggested that a large meteorite, roughly 8 kilometers in diameter, had smacked into Earth 65 million years ago, carrying with it some 200,000 tons of iridium.

But a key piece of evidence was still missing. If a huge bolide had indeed smacked into Earth, where was the hole? "The biggest string that was left untied for a long time was the question of where the crater is," says Walter Alvarez. It was a question, he notes, that had haunted him and his colleagues ever since they reported their initial findings in 1978.

Unknown to geologists, evidence had been slowly mounting for more than two decades that an area on the northern coast of Mexico's Yucatán Peninsula might contain the telltale crater. During the 1950s, the Mexican national petroleum company, Pemex, discovered an unusual layer of broken rocks and structure while drilling in the region. A magnetic survey commissioned by Pemex later revealed a large circular structure at the site, centered near the town of Chicxulub. In 1968 and 1972, two geologists suggested the region represented a large buried crater.

By 1991, several other researchers, including Kevin O. Pope of Geo Eco Arc Research in La Canada, Calif., Adriana Ocampo of JPL, and Charles Duller of NASA's Ames Research Center, had also published reports affirming the impact origin of the Chicxulub site. "Now there really is very little doubt that the impact site is in the Yucatán Peninsula," says Walter Alvarez. "I think that basically wraps up the biggest question about the whole thing."

cool and blacken. By the end of the first day, soot from charred debris would block out the Sun, plunging the world into darkness for as long as a year.

Six months to a decade: Sulfur ejected into the stratosphere by the impact and slowly converted to sulfuric acid, a highly efficient Sun blocker, may have prolonged the blackout for as long as a decade.

"The fires themselves wouldn't have killed ocean life, but if you turn the lights out, phytoplankton stops reproducing and you trigger the collapse of the ocean ecosystem," Toon notes.

Without sunlight, temperatures on land would take a nosedive — a far larger drop than Earth endured during the Ice Age. Global cooling may have lasted for a decade.

A decade after the hit: Following the cooling came the warming. The collision may have squeezed carbon-containing material out of rock at the impact site, creating a blanket of carbon dioxide in the atmosphere. This greenhouse gas would have heated Earth for 50 to 100 years, according to Toon. At first, he notes, the oceans would have soaked up some of the excess carbon dioxide, but their capacity would soon have been overwhelmed by the concentration in the atmosphere. Temperatures would rise for up to a century, Toon conjectures. "Temperatures don't just go back to normal after the global cooling but rise considerably higher," notes Sharpton. "There's a shift in temperature and it stresses the heck out of things. You might just be getting used to the cold and all of a sudden you've got this hot dry spell."

However, not all scientists agree on this warming. Kevin Pope of Geo Eco Arc Research in La Canãda, Calif., feels the warming would be negligible. "Recent research has shown that climate warming due to greenhouse effects from carbon dioxide released by the impact was extremely minor," says Pope. He is not alone: Studies by Thomas Ahrens of Caltech suggest the same result.

The Rock Hits the Powder Keg

The sheer size of the impactor — estimated to be 10 km in diameter — can't by itself account for the devastation. Haraldur Sigurdsson of the University of Rhode Island and other researchers believe that the impact sounded the death knell for so many species because of the unique composition of the crash site.

Several lines of evidence reveal that the bolide hit a sulfur-rich region of the Yucatán Peninsula, kicking up billions of tons of the element. "We think the severity of the event relates to the unusual chemistry of the terrain where the impact occurred," notes Sigurdsson. "The bolide hit the powder keg when it struck this chemical sediment."

He and his colleagues had found that the target rock in the Yucatán includes a 3-km thick sequence of carbonites and evaporites. Evaporites include gypsum and anhydrites, forms of calcium sulfate that contain large concentrations of sulfur.

Unlike soot and smoke, which probably washed out of the lower atmosphere in six months, the sulfur dioxide spewed into the upper atmosphere and lingered there for several years. The globally dispersed sulfur dioxide does not readily undergo chemical reactions and, because it is a gas, it does not settle out quickly like soot and dust.

By itself, sulfur dioxide doesn't do much damage. But ultraviolet sunlight and the presence of water eventually transform it into sulfuric acid. This falls to Earth as a toxic rain as corrosive as battery acid. "Unlike the aftermath of typical impacts, the skies remained murky for at least a decade due to chemically generated clouds of sulfuric acid high in the stratosphere," says Kevin Baines of NASA's Jet Propulsion Laboratory.

Baines and other team members led by Kevin Pope comprise an interdisciplinary group that is combining computer models of the Chicxulub impact and its atmospheric effects with geological studies of the crater.

Baines notes these same atmospheric conditions occur on Venus, which is perpetually cloaked in sulfur clouds. "The entire ecosystem of Earth, including plants and animals, was subjected to extreme environmental conditions for more than a decade. This Mother of all Environmental Disasters was simply too much for a large number of well-established species, such as the dinosaurs, to cope with," he says.

But Toon strikes a cautionary note. "The soot is in the K-T boundary layer, so you know that happened; we had to burn the entire world's biomass to have the amount of soot observed in the boundary layer," he says. In contrast, "We don't actually have any direct evidence that the sulfur did anything in the boundary layer."

"I would not bet my life on the contribution of sulfur to this whole business," says Peter Ward. "Sulfur is the least reliable kill mechanism we have in the sense that there is so much leeway, a lot of sway in the numbers. Dust going up in the atmosphere could have been every bit as catastrophic as the sulfur emissions. Several analyses suggest that when dust goes up, it perturbs rainfall cycles. Immediately after the K-T impact, places that had been previously wet became dry and those that were very dry became wet."

Despite the feelings of Toon and Ward, many scientists remain skeptical of the sulfur story and research continues, but evidence is mounting that a decade of cooling may have contributed to the dinosaurs' demise.

To Kill the Dinos or Not to Kill?

Some researchers consider micropaleontologist Gerta Keller of Princeton University a stubborn holdout. She has coined her own phrase for the true believers in the K-T catastrophe: "impact diehards."

It's largely physicists, astrophysicists, and chemists, she contends, who support the theory. "Relatively few paleontologsts believe the impact can snuff out life; they just don't see that in the data. Dinosaurs were already in decline the last few million years of the Cretaceous period," Keller adds.

32. Day the Dinosaurs Died

Keller says she does agree with other scientists that because dinosaur bones are scarce these fossils can't provide a detailed record of what happened 65 million years ago. "The best information is from the bottom of the food chain," she says, in part because such primitive organisms as pollen grains and fungal spores are plentiful and well preserved in sedimentary rocks.

Among the most abundant inhabitants of the sea is foraminifera, a type of plankton supposed to have suffered nearly complete extinction according to the impact theory, Keller says. These single-celled organisms live in the upper 100 to 400 meters of the ocean, and in the tropics they did indeed disappear quickly at the K-T boundary, she adds. But at more northerly and more southerly latitudes the die-off is considerably less severe. "Basically what paleontologists see is a differential pattern of extinction," Keller notes.

"The impact diehards don't like this," she asserts. "They predict that effects of such an impact must be global, extinction must be global, and there must be no differential across latitudes."

Pope, however, replies, "The fact is, such a latitudinal gradient, with most extinctions occurring in the tropics, is exactly what is to be expected from impact-induced global cooling and blackout because high-latitude species are better adapted to cold and low light levels."

Paleontologist Ward concurs that extinctions were not as great at the poles. "The high latitudes in all probability served as a refuge; plant extinctions were definitely not as severe at high latitudes." However, he adds, such limited protection in no way mitigates the global influence of the K-T impact.

Says Ward: "Even in a severe forest fire, there are always pockets of unburned material and pockets of survivors; it's very difficult to eradicate everything. But all the evidence points to a catastrophe."

Ward notes that for several years he, too, didn't buy the impact story. After all, the marine mollusks he studied appeared to have died off gradually rather than suddenly, in direct contrast to the effect of a large impact. Working with Charles Marshall of the University of California, Los Angeles, Ward subjected fossils found at various layers in the Cretaceous and Tertiary periods to a new statistical analysis. The researchers found that "in almost all cases, the big fossils all show a sudden extinction at the K-T boundary — even though the actual pattern in the rock record looks anything but sudden," says Ward.

"The bigger the fossil, the rarer they are, and the less chance you have of finding the last one right at the K-T boundary. So the dieoffs look gradual, but they're really not. A catastrophic extinction will always look gradual for big fossils because there are so few of them and because paleontologists have trouble finding them."

Ward says that several paleontological mysteries remain. "We want to know why some plankton go extinct and why others do not. The easy answer is that deep plankton survive the blast best, overwintering deeper in the ocean until the atmosphere warms."'

Subtle Evidence for the Big Smack

But that type of easy answer isn't what Jablonski and University of Chicago paleontologist David Raup discovered when they examined the survival of 350 evolutionary lineages of marine bivalves—clams and other two-shelled ocean dwellers—at the time of the Cretaceous period. Based on the rich, well-preserved fossil record of these at the K-T boundary, the researchers find that the normal rules of evolution simply didn't apply. Mass extinctions wipe out all types of bivalves, without regard to special survival strategies developed during other, less violent times, the paleontologists conclude.

Raup and Jablonski also uncovered one factor that seemed to enhance survival. Those organisms distributed over many continents had a higher survival rate than those that had a narrow geographic existence. One intriguing notion, he muses, is that "maybe this change in the rules for extinction or survival helps explain why the dinosaurs are gone and mammals weathered the extinction at the end of the Cretaceous." This could "open the door to a whole new way of looking at mass extinction as an evolutionary force."

Comet or Asteroid?

Scientists continue to debate whether the huge impactor was an asteroid or a comet. Sharpton argues that it was probably a comet. "The materials in the K-T boundary suggest that the impactor is primitive, undifferentiated material," Sharpton says. This suggests the possibility that it was a comet." He adds that the estimated size of the bolide, about 10 km, also makes a cometary source more likely.

On the other hand, he notes, "we have a lot of these objects in our vicinity called near-earth asteroids. So we can't eliminate the possibility that it was an asteroid." (See "Far Journey to a NEAR Asteroid," March 1996.) Geologist Eugene Shoemaker of Lowell Observatory in Flagstaff, Arizona, says he's puzzled by another mystery. According to Shoemaker, two distinct layers of material are associated with the K-T impact — created apparently by two distinct episodes of deposition, separated perhaps by a growing season.

"The serious problem is that we do not understand how those two layers formed. Along with shocked quartz, there is shocked zircon in the upper layer. Other people have glibly said, 'Oh well, the lower layer is the fallout layer, the upper layer is the fireball layer.' Well, that's bull. The lower layer is just as much the fireball layer as the upper."

Says Pope: "The best explanation for the two-part stratigraphy is that the lower unit, which contains very little shocked quartz (tiny crystal grains fractured by the impact) and lots of microtektites, was deposited ballistically on relatively short trajectories that did not leave the atmosphere. The upper unit, with abundant shocked quartz, as well as most of the iridium, contains material that was initially ejected much higher, perhaps above the atmosphere, and settled to the ground more slowly, thus landing on top of the tektite layer."

Additionally, Adriana Ocampo of JPL and Pope recently discovered a Chicxulub ejecta deposit in Belize, only 360 km from the center of the impact. This deposit also contains two layers, the lower one with tektites and carbonate spherules, and the upper one with boulders up to 7 m across floating in mud.

And then there's a more basic puzzle. After several years of taking gravity maps, drilling at the site, and evaluating seismic data from Chicxulub, geologists still don't agree on the size of the crater. Sharpton and his colleagues assert that it has a diameter of about 280 km. "Any estimate less than 200 kilometers is bordering on nonsense," Sharpton declares. A recent study of the topography of the northern Yucatán by Pope and his colleagues found several concentric troughs and ridges that also indicate that the crater is larger than 200 km in diameter (they suggest 240 km).

Yet Alan Hildebrand of the Commission Geologique du Canada and his team claim just as loudly that the crater is no more than 180 km across. At stake are estimates of the energy and mass of the impactor, properties directly linked to the diameter of the crater it gouges.

"It may seem strange that experts cannot agree on something so basic as the size of an impact crater," concedes Melosh in a commentary in the August 3 *Nature*. However, he notes, "the problem is that the structure of all craters is not the same and the relation of the final crater form to the impact that created it is often unclear."

The initial cavity gouged by an impactor, dubbed the "transient crater," collapses immediately under the influence of gravity to form one of a variety of crater types. And therein lies the challenge. "It is really the transient crater size that matters in deducing impact energy, and so the problem is to relate the measures of crater diameter to the crater size," Melosh says.

Determining the diameter of the Chicxulub crater poses further difficulties, he notes, because the crater lies buried under as much as a kilometer of sediment.

In recent studies, Sharpton and his colleagues have detected a ring-shaped feature surrounding the Chicxulub site that has a diameter of 280 km. Sharpton maintains that the ring indicates the edge of the crater and deduces that the transient crater must have had a diameter between 145 and 205 km. But using the same data set, supplemented by five new gravity maps, Hildebrand and his colleagues see no hint of Sharpton's outer ring.

With the debate about the gravity data likely to continue, Melosh suggests that geologists might rely on an alternative way of estimating the diameter of Chicxulub. In this method, researchers use the quantity of iridium scattered over Earth's surface by the impactor to determine the diameter of the body that slammed into Earth. Current iridium estimates indicate a 10-km projectile, which would gouge a transient cavity of about 70 km in diameter.

"Many opinions have been changed by the impact of Comet Shoemaker-Levy 9 on Jupiter, especially as it now seems that the largest individual fragments were only 700 meters in diameter," says Melosh. Although Jupiter's gravity caused the fragments to slam into the giant planet with a velocity of 60 km per second — much higher than in the terrestrial collision — the smaller size of the Shoemaker-Levy 9 projectiles means that the fragments carried at most one-hundredth the energy of the K-T impactor.

"Was the impact of an asteroid or comet big enough to gouge a 180-kilometer-diameter crater in the Earth also big enough to wipe out the dinosaurs?," asks Melosh. "Many people would now answer with a resounding 'yes'."

Ron Cowen is a frequent contributor to ASTRONOMY. *He is an editor at Science News.*

Escaping the Ultimate Disaster

A Cosmic Collision

Large celestial impacts don't happen often, but they can wipe out entire civilizations at once. An expert tells how we could protect ourselves—and even benefit—from these space invaders.

By John S. Lewis

There is something especially horrifying about unanticipated, rare, or unfamiliar lethal hazards, such as the possibility of an asteroid hitting Earth and wiping us out.

Americans know that about 50,000 people will be killed on their highways each year, though usually only one or two at a time. Yet this knowledge does not deter them from driving vast numbers of miles each year. In contrast, a single fatal airline accident involving 100 people takes on a spectacular aspect precisely because the rare fatalities, when they do occur, involve 100 or more deaths. Thus a single airline crash that kills 100 people is 100 times as visible as 10,000 highway crashes that kill 10,000 people.

So it is with the prospect of cosmic collisions, which may kill thousands, millions, and even billions of people in a single event.

Clearly, the actual importance of many threats differs radically from their perceived danger. How, then, would people react to a normal hazard that affects the entire planet, but occurs only very infrequently? Among forest fires and brushfires, earthquakes, lightning, landslides, floods, coastal storms, tornadoes, tidal waves, and impact events (i.e., asteroid, meteorite, or comet collisions), the *only* hazard that could cause the destruction of human civilization or the extinction of the human species is a large impact.

There is evidence that such catastrophic impacts have happened before on Earth. Paleontologists studying the pattern of appearance and disappearance of species in the fossil record have long been aware of abrupt, devastating global extinction events occurring at the ends of geological ages. In 1981, researchers discovered that a thin, global sediment layer that separates the end of the Cretaceous era (the last period of the age of dinosaurs) from the beginning of the Tertiary era (the start of the age of mammals) contained the unmistakable signature of an asteroid or comet impact.

And what actually happened 8,000–10,000 years ago to end the hunter-gatherer chapter in human history? Quite suddenly, agriculture became common, specialized occupations arose, and cities appeared. Writing was invented, giving rise to record keeping and literacy. And the earliest human records all relate stories of floods that devastated civilization.... What did happen then? Was the clock of human history reset to zero by an event (or more than one) that devastated civilization?

The Dangers of Impacts

There is a wide range of lethal consequences of asteroid and comet collisions: The death or injury of individuals struck by a falling meteorite affects probably one to 10 people per century. Villages or cities can be struck by showers of meteorites from high-altitude airbursts about once per century. Also about once per century, iron or other physically strong meteorites may resist atmospheric breakup to strike the surface as a single crater-forming body or as a compact shower of iron shrapnel. And low-altitude megaton air-

3. THE SOLAR SYSTEM: Asteroids

bursts should also strike at populated areas every century or so, setting fires, shattering windows, and demolishing buildings over an area of hundreds to thousands of square kilometers.

About half of impact fatalities are caused by the smaller, more frequent, localized events. About a quarter of the total deaths arise from tsunamis caused by impacts (once every 10,000 years), and another quarter from continental cratering events and low airbursts.

Every 70,000 to 1 million years, a global billion-casualty killer will strike Earth: Collisions of 10-gigaton objects may occur about every 70,000 years; 100-gigaton explosions occur about once per 250,000 years; and 1,000-gigaton events occur a little more than once per 1 million years. If your projected life-span is about 75 years, that means the probability that *you* will be killed in a global impact event is between 0.01% and 0.1%. By comparison, the probability that you will be killed in a civil airliner crash is 0.005%.

The long-term average death rate from impacts is 4 billion people per million years, or 4,000 people per year worldwide. The people of the United States make up about 5% of the global population, so the average American death rate from global-scale impacts is about 200 per year. By comparison, the death rate of American citizens from commercial aircraft crashes is 100 people a year.

As we have found with hurricanes, predicting impact events could eliminate much of the horror and lethality associated with them. Cataloging the orbits and properties of Earth-crossing objects is already in progress and could be scaled up at modest expense. If we had a discovery and tracking capability, areas threatened by gigaton impactors could at the very least be evacuated. This would be ineffectual at reducing the cost of physical damage and economic dislocation, but would at least reduce the death toll to near zero.

The problem with finding and tracking these very large bodies is that evacuation does not work; the effects of the disaster are global. The leading cause of death would probably be famine induced by climate change. If such a body hits Earth, there are no places to which refugees can be relocated. Moving away from the computed impact area means selecting a slow death over a quick one, since Earth's ability to support life would be universally diminished.

Finding, tracking, and predicting the orbits of kilometer-sized bodies is neither technically demanding nor fiscally draining; rather, the problem arises when we ask what we would do with the knowledge. We can in fact do nothing meaningful to avoid this threat unless we use space technology to divert or destroy the threatening objects. The prospect of letting one hit our densely populated planet is unacceptable.

What Can We Do?

A search-and-tracking system to find bodies down to 250 meters in diameter in near-Earth space would likely find a number of objects in threatening orbits. With probable lead times of centuries, evacuation plans can easily be made and executed. The impact could then be used by fiscally conservative governments as a sort of instant urban renewal program.

But with so much lead time, given the rate of advance of technology, might it not prove much less expensive and inconvenient to do something to avert the impact of a threatening object? And what exactly should we do?

Why not blow it up?

Bad idea. If we split an approaching one-gigaton object into 10 equal pieces of 100 megatons of energy each, they'd strike Earth like a giant shotgun pattern. The main effect of breaking up the threatening impactor would be to *increase* the damage done. The disruption of a threatening impactor is clearly not a sensible option unless we are certain that almost all of its fragments can be diverted so as to miss Earth.

But if we have the ability to divert dozens of pieces, why not elect the simpler option of gently diverting the whole thing?

The idea of diverting the course of an asteroid that is several hundred meters in diameter seems breathtakingly ambitious. Yet, human mining activities routinely crush, excavate, and move comparable volumes of rock. There is an important factor that makes this scenario much less daunting: We are merely trying to avoid a single predicted impact with Earth.

Suppose our asteroid-search team finds a 250-meter body that is due to hit Earth dead center a few hundred years from now. This same body has probably been crossing Earth's orbit for 10 million to 100 million years without an impact. If we can just ease it by Earth without an impact on this one occasion, we may well buy ourselves another 30 million years to figure out what to do the next time it threatens us. So the real problem is not to devise a permanent fix; it is to avoid a specific near-term event.

> "Remember that we will almost certainly have hundreds to thousands of years of warning time before a threatening global-scale impact. We need not be driven to rash and risky actions taken precipitously under threat of death."

We might give the asteroid a small sideways nudge so that, when it reaches Earth, it will skim by to one side of the planet rather than strike it directly. Or we could accelerate or decelerate the asteroid along its direction of orbital motion so as to change its orbital period slightly. This would cause the asteroid to cross Earth's orbit a little ahead of or behind the impact schedule it was following, and hence cross Earth's orbit at a point ahead of or behind Earth.

There are many methods available for making such small changes in the velocity of an asteroid. One of the favorite techniques proposed by military experts is to explode a small nuclear warhead well clear of the surface of the asteroid. But simply launching an existing intercontinental ballistic missile at the asteroid would not work: Such vehicles cannot achieve escape velocity to reach an asteroid on its orbit around the Sun. Further, missile guidance systems are designed to operate for the half-hour of an intercontinental trip—not the weeks or months required for the trip to an asteroid. The mission would have to be accomplished by a military warhead combined with a NASA planetary spacecraft bus that provides guidance and power.

Readers concerned about the environmental impact of such an explosion should realize that the asteroid would not be contaminated to any significant degree by radioactive bomb debris, since the surface layer would be boiled off by the blast. The bomb vapor would be swept out of the solar system by the solar wind at a speed of about 600 kilometers per second. The net result of the asteroid deflection is really a twofold benefit to Earth: A devastating impact would be avoided, and there would be one less nuclear warhead on Earth.

Other options for deflecting asteroids and comets include chemical propulsion, electrical propulsion, nuclear thermal propulsion, solar thermal propulsion, and solar sailing.

The choice of flight hardware for an asteroid-deflection mission clearly depends on what technological options are available at the time the problem arises. In general, of course, the more different technologies we have to choose from, the more likely we are to have a good choice. It is quite impossible to guess what the preferred solution will be in the year 2010 or 2050, let alone 2300 or 5875. But if a threatening asteroid were discovered this year and action had to be taken in the next 10 to 20 years, we would be forced to choose quickly. We almost certainly would have to make do with existing, tested technology, such as the nuclear warhead option.

Don't Panic

Our central conclusion is that there is no reason to panic. There is a real threat, confirmed by a wide range of evidence, but we are not helpless in the face of cosmic bombardment. We certainly have no reason to ignore the impact hazard. Once we understand that the threat exists, and once we begin to collect the information we need to deal with the threat intelligently, there is no longer any need to retreat into denial.

First, we have the technical ability to discover and track almost every body that poses a global threat. Within a few decades, a nearly complete catalog of the larger near-Earth asteroids and short-period comets could be compiled for modest cost with known technology.

Second, massive objects found to be on threatening orbits can be deflected using techniques and hardware developed by NASA and the Defense Department.

There is another way of looking at near-Earth bodies: as opportunities rather than threats. A large proportion of the most-threatening objects are also the most-accessible bodies in the solar system for spacecraft missions from Earth. These bodies are the most-promising sources of raw materials for a wide range of future space activities. They may provide the propellants for future interplanetary expeditions, the metals for construction of solar power satellites to meet Earth's energy needs in the third millennium, the life-support materials and radiation shielding to protect space colonies, and the precious and strategic metals needed by Earth's industries.

For instance, the *smallest* known metallic asteroid, 3554 Amun, contains over $1 trillion worth of cobalt, $1 trillion worth of nickel, $800 billion worth of iron, and $700 billion worth of platinum. The total value of this single small asteroid is approximately equal to the entire national debt of the United States. By comparison, the uncontrolled impact of Amun with Earth would deliver a devastating 7-million-megaton blow to the biosphere, killing billions of people and doing hundreds of trillions of dollars worth of damage.

Thus we come to our final, and most startling, discovery: The stick that threatens Earth is also a carrot. Every negative incentive we have to master the impact hazard has a corresponding positive incentive to reap the bounty of mineral wealth in the would-be impactors by crushing them and bringing them back in tiny, safe packages, a few hundred tons at a time, for use both in space and on Earth.

Remember that we will almost certainly have hundreds to thousands of years of warning time before a threatening global-scale impact. We need not be driven to rash and risky actions taken precipitously under threat of death. We will almost certainly have plenty of time to deal with the problem.

Dealing with near-Earth objects should not be viewed grudgingly as a necessary expense: It is an enormously profitable investment in a limitless future. It is a liberation from resource shortages and limits to growth. It is an open door into the solar system—and beyond.

About the Author
John S. Lewis is co-director of the NASA/University of Arizona Space Engineering Research Center and Commissioner of the Arizona State Space Commission. His address is Lunar and Planetary Laboratory, University of Arizona, Tucson, Arizona 85721. E-mail jsl@u.arizona.edu.

He is the author of *Rain of Iron and Ice: The Very Real Threat of Comet and Asteroid Bombardment* (Addison-Wesley, 1996), on which this article is based. His latest book is *Mining the Sky: Untold Riches from the Asteroids, Comets, and Planets* (Addison-Wesley, 1996).

The Universe

Stars (Articles 34 and 35)
Stellar Evolution (Articles 36 and 37)
Types of Stars (Article 38)
Black Holes (Article 39)
The Milky Way Galaxy (Article 40)
Dark Matter (Articles 41 and 42)
Galaxy Structure and Evolution (Articles 43-46)
Quasars (Articles 47 and 48)

The Stellar Magnitude System" should be read in conjunction with "The Spectral Types of Stars" found in Unit 1. Magnitude is one of the most important properties of stars and an understanding of the various types of magnitude is vital to an understanding of astronomy. Robert Naeye's report, "The Strange New Planetary Zoo," on the discovery of planets orbiting other stars, confirms what has been known all along—our Sun is not the only star with planets.

Stellar evolution is important to cosmologists because it provides one method of determining the age of the universe. (see Unit 5 subsection *The Age Paradox*) Ken Croswell's article on the life and times of a star shows how the Hertzsprung-Russell diagram can be used to show the evolution of a star. The essay "Extreme Stars: At the Edge of Stellar Behavior" discusses stars that lie outside the normal main sequence of stellar behavior.

Supernovae are the most spectacular end of certain stars. The essay "Ka-Boom! How Stars Explode" by Robert Naeye explains how computers are allowing astronomers to understand the dynamics of this process.

The most mysterious of all celestial objects, enigmatic black holes, seem to occur at the heart of many galaxies. John Wilford examines the evidence that has emerged on this subject in "New Findings Suggest Massive Black Holes Lurk in the Hearts of Many Galaxies."

Our galaxy is providing a lot of information to astronomers as they map the distribution of stars within it. Studying our galaxy is one source of information for explaining galaxy evolution. Ken Croswell's article, "The Milky Way," describes the mapping of the galaxy, the plotting of star distribution within the galaxy, and the classification of the types of stars that occur in the galaxy.

The next two articles, by Ken Croswell and James Glanz (see "The Dark Side of the Galaxy" and "Is the Dark Matter Mystery Solved?"), deal with the topic of dark matter. Evidence shows that galaxies must have more matter in them than is visible to us. This "dark matter" emits virtually no light, which makes it extremely difficult to study. Two possibilities exist for dark matter: either it is MACHOs (Massive Compact Halo Objects), a group of objects that includes such things as black dwarfs and brown dwarfs, or WIMPs (Weakly Interacting Massive Particles), a class of exotic particles that includes the neutrino. Evidence supports the first view, but the debate continues.

The next series of articles deal with galaxy evolution. The process by which galaxies came into being is still not completely understood, and their subsequent evolution is also open to interpretation. Images from the Hubble Space Telescope have allowed astronomers to more accurately classify galaxies by size, type, and magnitude. Knowing the types of galaxies and their variation allows a more accurate picture of galaxy evolution.

The final two articles in this section deal with quasars. Extraordinarily distant and bright, they have proved difficult to study. Radio astronomy observations of quasars indicate that "we are seeing the epoch when quasars become active."

Looking Ahead: Challenge Questions

What is the difference between visual magnitude and absolute magnitude?

How are extrasolar planets grouped?

What evidence is used to show that black holes occur at the hearts of galaxies?

How is the Hertzsprung-Russell diagram used in describing stellar evolution?

Describe four characteristics of the Milky Way galaxy.

Identify MACHOs and WIMPs and the part they play in the mysteries surrounding "dark matter."

After reading Marcia Bartusiak's article "What Makes Galaxies Change?" try to answer the question.

What are quasars? Why are they so difficult to study?

UNIT 4

Backyard Astronomy
Edited by Alan M. MacRobert

The Stellar Magnitude System

MOST WAYS of counting and measuring things work logically. When the thing you're measuring increases, the numbers get bigger. When you gain weight, the scale doesn't tell you a *smaller* number of pounds. But things are not so sensible in astronomy, at least not when it comes to the brightnesses of stars.

Star magnitudes do count backward, the result of an ancient fluke that seemed to make sense at the time. Since then the history of the magnitude scale has been, like so much else in astronomy, the history of increasing scientific precision being built on an ungainly historical foundation that was too deeply rooted for anyone to bulldoze it down and start fresh.

The story begins around 129 B.C., when the Greek astronomer Hipparchus produced the first well-known star catalog. Hipparchus ranked his stars in a simple way. He called the brightest ones "of the first magnitude," simply meaning "the biggest." Stars not so bright he called "second magnitude," second biggest. The faintest stars he could see he referred to as "sixth magnitude." This system was copied by Claudius Ptolemy in his own list of stars around A.D. 140. Sometimes Ptolemy added the words "greater" or "smaller" to distinguish between stars within a magnitude class. Ptolemy's works remained the basic astronomy texts for the next 1,400 years, so everyone used the system of first to sixth magnitudes. It worked just fine.

Galileo forced the first change. On turning his newly made telescopes to the sky, Galileo discovered that stars existed that were fainter than Ptolemy's sixth magnitude. "Indeed, with the glass you will detect below stars of the sixth magnitude such a crowd of others that escape natural sight that it is hardly believable," he exulted in his 1610 tract, *Sidereus Nuncius*. "The largest of these . . . we may designate as of the seventh magnitude. . . ." Thus did a new term enter the astronomical language, and the magnitude scale became open-ended. Now there could be no turning back.

As telescopes got bigger and better, astronomers kept adding more magnitudes to the bottom of the scale. Today a pair of 50-millimeter binoculars will show stars of about 9th magnitude, a 6-inch amateur telescope will reach to 13th, and the Hubble Space Telescope has seen sources as faint as 30th magnitude.

By the middle of the 19th century astronomers realized there was a pressing need to define the entire magnitude scale, both telescopic and naked-eye, more precisely than by eyeball judgment. They had already determined that a 1st-magnitude star shines with about 100 times the light of a 6th-magnitude star. Accordingly, in 1856 the Oxford astronomer Norman R. Pogson proposed that a difference of five magnitudes be defined as a brightness ratio of exactly 100 to 1. This convenient rule was quickly adopted. One magnitude thus corresponds to a brightness difference of exactly $\sqrt[5]{100}$, or very close to 2.512 — a value known as the Pogson ratio.

The resulting magnitude scale is logarithmic, in neat agreement with the 1850s belief that all human senses are logarithmic in their response to stimuli. (The decibel scale for rating loudness was likewise made logarithmic.) Alas, it's not quite so, not for brightness, sound, or anything else. Our perceptions of the world follow power-law curves, not logarithmic ones. Thus a star of magnitude 3.0 does not in fact look exactly halfway in brightness between 2.0 and 4.0. It looks a little fainter than that. The star that *looks* halfway between 2.0 and 4.0 will be about magnitude 2.8. The wider the magnitude gap, the greater this discrepancy. Accordingly, *Sky & Telescope*'s computer-drawn maps use star dots that are sized according to a power-law rela-

34. Stellar Magnitude System

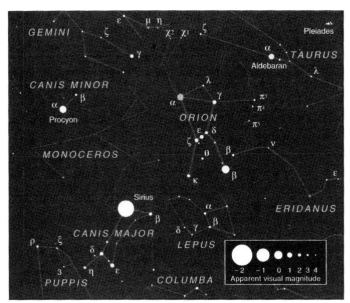
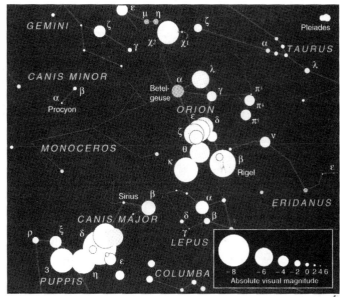

Left: On this map of the Orion–Canis Major area, star dots are sized proportionally to the stars' actual brightnesses. That is, a star 100 times (5 magnitudes) brighter than another will have 100 times the area. Note that the actual range of stars' apparent brightnesses is greater than maps usually suggest. *Right:* An absolute-magnitude map of the same stars, showing their actual luminosities — how they would look if they were all placed 32.6 light-years (10 parsecs) away. The dots here are not sized to fully represent the extreme differences in light output. If they were, and if Epsilon Eridani stayed the same size, the dots for the brightest supergiants would be larger than the entire map!

tion (see the March 1990 issue, page 311).

But the scientific world in the 1850s was gaga for logarithms, so now they are locked into the magnitude system as firmly as Hipparchus's backward numbering.

Now that star magnitudes were ranked on a precise scale, however ill-fitting a one, another problem became unavoidable. Some "1st-magnitude" stars were in fact a lot brighter than others. Astronomers had no choice but to extend the scale out to brighter values as well as fainter ones. Thus Rigel, Capella, Arcturus, and Vega are magnitude 0 — an awkward expression that might sound like they have no brightness at all. But it was too late to start over. The magnitude scale extends farther down into negative numbers: Sirius shines at magnitude –1.5, Venus reaches –4.4, the full Moon is about –12.5, and the Sun blazes at magnitude –26.7.

OTHER COLORS, OTHER MAGNITUDES

By the late 19th century astronomers were using photography to record the sky and measure star brightnesses, and a new problem cropped up. Some stars having the same brightness to the eye showed different brightnesses on film, and vice versa. Compared to the eye, photographic emulsions were more sensitive to blue light and less so to red light.

THE MEANING OF MAGNITUDES

This difference in magnitude . . .	. . . means this ratio in brightness
0	1 to 1
0.1	1.1 to 1
0.2	1.2 to 1
0.3	1.3 to 1
0.4	1.4 to 1
0.5	1.6 to 1
0.6	1.7 to 1
0.8	2.1 to 1
1.0	2.5 to 1
1.5	4.0 to 1
2	6.3 to 1
2.5	10 to 1
3	16 to 1
4	40 to 1
5	100 to 1
6	251 to 1
7.5	1,000 to 1
10	10,000 to 1
15	1,000,000 to 1

Accordingly, two separate scales were devised. *Visual magnitude,* or m_{vis}, described how a star looked to the eye. *Photographic magnitude,* or m_{pg}, referred to star images on blue-sensitive black-and-white film. These are now abbreviated m_v and m_p.

This complication turned out to be a blessing in disguise. The difference between photographic and visual magnitudes was a convenient measure of a star's color. The difference between the two kinds of magnitude was named the "color index." Its value is increasingly positive for yellow, orange, and red stars, and negative for blue ones.

But different photographic emulsions have different spectral responses! And people's eyes differ too. For one thing, your eye lenses yellow with age; old people see the world through yellow filters (*S&T:* September 1991, page 254). Magnitude systems designed for different wavelength ranges had to be more firmly grounded than this.

Today, precise magnitudes are specified by what a standard photoelectric photometer sees through standard color filters, which are depicted above. Several photometric systems have been devised; the most familiar is called UBV after the three filters most commonly used. U encompasses the near-ultraviolet, B is blue, and V corresponds fairly closely to the old visual magnitude; its wide peak is in the yellow-green band, where the eye is most sensitive.

Color index is now defined as the B magnitude minus the V magnitude. A pure white star has a B–V of about 0.2, our yellow Sun is 0.63, orange-red Betel-

4. THE UNIVERSE: Stars

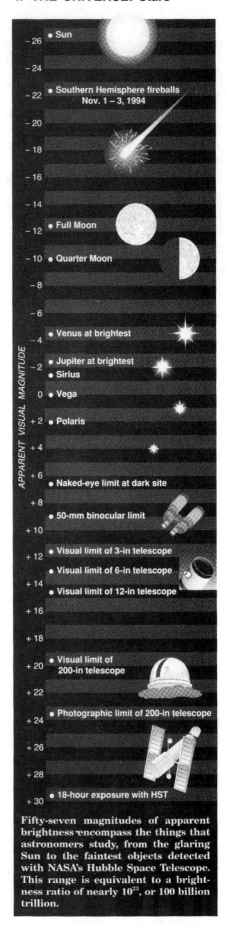

Fifty-seven magnitudes of apparent brightness encompass the things that astronomers study, from the glaring Sun to the faintest objects detected with NASA's Hubble Space Telescope. This range is equivalent to a brightness ratio of nearly 10^{23}, or 100 billion trillion.

geuse is 1.85, and the bluest star believed possible is −0.4, pale blue-white (see "The Truth About Star Colors," *S&T:* September 1992, page 266).

So successful was the UBV system that it was extended redward with R and I filters to define standard red and near-infrared magnitudes. Hence it is sometimes called UBVRI. Infrared astronomers have carried it to still longer wavelengths, picking up alphabetically after I to define the J, K, L, M, N, and Q bands (*S&T:* June 1995, page 23). These were chosen to match the wavelengths of infrared "windows" in the atmosphere where absorption by water vapor does not entirely block the view.

APPEARANCE AND REALITY

What, then, is an object's *real* brightness? How much total energy is it sending to us at all wavelengths combined, visible and invisible?

The answer is called the *bolometric magnitude,* m_{bol}, because total radiation was once measured with a device called a bolometer. The bolometric magnitude has been called the God's-eye view of an object's true luster. Astrophysicists value it as the true measure of energy emission as seen from the location of Earth. The *bolometric correction* tells how much brighter the bolometric magnitude is than the V magnitude. Its value is always negative, because any star or object emits at least some radiation outside the visual range.

Up to now we've been dealing only with *apparent* magnitudes — how bright things look from Earth. We don't know how intrinsically bright an object really is until we take its distance into account. Thus astronomers created the *absolute magnitude* scale. An object's absolute magnitude is simply how bright it would appear if placed at a standard distance of 10 parsecs (32.6 light-years).

Seen from this distance, the Sun would shine at an unimpressive visual magnitude 4.85. Rigel would blaze at a dazzling −8, nearly as bright as the quarter Moon. The red dwarf Proxima Centauri, the closest star to the solar system, would appear to be magnitude 15.6, the tiniest little glimmer visible in a 16-inch telescope! Knowing absolute magnitudes makes plain how vastly diverse are the objects that we lump together under the word "star." The maps of Orion and Canis Major on the previous page illustrate differences between the apparent and absolute magnitudes of some familiar naked-eye stars.

Absolute magnitudes are always written with a capital *M,* apparent magnitudes with a lowercase *m*. Any type of apparent magnitude — photographic, bolometric, or whatever — can be converted to absolute.

Lastly, for comets and asteroids a very different "absolute magnitude" is used. It tells how bright they would appear to an observer standing on the Sun if the object were one astronomical unit away.

So, are magnitudes too complicated? Not at all. They're as simple as they can be, considering their historical roots and what they have to describe today. Hipparchus would be enchanted.

A. M.

The Strange New PLANETARY ZOO

The dozen or so planets found orbiting solar-type stars have turned planet formation theory on its head.

by Robert Naeye

For centuries, the question of planets outside our solar system was rife with speculation. Astronomers could cite any number of indirect pieces of evidence that planets are sprinkled all over the Galaxy. But without any hard evidence, there was always a nagging element of doubt.

In 1991, radio astronomers Alex Wolszczan and Dale Frail discovered planets orbiting a type of collapsed star known as a pulsar. But that still left open the question of whether planets commonly orbit stars similar to the sun — a question of paramount importance to those interested in extraterrestrial life.

Then on October 6, 1995, everything changed. On that date Michel Mayor and Didier Queloz of the Geneva Observatory in Switzerland announced the discovery of a planet orbiting the solar-type star 51 Pegasi. Just three months later, Geoff Marcy of San Francisco State University and colleague Paul Butler of the University of California at Berkeley announced two planets of their own. Knowing that he and Marcy had just started analyzing their mountain of data, and that other groups were closing in as well, Butler confidently predicted: "Within one year, more extrasolar planets will be known than there are planets in our solar system."

Butler's prediction turned out to be remarkably prophetic. One year later, astronomers have detected nearly a dozen planets orbiting other solar-type stars, the exact number depending upon which objects one chooses to count as planets. But astronomers are not content to rest on their laurels. "Nature is very rich, much richer than we had thought at the beginning," says Queloz. "We have to find more objects to better understand where we are."

The Hot Jupiters

Although each extrasolar planet has its own unique characteristics, the ones discovered to date orbiting solar-type stars can be conveniently lumped into three families. The 51 Pegasi planet belongs to a family that stunned the astronomical community: Jupiter-mass planets situated so close to their host stars that they take only a few days to complete an orbit.

When Mayor and Queloz announced the 51 Pegasi planet, most astronomers assumed it was a freak of nature. Here was a planet that was at least half the mass of Jupiter orbiting its host star in a ridiculously short 4.2 days. At its 5-million-mile distance from the star, a mere one-eighth the Mercury-Sun distance, this planet would be baked to a temperature of about 1000° C. How could such a planet form? How could such a planet survive?

Shortly after the planet's discovery, calculations by Adam Burrows and colleagues at the University of Arizona showed that it was massive enough to hold onto its atmosphere despite its proximity to the stellar inferno. And then in April 1996, Marcy and Butler found a second example of a 51 Peg-type planet orbiting the star Rho[1] Cancri. Within three months Marcy and Butler found two more orbiting the stars Tau Bootis and Upsilon Andromedae. The evidence was undeniable: Roughly 5 percent of solar-type stars have these so-called "hot Jupiters." As Marcy says, "Maybe they think we're weird!"

A year ago, astronomers would have told you that finding four hot Jupiters was about as likely as Madonna taking an oath of celibacy. "We *knew* that you could not generate a giant planet closer in than the asteroid belt," says planet hunter George Gatewood of the University of Pittsburgh.

Because nobody has been able to image or take spectra of these hot Jupiters — they were discovered indirectly through their gravitational pull on their host stars — nobody knows exactly what these planets are like. Because there shouldn't be enough rocky material close to a star to form a giant planet, most astronomers think these planets are enormous gas bags similar in composition to Jupiter and Saturn.

According to prevailing theories, a planetary system forms from a disk of gas and dust that surrounds a newborn star. A giant planet forms from a core of mostly ice, with some rock thrown in for good measure. Once the core attains the mass of several Earths, its gravity gobbles up huge quantities of gas from the surrounding protoplanetary disk. But the young star's intense heat and wind prevents ice from condensing in the inner disk. In addition, the heat and wind drive the gas out of the inner disk, preventing gaseous planets from forming there. Gas giants, so the story goes, can only form in the frigid outer disk, at a distance of at least five times the Earth-Sun distance.

So how does one explain the existence of four Jupiter-mass planets orbiting stars well inside the Mercury-Sun distance? A decade before the hot Jupiters were discovered, planet formation models predicted that under certain conditions, gas giant planets would experience drag forces with their disks that would cause them to spiral inward from their birthplace. But these models had a sad ending: The planets would spiral all the way into the star, like a prisoner walking the plank.

"What 51 Pegasi did was make theorists realize that nature has found a way to save the planets," says Alan Boss of the Carnegie Institution in Washington, D.C., who is himself a planet formation theorist. Doug Lin and Peter Bodenheimer of the University of California at Santa Cruz and Derek Richardson of the Canadian Institute for Theoretical Astrophysics developed a theory to explain how a planet could spiral toward a star without plunging all the way in. According to their scenario, when the planet gets very close to the star,

from the system. "A system like our solar system, which has only one dominant massive planet, Jupiter, and is stable over billions of years, may be rare," says Rasio. "The only reason why we happen to live in such a rare system is that long-term stability may be necessary for the development of intelligent life"

The Eccentric Planets

Most planets both inside and outside the solar system, including the hot Jupiters, have orbits that are near-perfect circles. But three of the newly discovered planets have highly elliptical orbits, and thus belong to the family of "eccentric planets."

The first eccentric planet was discovered in 1988 by Harvard University astronomer David Latham's group. The object has a mass of at least 9 Jupiters and orbits the star HD 114762 (remember this name, there will be a quiz at the end of the article). The planet's eccentric orbit brings it as close as a 51-Pegasi-like 0.22 AU to the star and as far as a Mercury-like 0.46 AU (the Earth-Sun distance is 1 AU).

Because the object is much more massive than Jupiter and has such an eccentric orbit, astronomers have been reluctant to classify it as a planet. This explains why Latham was not credited with discovering the first planet orbiting another star.

But if the HD 114762 object isn't a planet, then what is it? For the time being, many astronomers are content to call it a brown dwarf. Brown dwarfs are similar

"Maybe they think we're weird!"

tidal interactions transfer some of the momentum from the young star's rapid spin to the planet's orbital motion. This gives the planet a slight outward boost, thus preventing its death plunge into the star.

This theory works, but it doesn't explain why some planets march inward, while others, such as those in our own solar system, presumably moved very little from their birthplaces. "My guess is that it has something to do with the lifetime of the disk," says Boss. "Maybe our solar system had a very short-lived disk. In the case of 51 Pegasi, the disk was long-lived, so the system had plenty of time to self-destruct."

While the observations don't rule out the possibility of terrestrial planets in these hot Jupiter systems, they're not likely to be there if the spiral mechanism is correct. First generation terrestrial planets would either spiral all the way to their doom or the giant planet's gravity would eject them from system, taking any chance for life with them. The only hope for life in these systems is if the disk manages to form second generation planets.

Another theory, recently developed by Fred Rasio and Eric Ford of MIT, predicts that if two Jupiter-mass planets form around a star, gravitational interactions will eject one planet from the system while the other will be launched inward toward the star, where it will settle into a tight orbit. This theory is also bad news for life, since any terrestrial planets would be ejected

in composition to gas giant planets such as Jupiter. But instead of forming from a disk like planets, brown dwarfs form from collapsing gas clouds, like stars. Brown dwarfs don't shine like stars because they lack the 80 Jupiters' worth of mass necessary to ignite nuclear reactions in their cores. So in a sense, brown dwarfs are transitional objects that bridge the gap in mass between stars and planets.

For eight years, the HD 114762 object was the only example of its type known to science. But in January 1996, Marcy and Butler announced the discovery of a near-twin, an object with a minimum mass of 6.5 Jupiters orbiting the star 70 Virginis. This object's orbit takes it from 0.27 AU to 0.59 AU, almost the same as HD 114762's companion.

Determining whether these objects are planets or brown dwarfs will be difficult. The mass alone doesn't resolve the issue because there's no theoretical reason why the heaviest planets couldn't be more massive than the lightest brown dwarfs. Even a spectrum — which would reveal the chemical composition of their outer atmospheres — probably wouldn't be enough because both brown dwarfs and gas giant planets are composed primarily of hydrogen and helium.

The critical difference lies deep down in the core. Because a brown dwarf forms from a collapsing gas cloud, like a star, it maintains a gaseous composition all the way to the very center. A planet, on the other hand, forms from accreting material in a disk, and will have a core made of ice and rock. But because we're looking at these objects from enormous distances, it will be virtually impossible to determine their interior structures, which may be the key that will unlock the mystery of how they formed.

Some astronomers argue that eccentricity can tell the tale. Mayor and Queloz have recently discovered 10 brown dwarfs orbiting solar-type stars, and most of them have highly eccentric orbits. Stars in binary systems usually orbit each other in highly eccentric orbits too, so the high eccentricity of the HD 114762 and 70 Virginis objects might be telling us that they formed in the way stars do, making them brown dwarfs.

are driven by its distance from the star rather than the way our seasons are driven, by the tilt of the axis." Cochran and his colleague Artie Hatzes independently discovered this planet with Marcy and Butler.

The planet orbits 16 Cygni B, a virtual solar twin that belongs to a triple star system. 16 Cygni B and its slightly larger companion, 16 Cygni A, leisurely orbit each other in a cigar-shaped ellipse that takes at least 125,000 years to trace. The third member of the system, 16 Cygni C, is a red dwarf star roughly half the size of the sun that orbits the main pair at a distance of at least 100,000 AU.

No one knows for sure how this planet ended up with such an eccentric orbit. It could have formed via

> *"You really can't understand a system when you only have one of them to study."*

Marcy and Butler counter that the least massive of Mayor and Queloz's 10 new brown dwarfs has a minimum mass of 17 Jupiters, significantly heavier than the HD 114762 and 70 Virginis companions. Marcy and Butler contend that these latter two objects are indeed planets, or that they belong to a separate population. "Maybe they represent a new class that theorists haven't envisioned yet," says Marcy.

Doug Lin and his colleague Shigeru Ida have even developed a model that explains how these objects could have formed from a protoplanetary disk. If two or more massive planets form in circular orbits around a young star, gravitational interactions will stretch out their orbits into ellipses. After tens of thousands of years, the planets will start crossing one another's orbits. Eventually they will collide and merge into a superplanet. Lin and Ida's model yields a single supermassive planet in an eccentric orbit similar to those of the HD 114762 and 70 Virginis companions. The Rasio and Ford model, which is similar to the Lin and Shigeru model, also produces supermassive planets in eccentric orbits. In other words, the hot Jupiters and eccentric planets could be telling astronomers that planet formation can be a much more violent and dynamic process than they had ever imagined.

The Oddball Eccentric

The third eccentric planet is quite different from the other two. Its minimum mass of 1.5 Jupiters is much lower, and it has by far and away the most eccentric orbit of any planet discovered to date. The planet comes as close as 0.6 AU from the star (which would put it inside Venus) and ventures as far away as 2.8 AU (the asteroid belt). "This planet has wild seasonal variation," says co-discoverer William Cochran of the University of Texas at Austin. "But here the seasons

the colliding planets mechanism suggested by Lin and Ida. Or perhaps when the system formed, gravitational encounters between the three stars stretched the planet's orbit into an ellipse. The simplest explanation is that the planet formed in a circular orbit, but each time 16 Cygni A comes in for a close approach, its gravity tugs on the planet, gradually yanking the orbit into its current elongated shape.

The objects orbiting HD 114762, 70 Virginis, and 16 Cygni B are so massive and have such eccentric orbits that other planets in the stars' "life zones" would either be kicked out of their systems entirely, or would crash into the big planets. If one is prospecting for life, one would be advised to look elsewhere.

The Jupiter-like Planets

Most astronomers believe that planetary systems like our own, with small, rocky, inner planets and giant, gaseous, outer planets, offer the best prospects for life. But none of the planets discussed so far even remotely resembles the retinue of planets in our solar system. So finding a planet with the same mass and orbit as Jupiter, a true member of the Jupiter family, has become one of the holy grails of extrasolar planet detection. Astronomers have yet to turn up a Jupiter twin, but they have discovered a few planets that sound a note of familiarity.

One of Marcy and Butler's first two planets was a world with a mass of at least 2.3 Jupiters orbiting the solar-type star 47 Ursae Majoris. The planet orbits at a distance of 2.1 AU, which would put it on the inner edge of the asteroid belt if it were in our solar system. It's not exactly Jupiter's twin, but it could be a cousin.

Perhaps most intriguing of all, George Gatewood has possibly found two Jupiter-like planets orbiting Lalande 21185, a red dwarf star only 8.25 light-years

4. THE UNIVERSE: Stars

BIZARRE DENIZENS OF THE PLANETARY ZOO

Star	Spectral Type of Star	Planet Mass (Jupiter=1)	Average Distance from Star (AU)	Orbital Period	Eccentricity	Discoverers
HOT JUPITERS						
51 Pegasi	G3	*0.46	0.05	4.2 days	0.00	Mayor/Queloz
Rho¹ Cancri	G8	*0.84	0.11	14.7 days	0.05	Marcy/Butler
Tau Bootis	F6	*3.87	0.05	3.3 days	0.02	Marcy/Butler
Upsilon Andromedae	F7	*0.68	0.06	4.6 days	0.11	Marcy/Butler
ECCENTRIC PLANETS						
HD 114762	F9	*9.0	0.34	84.0 days	0.35	Latham, et al.
70 Virginis	G5	*6.5	0.43	116.6 days	0.38	Marcy/Butler
16 Cygni B	G2	*1.5	1.72	2.2 years	0.67	Cochran/Hatzes & Butler/Marcy
JUPITER-LIKE PLANETS						
Jupiter	G2	1.00	5.2	11.9 years	0.05	Prehistory
Saturn	G2	0.30	9.5	29.5 years	0.06	Prehistory
47 Ursae Majoris	G0	*2.3	2.1	3.0 years	0.03	Marcy/Butler
Lalande 21185	M2	~0.9	~2.2	~5.8 years	0	Gatewood
Lalande 21185	M2	~1.1	~11	~30 years	0	Gatewood
UNCONFIRMED PLANETS						
Rho¹ Cancri	G8	~5	~5	~20 years	?	Marcy/Butler
Upsilon Andromedae	F7	?	?	~2 years	?	Marcy/Butler
Lalande 21185	M2	?	>11	>30 years	?	Gatewood
CM Draconis	M4.5 & M.4.5	?	?	?	?	several teams
PULSAR PLANETS						
B1257+12	Pulsar	*0.015 (Earth)	0.19	23 days	0.00	Wolszczan
B1257+12	Pulsar	*3.4 (Earths)	0.36	66 days	0.02	Wolszczan/Frail
B1257+12	Pulsar	*2.8 (Earths)	0.47	95 days	0.3	Wolszczan/Frail
B1620-26	Pulsar	5 to 15 (Jupiters)	~20	~100 years	?	Rasio, et al.

*minimum mass

> *"I'm very happy Columbus has landed so we can get down to the science of the project."*

away — five times closer than than the next closest known extrasolar planet. Although each of the dozen or so planets found to date can be considered somewhat uncertain because all of them have been detected through indirect means, Lalande 21185 is the most iffy of them all. Gatewood uses a different technique than the other planet hunters, one that doesn't immediately yield a mass and orbit. He will need to observe the star for several more years before he can pin down specific masses and orbits. All Gatewood can say for now is that Lalande 21185 appears to be orbited by at least two Jupiter-mass planets at distances of about 2.2 AU (the inner asteroid belt) and 11 AU (a little beyond Saturn) respectively. There are also hints that a third planet orbits much farther out.

Of all the planets found so far, Lalande 21185's purported planets most closely resemble Jupiter in terms of mass and orbit. "I was tremendously reassured when Gatewood came up with the Lalande 21185 detection because we didn't have anything that looked like Jupiter or Saturn," says Boss. "We might very well have twisted in the wind for quite some time before we found anything that looked like a 'normal' system." Because Lalande 21185 is practically a next-door neighbor, this adds to the excitement as well. If one of the sun's closest neighbors has a similar planetary system, it's not hard to imagine that systems resembling our own are sprinkled throughout the galaxy.

The Next Step

It's been a wacky and wonderful year. In September 1995 astronomers were wondering if Jupiter-mass planets were as rare as gemstones. Now, the number of planets orbiting solar-type stars has jumped from zero (or one, if you consider the HD 114762 object a planet) to a dozen or so. "Things are very encouraging now," says Cochran. "A year ago, everyone in this game was sort of morose. We weren't seeing things, people were trying to explain why this was. Now we're seeing that there's a lot of things out there."

The newly discovered planets have turned theory on its head, demonstrating that planet formation is a much more diverse process than anyone had imagined. Theorists can hardly be blamed for their lack of foresight, however, because until recently they only had one example of a planetary system to study: ours. Because their ideas so beautifully described the architecture of our solar system, it's no wonder they expected to find carbon copies. As Gatewood says, "You really can't understand a system when you only have one of them to study." Now with other systems turning up, and with a diversity exceeding all expectations, the creativity of theorists will be unleashed on the universe.

But before theorists go overboard, one must consider that the searches are most sensitive to massive planets orbiting close to their host stars, which is exactly what astronomers are finding. "The planets we found are the easiest ones to find, and therefore might not be representative of what's out there," says Cochran. "We have to be very careful about making any broad, sweeping interpretations based on what's been found so far." Butler adds, "We're just beginning to get a sense of how bizarre the planetary zoo is."

To really understand the whole story of planets and planet formation, astronomers need to identify entire systems, like the one that might be orbiting Lalande 21185. Marcy and Butler have compelling evidence for second planets orbiting Rho[1] Cancri and Upsilon Andromedae, demonstrating that astronomers are rapidly making progress in this arena. "We're moving from planet detection to planetary system characterization," says Gatewood. "It's fine to go out and find new planets, we need to classify these things. But this is all about relating these back to our own solar system. We're going to say, 'Yes, you've found a planet around these stars. See if you can find a second one or a third one.' To say that there's a world here and a world there is more exploration than science. I'm very happy Columbus has landed so we can get down to the science of the project."

With instrumentation and telescopes improving rapidly, and with more teams looking at more stars, the future will be more exciting than the past. Butler offers the following prognostication: "Within the next couple of years, I expect the discovery of systems with multiple planets and the discovery of a planet that is as small as 10 Earth masses. We've literally just started scratching the surface, and we're going to find many things that we didn't expect to find."

> *"We're just beginning to get a sense of how bizarre the planetary zoo is."*

Associate editor Robert Naeye (rnaeye@astronomy.com) thinks these new planets are really cool, but his favorite planet is still Earth.

Article 36

At first glance, all stars may look the same. But stars actually differ greatly from one another. By understanding and classifying these differences, astronomers have been able to learn the secrets of the stars and what makes them tick

LIFE AND TIMES OF A STAR

Ken Croswell

STARS add beauty to the sky and are the building blocks of our Galaxy, the Milky Way. But they are also essential for life on Earth. Many of the atoms in our bodies were forged inside stars, and one star, the Sun, sustains us all. The Sun looks different from other stars simply because it is so much closer. Sunlight is really just very bright starlight, and if our star stopped shining, all life on Earth would perish.

In addition to the Sun, our Galaxy harbours hundreds of billions of other stars. But like snowflakes, no two stars are the same. Some stars are bright, others faint; some are blue, others white, yellow, orange or red; some are enormous, others tiny; some are young, while others are old and dying.

To make sense of the tremendous diversity of the stars, astronomers use the **Hertzsprung-Russell diagram**, a method of studying the differences between stars which was first published by the Danish astronomer Ejnar Hertzsprung in 1911 and independently by the American astronomer Henry Norris Russell in 1913. The H-R diagram is central to the study of stars and stellar evolution. Just as the periodic table allows chemists to sort the elements by their fundamental characters, so the H-R diagram allows us to distinguish stars by their main features.

Luminosity v. colour
A powerful diagram

THE H-R diagram is so powerful for astronomers because it plots two basic stellar properties: luminosity and colour.

Stellar luminosity is the amount of light that a star emits. The most luminous stars are a million million times more luminous than the least luminous. This means that the most luminous star in the Galaxy sends out more light in a single second than the least luminous does in a hundred centuries. If the most luminous star in the Galaxy replaced the Sun, the Earth would become so hot that its oceans would boil and its rocks melt. Conversely, if the least luminous star replaced the Sun, daytime would be darker than a moonlit night and Earth's oceans would freeze.

The Sun's luminosity falls midway between these two extremes. But the Sun is by no means just an "average star", because most stars are in fact much dimmer.

On the H-R diagram, the most luminous stars appear at the top

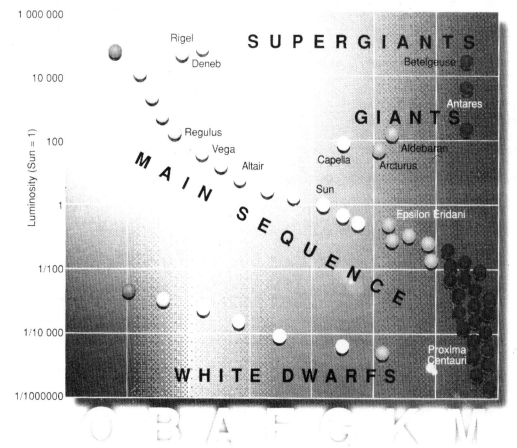

Figure 1 The Rosetta stone of the sky: the H-R diagram plots stellar luminosity against stellar colour and reveals three types of stars: main-sequence stars (diagonal band); giants and supergiants (upper right); and white dwarfs (line below main sequence)

36. Life and Times of a Star

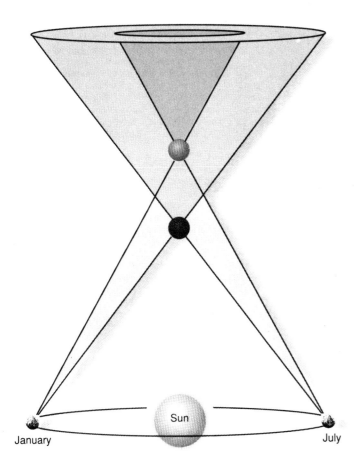

Figure 2 Shifting stars: as the Earth moves round the Sun, the apparent position of a star in the sky changes slightly. The greater this change, or parallax, the closer the star. Here the bottom star is closer to Earth than the top star and so shows a larger parallax

and the least luminous at the bottom. Because its luminosity is in the middle of this range, the Sun appears about halfway down on the H-R diagram.

Colour is the other stellar property used in the H-R diagram. To the untrained eye, all stars may look white or yellow. But in fact, stars range in colour from blue and white to yellow, orange and red. The Sun is yellow.

Colour tells us how hot or cool a star is. The blue and the white stars are hot (most of them are between 7500 °C and 50 000 °C), yellow stars are warm (between 5000 °C and 7500 °C), and orange and red stars are cool (between 2000 °C and 5000 °C). In the same way, a poker placed in a fire will first glow red and then, as it becomes hotter, and hotter, will glow orange, yellow and white.

On the H-R diagram, the hot blue stars appear on the left hand side, the warm yellow stars in the middle and the cool red stars on the right. Because the Sun is yellow, it again lies near the middle of the diagram.

Strength in numbers
The main sequence

WHEN Hertzsprung and Russell first plotted the H-R diagram, they were astonished to find that stars did not scatter randomly over it. Instead, most stars lay in a band that stretched diagonally from the upper left (bright and blue) to the lower right (faint and red). This diagonal band included the Sun. Astronomers now know that 90 per cent of all stars fall in this band. It is, therefore, called the **main sequence**.

Main-sequence stars obey a law: the brighter the star, the bluer it is. This law arises because every main-sequence star generates energy the same way. Every such star fuses hydrogen nuclei into helium nuclei at its centre.

The reason the luminosities and colours of main-sequence stars differ is because the stars differ in **mass**—the total quantity of material in the star. The greater the mass of a main-sequence star, the hotter the star's centre and the faster the hydrogen fuses, so the hotter and brighter the star.

Because of this dependence on mass, the main sequence is really a mass sequence. Blue and white main-sequence stars have more mass than the Sun; yellow main-sequence stars have about the same mass as the Sun; and orange and red main-sequence stars have less mass than the Sun. The most massive stars have about 100 times the Sun's mass, whereas the least massive main-sequence stars have only 0·08 solar masses.

But not all main-sequence stars are equally common. Massive stars are rare, both because few such stars are formed and because they do not live for long. Consequently, only one star in more than a thousand is a blue main-sequence star. Yet many such stars are visible to the naked eye,

1: Magnitude, distance and luminosity

A GLANCE at the night sky reveals that some stars look bright and others faint. To quantify this impression, astronomers have for a long time used the concept of **apparent magnitude**. In about 120 BC, Hipparchus classified the stars into six groups by calling the brightest stars those of first magnitude down to the faintest stars of the sixth magnitude. The system was calibrated more exactly in the 1850s, and put on a logarithmic scale so that each magnitude corresponds to a factor of 2·5 in brightness. A star with an apparent magnitude of 1·00 is 2·5 times brighter than a star with an apparent magnitude of 2·00 and so on. Most of the brightest stars in the night sky are of the first magnitude; the faintest stars that the naked eye can see are sixth magnitude and look only about 1 per cent as bright as first-magnitude stars.

Telescopes penetrate to fainter apparent magnitudes. The largest telescope on Earth can detect stars as faint as 30th apparent magnitude. This is 4 billion times fainter than the naked eye can see.

But apparent magnitude alone does not reveal a star's luminosity. To calculate that, astronomers must also know the star's distance. A star that looks faint may have a low luminosity and be nearby, or it may be luminous and lie far away. For stars near the Sun, astronomers can measure the distance because they view the star from slightly different perspectives as the Earth circles the Sun. For example, the Earth is on one side of the Sun in January and on the opposite side in July. So, from January to July, every star's apparent position in the sky shifts slightly. The larger this shift, or **parallax**, the closer the star. Only nearby stars have parallaxes that are large enough for astronomers to measure.

Distances are often given in **light years**. One light year is the distance that light travels in a year, or 9·5 million million kilometres. This is an enormous distance: there are as many Earth-Sun distances in a light year as there are inches in a mile (more than 60 000). Yet even the nearest star to the Sun is 4·3 light years away, and most stars that you can see in the night sky are a few hundred light years away. Another unit of distance that astronomers use is the **parsec**. A parsec is the distance of a star whose parallax is one arcsecond or 1/3600th of a degree. One parsec is also 3·26 light years. The nearest star is, therefore, 1·3 parsecs away.

Once they know a star's distance, astronomers can calculate its luminosity from its apparent magnitude. They express this luminosity in terms of **absolute magnitude**. This is the apparent magnitude the star would have if it were 10 parsecs (32·6 light years) from Earth. For example, the Sun has an absolute magnitude of 4·83, which means that if we viewed it from a distance of 32·6 light years, the Sun would have an apparent magnitude of 4·83, barely visible to the naked eye.

But the absolute magnitudes of other stars run the gamut from bright to dim. The most luminous stars have absolute magnitudes around -10, which means that if such a star were 32·6 light years away it would be incredibly bright—about a tenth as bright as the full Moon. In contrast, the dimmest have absolute magnitudes around +20, which means that such a star 32·6 light years from Earth would look hundreds of times fainter than Pluto. The Sun, with an absolute magnitude around +5, is right in the middle of this range.

4. THE UNIVERSE: Stellar Evolution

because they are so luminous that they can be seen from great distances.

In contrast, less massive stars abound but are hard to see. The most abundant main-sequence stars are the red ones, which appear at the bottom right of the H-R diagram and are called red dwarfs. They outnumber all other stars put together, accounting for 70 per cent of the stars in the Galaxy. Yet they are so faint that not a single one is visible to the naked eye.

If a star has even less mass than a red dwarf, it never becomes hot enough to ignite its hydrogen and so never joins the main sequence. These stars, which astronomers have not yet definitely detected, are called **brown dwarfs**. The name is misleading, however, because if these stars exist, they will be faint and red and look like red dwarfs.

A star's mass dictates how long the star will live. Although stars with more mass have more hydrogen fuel, they "burn" it much faster and die sooner—just as a millionaire who spent a million pounds a day would go broke long before a poor person with a sensible budget.

The most massive stars use up the hydrogen fuel at their centres just a few million years after they are born. In contrast, the Sun has been burning hydrogen for 4·6 billion years and will continue to do so for billions of years more. Red dwarfs burn their fuel so slowly that some will remain on the main sequence for thousands of billions of years.

Nevertheless, every main-sequence star will someday use up the hydrogen at its centre. When that happens, the star begins to burn hydrogen outside its centre and so is no longer a main-sequence star. The star expands and cools, moving to the right on the H-R diagram, and it may also get brighter, in which case it moves up on the diagram. The star is now a giant or supergiant.

Big and bright
Giants and supergiants

AS THEIR name implies, giants and supergiants are much bigger than the stars on the main sequence. Aldebaran, an orange giant, is 40 times bigger than the Sun. If put in the Sun's place, Aldebaran would stretch halfway to the innermost planet Mercury. Supergiants are even bigger. Betelgeuse, a red supergiant, is so large that if it replaced the Sun it would engulf Mercury, Venus, Earth, Mars and part of the asteroid belt. The largest red supergiants would stretch all the way

2: Stellar colour and spectral type

COLOUR is such a basic property of stars that astronomers have developed a system for classifying it. Because stars of different colour have different temperature, and because temperature affects different types of atoms differently, astronomers can classify a star by studying the atoms and molecules that mark its spectrum. For example, white stars have strong spectral lines due to hydrogen, whereas yellow stars, like the Sun, have strong lines due to calcium.

Astronomers have used stellar spectra to divide the stars into seven main types, which can best be remembered with the mnemonic "Oh, Be A Fine Guy/Girl, Kiss Me!"

O: These stars are the hottest and bluest of all. But O stars are rare, and few appear in the night sky.

B: Somewhat cooler than O-type stars, stars of spectral type B are nevertheless hot and blue. Many bright stars in the night sky, such as Spica, Regulus and Rigel, are B-type stars.

A: A-type stars are white and contribute much to the light of our Galaxy. They include among their ranks Sirius, Vega and Altair, which are all main-sequence stars, and Deneb, which is a white supergiant.

F: The F stars are a bit hotter than the Sun and are yellow-white. Viewed from Earth, the two brightest F-type stars are Canopus and Procyon. Polaris, the Pole Star, is also an F star.

G: The Sun is a member of this spectral class. Like the Sun, G-type stars are warm and yellow. Other G stars are Alpha Centauri A, which is a main-sequence star, and Capella, which is a giant.

K: Cooler than the Sun are the orange K stars, which include giant stars, such as Arcturus and Aldebaran, and fainter main-sequence stars, such as Epsilon Eridani. K-type main-sequence stars are called orange dwarfs.

M: These stars are cool and red. Some M stars, such as Betelgeuse and Antares, are supergiants that shine thousands of times more brightly than the Sun, but most M stars are faint stars on the main sequence—red dwarfs.

---

to the planet Saturn.

Most giants and supergiants are warm or cool, being yellow, orange, or red, so on the H-R diagram they appear at the upper right (bright and red). A few are blue or white, such as Rigel, a blue supergiant, and Deneb, a white supergiant.

In general, supergiants have evolved from the hottest and bluest main-sequence stars, whereas giants have evolved from less massive main-sequence stars. Red dwarfs are so long-lived that none has yet had enough time to evolve to become a giant.

Because they are so big, giants and supergiants have a low average density, since their mass is spread throughout a large volume. The outer parts of a giant or supergiant are so tenuous that they would pass for a perfect vacuum here on Earth, so a rocket ship could fly through most of such a star unimpeded.

Because of their large size, giants and supergiants are luminous and outshine the Sun. When our Sun becomes a giant, it will be about 100 times brighter than it is now. Supergiants are even brighter, outshining the present Sun thousands of times over.

Despite their impressive size and luminosity, giants and supergiants are rare. In fact, they account for less than 1 per cent of all stars. They are prominent in the night sky simply because they are luminous and easy to see.

Giants and supergiants are rare because they don't last long. Stars with more than eight times the mass of the Sun have a dramatic but brief life after becoming supergiants. Having consumed all the hydrogen at their core, they soon start fusing helium, carbon, neon, oxygen, silicon and sulphur, the last two of which fuse into iron. But iron does not fuse into heavier elements, and at that point the star is doomed and explodes as a **supernova**.

Normally, the supernova will be classified **type II**, which means that hydrogen appears in the exploding star's spectrum. The hydrogen comes from the star's outer atmosphere. However, a massive star may lose this hydrogen atmosphere before exploding. If so, the star becomes a **type Ib** supernova, and no hydrogen appears in the spectrum. A star may even lose both its hydrogen and its helium—which lies deeper inside the star—in which case neither element appears in the exploding star's spectrum and the supernova is classified as **type Ic**.

After the supernova, the remains of the star collapse into a small but dense sphere. This may be a **neutron star**, so called because the star's protons and electrons smash together and become neutrons. A neutron star is the size of a large city, but typically has a mass 1·5 times that of the Sun. Or the collapsed star may become a **black hole**, which has such intense gravity that nothing—not even light, the fastest thing in the Universe—can escape it. Neither neutron stars nor black holes appear on the conventional H-R diagram, because they are dimmer than even the dimmest stars.

But most stars never go through the ordeal of a supernova, because most are born with less

Figure 3 Colourful stars: each spectral type corresponds to a different temperature [See Box 2 above.]

36. Life and Times of a Star

than eight solar masses. When such a star becomes a giant, it eventually burns helium into carbon and oxygen. During this time, instabilities develop in the star's outer atmosphere, which gets ejected into space. The star's hot core emerges and makes the ejected atmosphere glow. This glowing gas is a **planetary nebula**, not because it has anything to do with planets but because through a small telescope it may look like a planet's disc.

In only a few tens of thousands of years, the planetary nebula expands so much that it disappears from view. All that is left is a small but extremely hot star. The star is so hot that it does not appear on the conventional H-R diagram, since the star would be far left of even the hottest and bluest main-sequence stars.

But the star soon cools and fades, reappearing on the H-R diagram below the main sequence, because the star is fainter than a main-sequence star of the same colour. The star is now a **white dwarf**.

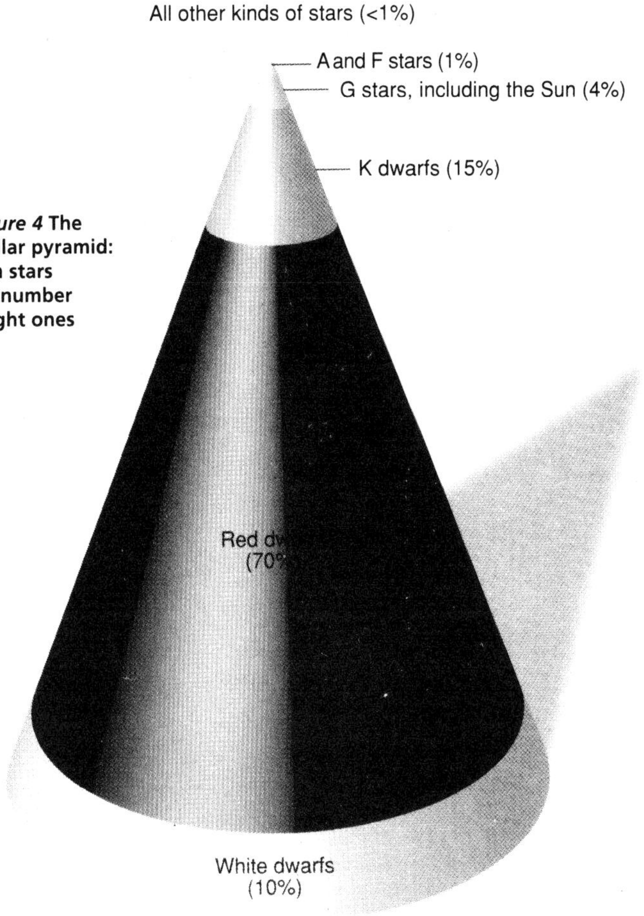

Figure 4 The stellar pyramid: dim stars outnumber bright ones

All other kinds of stars (<1%)
A and F stars (1%)
G stars, including the Sun (4%)
K dwarfs (15%)
Red dwarfs (70%)
White dwarfs (10%)

Fading stars
White dwarfs

THE FORMER core of a once-living star, a white dwarf is hot and dense. A typical white dwarf is little larger than the Earth but contains about 60 per cent of the mass of the Sun. A teaspoonful of white dwarf matter would weigh more than a tonne.

Because so many stars become white dwarfs, these objects are common, making up 10 per cent of all stars in the Galaxy. But they are so faint that all are invisible to the naked eye. A typical white dwarf leads a boring life. It no longer burns fuel; it shines simply because it contains a store of heat. As it radiates energy into space, the star fades and cools over billions of years. As it does so, it changes colour, so despite their name, white dwarfs can, in fact, be any colour. The newest members are hot and blue, while those that have been around a long time and lost most of their energy are orange or red. So, on the H-R diagram, white dwarfs form a sequence that is parallel to the main sequence. If enough time elapses, the white dwarf will fade completely and become a **black dwarf**. But no black dwarfs exist, because the Universe is not old enough for any white dwarfs to have faded from view.

On rare occasions, white dwarfs can create spectacles. If another star orbits the white dwarf and dumps material onto it, the material can explode. Astronomers then see a **nova**. During a nova, the star increases in brightness some 100 000 times. Violent though it may be, the explosion does not destroy either star.

However, if a companion star transfers too much mass, then the white dwarf can annihilate itself in a supernova. This is because the most mass that a white dwarf can have is 1·4 times that of the Sun: anything more than that and the star blows up.

This is now known as the **Chandrasekhar limit**, so named after the astronomer who discovered it in the 1930s. If a companion star forces the white dwarf over this limit, the white dwarf explodes. This supernova is classified as **type Ia**; it is the only type of supernova that does not arise from a massive star, the way that types Ib, Ic and II supernovae do.

3: The Sun: something special in the sky

TEXTBOOKS often malign our star, the Sun, by calling it "average". But it isn't, not by a long shot. The truth is, the Sun is much more luminous than most other stars.

Seventy per cent of all stars are red dwarfs. These stars are much dimmer than the Sun, and most emit less than 1 per cent of its light. Another 15 per cent of stars are orange dwarfs, which are also less luminous than the Sun. And another 10 per cent of stars are white dwarfs, which again are less luminous than the Sun. Thus, 95 per cent of the Galaxy's stars are fainter than the Sun. Or, to put it the other way, our supposedly "average" Sun is in the most luminous 5 per cent of all stars in the Milky Way.

The night sky is deceptive, however, as dim stars can't be seen with the naked eye, whereas luminous ones can be seen even if they are hundreds or thousands of light years away. Therefore, nearly all the stars visible to the naked eye are more luminous than the Sun. If you judged the Galaxy by these, you'd fooled into thinking the Sun was a dim star. Although it looks as if luminous stars outnumber dim ones, the reality is just the reverse.

Supernovae cast heavy elements, such as oxygen and iron, into the Galaxy. These elements were formed both during the star's life and in the explosion itself. This material, together with debris from planetary nebulae, eventually gathers in star-breeding areas of the Milky Way, where it will give birth to new stars and planets, some of which may one day support life. This is how the Sun and Earth formed 4·6 billion years ago. We are part of this legacy: our bodies contain atoms that were created by the stars.

FURTHER READING

Stars, by James Kaler (Freeman, 1992); *The Supernova Story*, by Laurence Marschall (Princeton, 1994). "Life of a star", Inside Science, number 10.

Ken Croswell is an astronomer in Berkeley, California, and author of *The Alchemy of the Heavens*, a popular-level book about the Milky Way to be published in 1995.
All graphics are by Nigel Hawtin.

Article 37

EXTREME STARS
At the Edge of Stellar Behavior

From tiny balls no larger than a city to swollen giants that would fill up the inner solar system, stars come in an amazing variety of sizes, temperatures, and luminosities.

by James B. Kaler

They shimmer around us, sparkling grains of quartz tossed against the nighttime sky. At first, they seem similar, tiny points displaying only differences in brightness and subtle variations in color. Over the past 200 years, however, astronomers have slowly revealed their true natures. Far more than a simple evening delight, the stars are perhaps the proprietors of other planetary systems and the keepers of life.

Along with their great importance go spectacular extremes of size, temperature, and luminosity. In one of nature's most ironic twists, one extreme may be followed by its opposite, the coolest stars becoming the hottest and the largest the smallest. By understanding these transformations astronomers have pieced together the story of how stars live and die.

Limits to Size

Let's begin with our Sun as a reference: 109 Earths could be stretched along its diameter and its volume could hold a million terrestrial worlds. The Sun's mass is 300,000 Earths and a thousand times that of the king of the planets, Jupiter. From its 6,000-kelvin surface, energy pours out at a rate of 4×10^{26} watts, the equivalent of over 10 trillion trillion standard 100-watt light bulbs. The cost of running the Sun for one second by your local power company would equal the gross national product of the United States accumulated over a million years. Moreover, this rate of energy production has been sustained for the 4.6 billion-year lifetime of the solar system. This remarkable energy output is produced by the fusion of hydrogen into helium deep in the Sun's hot (15 million-kelvin) core. Astronomers calculate that the Sun started with a 10-billion-year supply of hydrogen. The Sun is 4.6 billion years old, so it should last for another 5 or so billion years.

Though commonly touted as an "average" star, the Sun is considerably larger, hotter, and brighter than most. But as impressive as our star is, many are much grander in all their aspects. But what do we mean by "grand"? When we speak of impressive qualities, we commonly think in terms of the large. Small, however, can be impressive as well.

Stars are characterized chiefly by their diameters and their surface temperatures. Together these properties determine the third principal characteristic, luminosity. The range of stellar diameters is perhaps most astonishing. Pluck a few stars down from the nighttime sky, beginning with spring's Arcturus or winter's Aldebaran. These orangy "giant stars," not just a description but also a name for a type, are respectively 20 and 40 times larger than the Sun. Stars like the Sun are called "dwarfs," hardly an apt term, merely to distinguish them from these giants.

Arcturus and Aldebaran pale, however, beside larger but less-easily visible stars like reddish Mira in the heart of Cetus, the Whale. Mira is a long-period variable, typical of a yet-larger class of giants. The star is normally invisible to the naked eye, but every 11 months it brightens to roughly third magnitude, about as bright as the faintest star of the Big Dipper. Mira's variation in brightness is in part the result of expansion and contraction. If Mira were magically transported to the Sun's location, it would fill out the inner solar system beyond Earth even at its smallest size. At its largest, the star would approach the orbit of Mars and would be 150 times the solar diameter.

But we are not nearly at the limit. Two reddish jewels, Betelgeuse in winter's Orion and Antares in summer's Scorpius, belong to a different category altogether. Both of these "supergiants" would fill our solar system well past the orbit of Mars, with Betelgeuse (twice Antares' size) stretching to near the orbit of Jupiter.

And we are still not at the end. In Cepheus gleams Mu (μ) Cephei, impressively reddish through binoculars. One of the largest stars

37. Extreme Stars

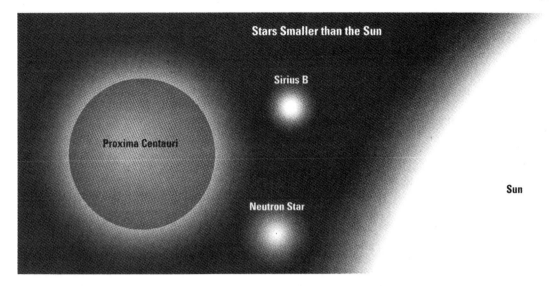

known, Mu Cephei's diameter is crudely comparable to the orbit of Saturn, over 1,000 times the diameter of the Sun. It could hold a billion Suns.

Now let's go in the other direction, from the large to the small. The nearest star — only 4.2 light-years away — is a red dwarf called Proxima Centauri. It's diameter is only 1/10 that of the Sun. Below Orion sparkles the night sky's brightest star, Sirius. Tucked next to it is a dim companion 10,000 times fainter than Sirius itself, each star in orbit around the other. The faint component, Sirius B, is the size of Earth and can be seen only in large amateur telescopes. The Sun could hold a million Sirius Bs, and Mu Cephei could contain a million billion. The first of this strange breed to be discovered was white in color, hence the whole class of such stars came to be known as "white dwarfs" (even though some are blue and others orange) to distinguish them from solar-type "dwarfs."

We can go yet smaller. Off the southern horn of Taurus is a dim cloud of gas, the Crab Nebula. Within it lies a strange star that produces 30 bursts of radiation per second at extraordinarily regular intervals. Precision timing suggests rotation, and fast rotation can only take place in something very small. This tiny star is beaming radiation like a lighthouse, and the beam sweeps past Earth every rotation. These small stars — called neutron stars — are densely packed balls made nearly exclusively of neutrons (the neutral particles that help bind atomic nuclei). Neutron stars are no more than 30 kilometers or so across, the size of a small city. We could line up 50,000 neutron stars across the solar diameter. The Sun could contain 100 quadrillion (10^{17}) of them and Mu Cephei an uncountable (10^{23}), which is more than all the stars in the visible universe!

Limits to Temperature and Luminosity

While the range of stellar surface temperatures is not so great as

4. THE UNIVERSE: Stellar Evolution

that of sizes, it is still impressive. The color of a star, or any other dense glowing body, correlates with temperature. As you heat a bar of iron (or if you could heat a star) it glows red at a temperature of 1500 K, turns orangy at about 4000 K, yellowish-white at 6000 K, and white at 10,000 K. Much hotter than that and the body takes on a bluish cast. Because the Sun's temperature is about 6000 K, sunlight has a soft yellow-white glow. Aldebaran and Arcturus — and the other ordinary giants — typically have relatively cool surface temperatures of 4000 K, giving them a yellow-orange cast. The reddish supergiants Betelgeuse and Antares go down to 3500 K. The lowest surface temperatures are reserved for the large giants of the Mira class, which can drop to 2000 K or lower.

In the opposite direction we find other dwarfs: white Sirius and summer's Vega, with surface temperatures of about 10,000 K, and Orion's Rigel — a blue supergiant — at 20,000 K. Higher yet is the brightest star in the center of the Orion Nebula, located at the center of the Hunter's sword, which tops 45,000 K. The high-temperature limit is reached in the central stars of the lovely gas clouds called planetary nebulae. They start at 25,000 K and climb to an observed maximum of nearly 250,000 K, over 40 times hotter than the Sun's surface.

The amount of energy pouring out of a body's surface hinges critically on its temperature. Hot bodies naturally radiate more energy than cool ones, the output power per square meter depending on the fourth power of the temperature: double it and the radiant luminosity goes up by a factor of $2 \times 2 \times 2 \times 2 = 16$. The luminosity of a body also depends on how big it is, or on the number of square meters that radiate light. The total luminosity therefore depends on the product of surface area and temperature. If we know temperature and diameter (from which we can calculate surface area), we can thus determine the luminosity. Conversely, if we know luminosity and temperature, we can calculate the diameter. Much of our knowledge of stellar sizes comes from just such a procedure.

The range of stellar luminosities is staggering. The red supergiant Mu Cephei is nearly a million times more luminous than our Sun. At the other end of the luminosity scale, we find red dwarfs a million times fainter than our Sun. These dim bulbs could be seen with the naked eye only if brought within 200 times Pluto's distance.

Origins

The origins of these variations in size, temperature, and luminosity lie in the combination of mass and age. The course of a star's life is set by the amount of matter given to it at birth. The wide range of physical characteristics among dwarfs, all of which have similar chemical compositions, is actually caused by their wide range of masses. Dwarf stars like the Sun are defined as those that are producing energy by the nuclear fusion of hydrogen into helium. Dwarf masses range from about 0.08 that of the Sun to about 100 times solar.

Stars at the upper end of this mass range have more gravitational energy, and are consequently hotter and denser inside. Higher temperature means a higher rate of energy production and thus greater luminosity. Double the mass of a star and it becomes roughly 10 times brighter. A 100 solar-mass dwarf is approximately a trillion times brighter than one of 0.08 solar masses.

All the other categories of stars arise from the aging of the dwarfs as they almost magically transform themselves under the inexorable force of gravity. When the giants and supergiants were first distinguished from the dwarfs in the early part of the century, it seemed only logical that stars ought to be born large and then squeeze down into the smaller varieties. Therefore giants and supergiants should be the youngest. The astronomical pioneers, however, had not realized that most of the gravitational squeeze takes place deep inside the star where it is not immediately visible.

Astronomers gradually worked out the story of stellar evolution over the latter part of this century. The longest-lived stars are the low-mass dwarfs. The lower the mass of a star, the slower the rate of nuclear fusion and the longer the star lives in spite of its smaller fuel supply. The lowest mass dwarfs will live for over a trillion years, far longer than the age of the Galaxy. All the low-mass dwarfs ever born are still shining.

The dwarfs of intermediate mass, those between just under a solar mass and about 8 times the mass of the Sun (the number is poorly known) perversely become the giants. When the hydrogen fuel runs out in such a star, its helium-rich interior shrinks. But the gravitational energy released heats the interior, pushing the outer envelope outward: A giant is born. The giant's core eventually gets hot enough to fuse helium into carbon, the star shrinks and stays alive for awhile. But eventually the helium, too, runs out and the star transforms itself again into an even larger giant like Mira. The intermediate dwarfs thus become the coolest stars.

But while these transformations are taking place, spanning roughly 10 percent the lifetime of the star, the huge star begins furiously losing mass. The outer envelope streams away in a powerful wind, gradually revealing more and more of the intensely hot core. In one of the more remarkable metamorphoses in nature, the coolest stars are thus transformed into the hottest. The cores illuminate the surrounding shells of gaseous mass given off by the previous stellar winds, and the giants become planetary nebulae with hot stellar cinders at their centers.

The central stars of the planetary nebulae continue to shrink, the surrounding gaseous clouds dissipate into space, and the hottest stars become small white dwarfs. Most stars are supported, that is, kept from collapsing under the inward pull of their own gravity, by the outward push of the

37. Extreme Stars

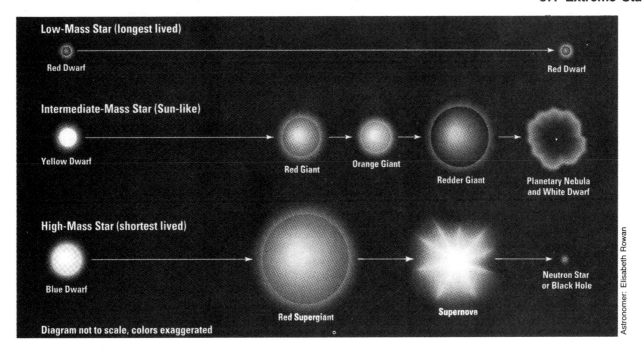

Diagram not to scale, colors exaggerated

Astronomer: Elisabeth Rowan

pressure provided by their hot gases. The white dwarfs, however, can't shrink further because they are supported by the outward push of densely packed electrons. Their destiny is to shrink and cool forever. But the cooling time is so long, longer than the lifetime of the Galaxy itself, that every white dwarf ever created still glows (see "White Dwarfs Confront the Universe," May 1996).

The high-mass stars, those above 10 solar-masses, have the shortest lives. At the upper limit of about 100 solar-masses, where the stars are also the luminosity record holders (see "Star on the Brink," page 46), fuel is consumed at such prodigious rates that the stars survive for only a few million years, a lifetime only a millionth that of the lowest-mass dwarfs. When their internal hydrogen is all gone, the stars balloon outward to become the huge supergiants. As a result of their great masses, supergiant interiors become so hot that they can carry nuclear fusion past the conversion of helium into carbon, and can "burn" carbon to magnesium, magnesium to silicon, and finally silicon to iron. Iron is the end of the road. No energy can be generated by additional fusion. Instead it takes an input of energy to fuse iron.

Once the iron core develops, fusion stops. The outward push of hot gases from the interior can no longer counteract the inward pull of gravity, so gravity wins in the end. The core collapses, releasing a vast amount of gravitational energy in the blink of an eye. The iron atoms break down ultimately to protons and electrons, which merge, producing a ball of neutrons that squeezes to 30 kilometers across: a neutron star. As the core collapses, it first rebounds, sending a shock wave out into the surrounding envelope. The shock wave and a huge number of nearly massless particles called neutrinos push the envelope outward, creating a cataclysmic explosion — a supernova — that can be seen across vast reaches of the universe.

Dozens of supernova remnants, some in the form of rapidly expanding gaseous ejecta like the Crab Nebula, others blast waves sweeping up interstellar gas, dot the Galaxy. Nuclear reactions gone berserk (as well as those generated in giants) have produced all the atoms heavier than lithium in the universe. As the coolest stars (the giants) transform themselves into the hottest stars (the nuclei of the planetary nebulae), the largest stars (the supergiants) transform themselves into the smallest stars, the neutron stars. Thus we have the seemingly paradoxical situation whereby the larger the star becomes after its initial dwarf state, the smaller it becomes at the end, the result of gravity's great crunch.

Stars can squeeze down even further. The pressure of neutrons in a neutron star can support the star only if the mass is less than two or three times that of the Sun. If the resulting neutron star has a greater mass, it collapses further, in a sense into a point. Ultimately, gravity becomes so powerful that light cannot escape and the "star" — no longer recognizable as such — disappears from view: a black hole. Some of the vast supergiants might thereby plummet into the ultimate in smallness.

The variety of stars in our Galaxy is remarkable, and we might stand in awe of what we see. But if we are to speak of awe, should we not also factor in ourselves? Perhaps more remarkable is that we have the ability to look back upon the stars and understand them and the enormous changes wrought by age.

Astronomer James B. Kaler (kaler@astro.uiuc.edu) studies stellar death at the University of Illinois at Urbana-Champaign. His new book The Ever-Changing Sky *is published by Cambridge University Press.*

KA-BOOM!

How Stars Explode

Some of the wimpiest particles in the universe power the awesome destructive fury of an exploding star.

by Robert Naeye

For millions of years, the massive star Sanduleak −69°202 was the master of its domain. At the height of its powers, the star was so big that 170,000 suns would have fit inside. Weighing in at 20 solar masses, Sanduleak −69°202's nuclear furnace pumped out so much energy that the star shone with the brightness of 20,000 suns, dominating its region of the Large Magellanic Cloud. But like many Egyptian pharaohs, this mighty star's glory would prove to be short-lived.

One day, its nuclear furnace ran out of fuel. In a fraction of a second, the core collapsed, triggering a shock wave. The shock wave ripped through the outer layers, blowing the star to kingdom come. Poor Sanduleak −69°202 was no more.

For 167,000 years, news of the star's catastrophic demise traversed space at the speed of light, until on February 23, 1987, it reached a small, rocky planet called Earth. Astronomers on Earth have combed each photon of light they could get their hands on to help them better understand the story of stellar death.

The fact that massive stars end their lives in titanic blasts called supernovae is old news. Astronomers have known that since the 1930s. But how do stars explode? Surprisingly, scientists are just now gaining a detailed understanding of this process. The key development has been high speed supercomputers, which can unveil the immensely complicated and messy goings-on that occur deep inside an exploding star.

Despite their immense size, stars are rather simple objects. A star, no matter how big or how small, lives its life on a high wire, carrying out a delicate balancing act. It generates energy from nuclear reactions in the core. This energy, in the form of heat, exerts a gas pressure that pushes outward while gravity tries to pull the star inward.

No matter how big the star, it spends most of its life fusing hydrogen atoms into helium deep down in its core. This reaction, similar to the one that powers a hydrogen bomb, generates enormous amounts of energy and helps prevent the star from contracting. But eventually the star will run out of its hydrogen fuel reserves. For the star, this is equivalent to a near-death experience.

But stars are feisty creatures with a built-in safety mechanism, so they always manage to survive this crisis. When a star runs out of hydrogen, gravity causes the core to contract. The contraction heats the star's interior, which increases the outward push of gas pressure. This causes the star's outer envelope to expand and cool: The star becomes a bloated red giant. The contracting core becomes hot enough to begin fusing helium into carbon and oxygen. This round of fusion gives the star a new lease on life.

38. Ka-Boom! How Stars Explode

Ten years ago, Supernova 1987A graced the skies of the Southern Hemisphere. The explosion occurred 167,000 light-years away in the Large Magellanic Cloud. It was the first supernova visible to the naked eye since 1604. Despite its dazzling appearance, SN1987A was a somewhat fainter-than-normal supernova. These stellar cataclysms can sometimes briefly outshine an entire galaxy in visible light.

Low-mass stars like the sun end their lives in a whimper. After the core runs out of helium, the core contracts and heats up again. But the temperature never gets hot enough to ignite another round of fusion. The core continues to contract until gravity scrunches the atoms together to the point where electrons exert their own pressure, preventing the core from collapsing any further. The core is a stellar cinder called a white dwarf. A white dwarf, about the size of Earth, is destined to slowly cool and fade for billions of years. Before dying, the bloated star sheds its outer layers of gas, which briefly form a brightly glowing planetary nebula around the white dwarf.

Stars that start out with about 11 or more solar masses end their lives in a blaze of glory. When the core runs out of helium, it shrinks. But the massive overlying layers compress the core to a density where it can become hot enough to fuse carbon, oxygen, and even heavier elements in succession, until the core is made up of iron. At this point, the star's internal structure resembles an onion: An iron core is surrounded by a layer of silicon, which in turn is surrounded by shells of increasingly lighter elements.

Iron is like a stop sign. It has the most stable nucleus of any element — it takes more energy to fuse iron than the reaction gives off. So when the core is made of iron, fusion reactions come to a halt, and the star is ready to meet its doom.

If the supernova is asymmetric, a slight excess of neutrinos flying off in one direction can give a "kick" in the opposite direction to the resulting pulsar or black hole. Astronomers have in fact observed pulsars racing through the galaxy like bullets. Here we see the fastest-moving pulsar known, which is zooming through the galaxy at 1,000 miles per second. The pulsar is moving so fast that it will easily escape the Milky Way's gravitational clutch. The pulsar plows through interstellar gas, forming the so-called "Guitar Nebula."

When an iron core exceeds the mass of 1.44 suns — the maximum mass of a white dwarf — gravity takes over and initiates a catastrophic collapse. In less than a second, the core shrinks from a sphere about the size of Earth to a sphere the size of a small city, attaining the density of an atomic nucleus. All the protons and electrons in the core squeeze together into neutrons. The resulting stellar cinder is called a neutron star. A sugar-cube-sized lump of neutron star material contains as much mass as the entire human population.

This time the neutrons bunch so closely together that they exert their own pressure, which prevents the core from

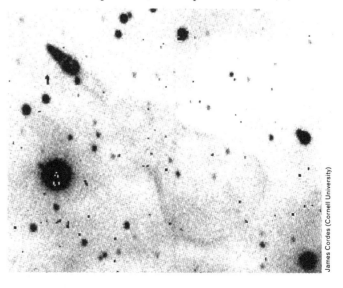

4. THE UNIVERSE: Types of Stars

1. The star's core sits in the lower-left corner, 0.3 second before it will explode. The colors represent different gas densities, pressures, and temperatures, all of which decrease the farther they are from the core. The density of material in the inner core is so great that each cubic micrometer of material (which is much smaller than a bacterium) has the energy equivalent of a hydrogen bomb.

2. A mere 0.02 second before the explosion, the core has collapsed and rebounded. The rebounding core has produced a shock wave that is moving outward. The region just outside the core is boiling. The shock wave is located at the boundary between the boiling region and the dark blue outer layer, which is comprised of silicon. The gas in the dark area is falling inward toward the core. The splotches inside the shock wave are hot, buoyant bubbles of gas that act like expanding balloons.... [d]enser material ... tends to sink. The dark buoyant bubbles are gas heated by neutrinos emanating from the core. As they rise upward, like hot water rising from a boiling pot, they push the shock wave ever outward.

3. The explosion is just under way. The region in the upper right is made of oxygen, both the oxygen and silicon layers are falling inward at about 10,000 km per second. In just a few milliseconds, they will collide with the shock wave, which is traveling outward at 30,000 km per second.

4. The buoyant plumes are created by instabilities that arise whenever a heavy gas overlies a lighter one. These instabilities drive convection: whereby hotter material rises and cooler material sinks. The convection thus pushes the shock wave ever outward. The oxygen and silicon layers continue to fall inward.

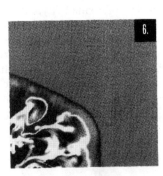

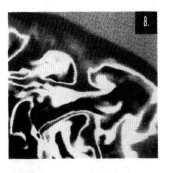

5. About 0.04 second after the explosion, the shock wave, which is now about 750 km from the core, has gobbled up the oxygen/silicon boundary. Huge plumes of hot material — heated by neutrinos emanating from the core — are pushing the shock wave outward.

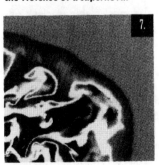

6. By this point, it's clear that the chaotic, turbulent motions below the shock wave are not driving it outward in a spherical manner. Small deviations from perfect symmetry grow with time, creating an asymmetry in the explosion. Many supernovae, including Supernova 1987A, spew their material out in clumps, perhaps leading to many asymmetrical supernova remnants.

7. About 0.08 second after the explosion, the shock wave has ripped through the inner part of the star. It's important to remember that this is just the central 2,500 km of a star hundreds of times larger than the sun. But in about a day, the shock wave will rip through the entire star. When the shock wave reaches the surface, the star might shine as bright as its host galaxy in visible light. Such is the violence of a supernova.

8. This is the central region 0.3 second after the core collapsed. The inexorable scrunch of gravity will cause the core to shrink even further. Eventually, it will form either a neutron star — a city-sized ball of superdense neutrons weighing more than the sun — or a black hole. Neutron stars are also called pulsars.

All images on this page: Adam Burrows (University of Arizona), John Hayes (University of Illinois), and Bruce Fryxell (NASA/GSFC)

collapsing any further. The core rebounds with 100 billion g's of acceleration (an F-16 fighter pilot can withstand about 10 g's at most). This violent action helps trigger a shock wave that rips through the star and tears it to shreds.

But things aren't quite that simple. Without an outward pressure emanating from the core, gravity causes the star's outer envelope — previously unaffected by the cataclysmic happenings in the core — to start falling inward. Until recently, astrophysicists who modeled supernova explosions couldn't get a star to explode in their computer models because infalling material would ram into and overpower the shock wave, causing it to peter out before breaking out of the core. Because nature has clearly found a way around this problem, getting a star to actually explode inside a computer was one of the biggest unsolved problems in astrophysics.

Now if you're an astrophysicist, subatomic particles called neutrinos come to the rescue. If you're a star, maybe you're not so crazy about these little critters. The breakdown of iron in a star's core produces 10 billion trillion trillion trillion trillion of these massless or virtually massless particles, which stream out of the core at near-light speed. These ghostly particles could pass through half a light-year of lead as if it weren't there, just like a stream of marbles could easily pass through a hula hoop.

While each individual neutrino carries only a negligible amount of energy, supercomputer simulations show that their enormous numbers enable them to reenergize the stalled shock wave by heating the material underneath it, similar to how a stove's hot burner boils water. The influx of energy creates buoyant plumes of hot gas, which push the shock wave outward, enabling it to rip through the entire star and blow it to smithereens. So ultimately, these incredibly ethereal particles are responsible for one of the most cataclysmic events in the universe.

The supercomputer simulations on the previous page show how a shock wave forms, is energized by neutrinos, and moves outward through a star. The images come from a group of researchers that includes Adam Burrows of the University of Arizona, John Hayes of the University of Illinois, and Bruce Fryxell of NASA's Goddard Space Flight Center in Maryland. The key to their work, and to other groups that do similar research, is the ability of modern-day supercomputers to model supernovae in two dimensions. Until recently, computer simulations only revealed what was happening in one dimension, essentially a straight line from the star's core to its surface. These models fell short because they couldn't fully account for all the turbulent motions going on deep in a star's bowels, which power the explosion.

For these two-dimensional models, supercomputers must perform 300 billion operations per second. The simulations run over a period of about 20 days and perform 1,000 trillion calculations. This gives astrophysicists the resolution required to see processes never before visible.

Even with the new computing power, astrophysicists still don't understand all the details of the turbulent motions inside an exploding star. Burrows's team and others will ultimately need to model supernovae in three dimensions, necessitating a hundredfold increase in computing power to achieve the same resolution they're getting now. "Over time we hope to develop knowledge, intuition, and insight into what's going on by ratcheting up the level of complexity," says Burrows. "Eventually, our graduate students' graduate students will be doing this in their classes."

During the few seconds that neutrinos stream out of a collapsing stellar core, more energy is released in the form of neutrinos than the energy radiated in visible light during the same amount of time by the trillions upon trillions of stars in the observable universe. Think about that next time you are out under the stars!

Robert Naeye (rnaeye@astronomy.com) is an associate editor of Astronomy *magazine. This article is partially based on his forthcoming book* Through the Eyes of Hubble: The Birth, Life, and Violent Death of Stars, *which will be published this fall by Kalmbach Publishing Co.*

New Findings Suggest Massive Black Holes Lurk in the Hearts of Many Galaxies

X-ray astronomy also confirms an elusive boundary of no return.

JOHN NOBLE WILFORD

TORONTO, Jan. 13—In the realm of extraordinary scientific concepts, black holes are the strangest and most awesome places in the universe, something out of the mind of Albert Einstein sharing spacetime with the imagination of Lewis Carroll.

A black hole, as conceived in interpretations of Einstein's theory of general relativity, is a region of space in which matter is so concentrated and the pull of its gravity is so powerful that nothing, not even light, can emerge from it. It is the ultimate point of no return. If Alice had ventured into this Wonderland, she would have found herself in a void created by this hidden mass and experienced time and space behaving in bizarre ways—but would have been unable to return or even report back what she had seen.

By definition, a black hole cannot be seen. Its presence must be detected through indirect evidence: the vast whirlpools of matter being sucked in by consuming gravity at ever increasing velocities. Recent observations, after three decades of theorizing and searching, have satisfied even the most cautious astrophysicists that black holes almost certainly exist.

Now scientists have found evidence that supermassive black holes probably lurk at the core of nearly all galaxies and that the mass of each one seems to be proportional to the mass of its host galaxy. This conclusion was drawn from the discovery by the Hubble Space Telescope and ground-based telescopes in Hawaii of three more black holes in relatively nearby ordinary galaxies, and from a wider survey of other possible black holes.

Scientists have also detected for the first time what they think is confirming evidence for the existence of the boundary of no return surrounding a black hole—an "event horizon" across which matter and energy pass in one direction only, falling in but never coming back out. The new findings, based on several telescope and spacecraft observations and innovative computer analysis, were reported here today at the 189th meeting of the American Astronomical Society.

Dr. Scott D. Tremaine of the Canadian Institute for Theoretical Astrophysics in Toronto said the new research seemed to show that in "two quite different contexts, black holes really do exist."

Dr. Douglas Richstone of the University of Michigan said he would still not bet everything he owned on the reality of black holes. "But I would be willing to bet my car on it, and it's a good car," he added.

Commenting on the detection of the event horizon, Dr. Ramesh Narayan, associate director of the Harvard-Smithsonian Center for Astrophysics in Cambridge, Mass., said, "This is the most direct evidence scientists have had that black holes are

39. New Findings Suggest Massive Black Holes Lurk

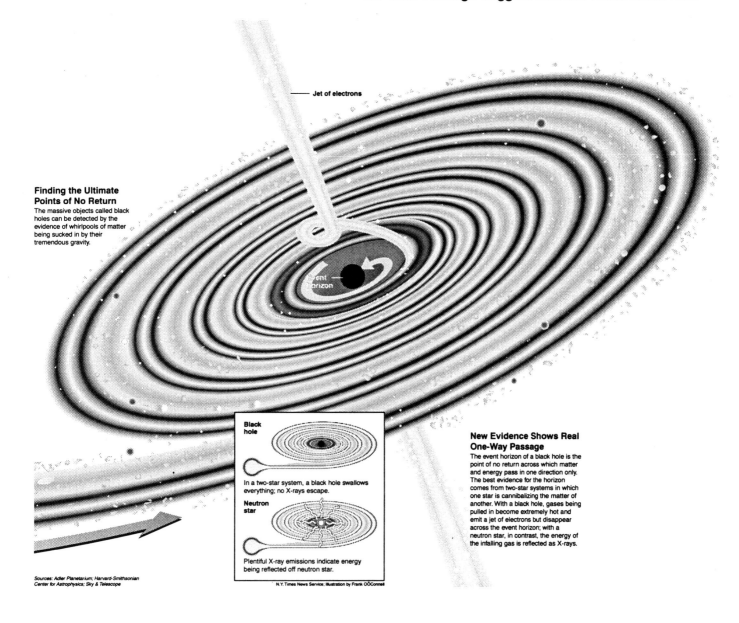

Finding the Ultimate Points of No Return
The massive objects called black holes can be detected by the evidence of whirlpools of matter being sucked in by their tremendous gravity.

New Evidence Shows Real One-Way Passage
The event horizon of a black hole is the point of no return across which matter and energy pass in one direction only. The best evidence for the horizon comes from two-star systems in which one star is cannibalizing the matter of another. With a black hole, gases being pulled in become extremely hot and emit a jet of electrons but disappear across the event horizon; with a neutron star, in contrast, the energy of the infalling gas is reflected as X-rays.

In a two-star system, a black hole swallows everything; no X-rays escape.

Plentiful X-ray emissions indicate energy being reflected off neutron star.

Sources: Adler Planetarium; Harvard-Smithsonian Center for Astrophysics; Sky & Telescope
N.Y. Times News Service; Illustration by Frank O'Connell

real." Dr. Narayan headed the team making the detection.

The group of astronomers predicting the prevalence of galactic black holes, led by Dr. Richstone, said it was still a challenging puzzle as to why black holes were so abundant, or why they should be proportional to a galaxy's mass. Whatever the explanation, it could yield insights about the formation and evolution of galaxies. The more astrophysicists can probe the mysteries of black holes, the more they feel they will learn about cosmic history.

In "Gravity's Fatal Attraction," published last year by the Scientific American Library, Dr. Mitchell Begelman of the University of Colorado and Sir Martin Rees, an astrophysicist at Cambridge University and Astronomer Royal of England, wrote that black holes "represent the ultimate triumph of gravity over all other forces." They concluded, "When we really understand black holes, we will understand the origin of the universe itself."

As scientists now recognize, black holes come in at least two varieties: stellar and galactic. The stellar ones, much smaller, are formed when massive stars, each weighing several times more than the Sun, run out of nuclear fuel and collapse into a compact object only a few miles in diameter. In most cases, the result is a neutron star, a dense stellar cinder. But the most massive stars are likely to collapse into black holes. Most of these probably exist alone in interstellar space, with not enough gas and other matter around them to be swallowed up at energies sufficient to produce detectable emissions of radiation; thus their presence remains beyond recognition.

But some stellar black holes make themselves known because they exist in a binary system as the companion to a normal star, which they are cannibalizing. X-ray emissions around a rotating object known as V404 Cyg revealed that its powerful

4. THE UNIVERSE: Black Holes

gravity was tearing gas away from its larger companion star, pulling it into a spiraling disk and finally consuming it. Studies of V404 Cyg and other binary systems that emit copious X-rays produced the new findings about the event horizon of a black hole.

The most formidable black holes are those at the hearts of galaxies. Some of these may have existed for 90 percent of cosmic history.

Over time, some runaway processes, as yet unexplained, have concentrated matter at galactic cores that can be equivalent to the mass of as many as three billion Suns, all squeezed into an area no larger than the solar system. Dr. John Archibald Wheeler, a Princeton University astrophysicist, had these in mind when he coined the term black hole.

The most compelling evidence for the reality of such a supermassive black hole, in the giant galaxy M87, was gathered three years ago by the Hubble telescope. The swirling gases around the black hole stretched across a distance of 500 light-years. Energy released from gas falling toward the hole produced a jet of electrons spiraling outward at nearly the speed of light. Evidence is also mounting for other galactic black holes, including a likely one at the center of Earth's Milky Way galaxy.

Two of the newly discovered black holes, reported today, are at the cores of the galaxies NGC 3379 and NGC 3377 in the constellation Leo. They weigh in at 50 million and 100 million solar masses, respectively. The third new galactic black hole is at the center of NGC 4486b, in the constellation Virgo.

An analysis of observations by the orbiting Hubble telescope showed a sharp rise in the velocities of stars near the centers of these galaxies, accelerations that would be produced by the gravity of mass concentrations on the order of black holes. The research was complemented by observations at the Canada-France-Hawaii telescope on Mauna Kea, Hawaii, conducted by Dr. John Kormendy of the University of Hawaii.

A Census of Galactic Black Holes

Only 11 of of the extremely dense objects called galactic black holes have been definitively detected to date.

Galaxy	Constellation	Distance in light-years	Mass in units of solar mass
Milky Way	Ñ	28,000	2 million
NGC 224 (M31)	Andromeda	2.3 million	30 million
NGC 221 (M32)	Andromeda	2.3 million	3 million
NGC 3115	Sextans	27 million	2 billion
NGC 4258	Canes Venatici	24 million	40 million
NGC 4261	Virgo	90 million	400 million
NGC 4486 (M87)	Virgo	50 million	3 billion
NGC 4594 (M104)	Virgo	30 million	1 billion
NEWLY IDENTIFIED			
NGC 3377	Leo	32 million	100 million
NGC 3379	Leo	32 million	50 million
NGC 4486b	Virgo	50 million	500 million

N.Y. Times News Service

From these data and other findings, astronomers have developed computer models of the velocity patterns apparently produced by galactic black holes. Using this information, they made a quick survey of 15 other nearby galaxies and found that 14 of them exhibited behavior consistent with the model for the gravitational influence of a black hole.

On the strength of this, Dr. Richstone predicted that nearly all galaxies probably contain black holes. And the estimated masses of these led to the conclusion that the strength of a black hole is related to the size of its galaxy. "This is probably the key evidence that a black hole's presence is intimately connected to the evolution of the galaxy," Dr. Richstone said at a news conference. This in turn, he said, "strongly supports the theory that black holes are the energy sources for quasars."

Quasars are the most distant, oldest and most luminous objects that can be seen. At an earlier time, when they were young, galaxies would probably have been visible only as brightly burning quasars, the result of tremendous energies produced by gas and stars plunging in toward black holes at their cores. Some quasars can still be seen, but the new evidence suggests that in galactic black holes, astronomers are detecting surviving "fossil quasars."

If galactic black holes may reveal the history of galaxies, the smaller stellar black holes should provide insights into the evolution and death of stars. As a result of the new research, said Dr. Tremaine, the evidence for the existence of stellar black holes is even stronger than that for galactic ones.

The detection of an event horizon, a kind of invisible one-way membrane that Einstein's theory predicted should surround a black hole, was made possible by measurements primarily from the Japanese-American X-ray astronomy satellite, ASCA. The measurements of radiations from two stars orbiting each other, one normal and the other extremely compact, agreed with the computer model of how gases accreting around a stellar black hole should behave.

In the model, gases pulled in by the compact star's gravity should become extremely hot but should radiate only a small fraction of the heat as radiation. When the superheated gases reach the center, they would disappear through the event horizon and carry with them their enormous thermal energy—if the compact object is a black hole. In that case, the object would appear very dim in the X-ray spectrum. But if the object is a neutron star, the energy from the infalling gases would be reflected by the star's solid surface, releasing a shower of X-rays.

Observations of two binary-star systems showed the difference, scientists said. Referring to V404 Cyg, Dr. Narayan said, "This star seems to be swallowing nearly a hundred times as much energy as it radiates, and the only way this can happen is if the star is a true black hole."

Two other reports on black hole research are being discussed at this week's astronomy conference. A team of American and German scientists announced observations of a new black-hole candidate, GRS 1915 + 105, made by the Rossi X-ray Timing Explorer satellite. The observations appeared to be consistent with predictions made by a Stanford University team about how the disk of dust and gas that forms around a black hole should vibrate in distinctive ways. Such a signature, astronomers said, should help determine the mass and rotation rate of black holes with greater accuracy.

Scientists at the National Radio Astronomy Observatory in Socorro, N.M., described new radiotelescope studies of water molecules in a disk around a suspected black hole at the center of the galaxy NGC 4258. They said the research made it possible to deduce the exact location of the object the disk is orbiting and measure the distance between the black hole and the innermost observable portions of high-speed jets of energy shooting out from the galactic core.

Article 40

Far larger and brighter than most other galaxies in the Universe, our Galaxy abounds with hundreds of billions of stars . By exploring our Galaxy, astronomers learn about the nature and evolution of galaxies in the cosmos

THE MILKY WAY

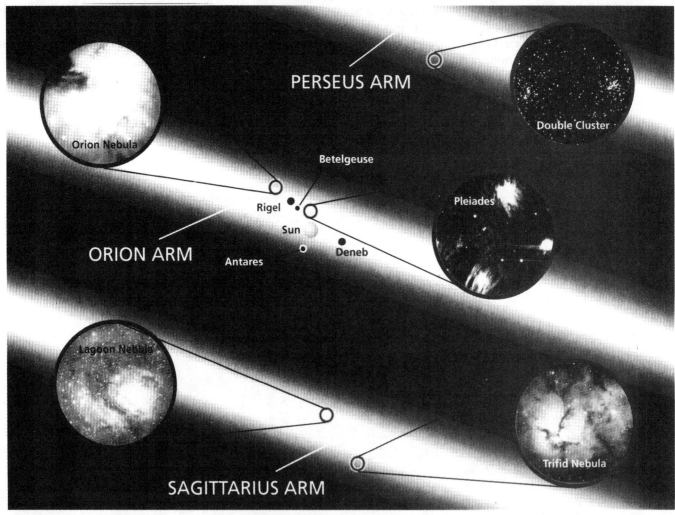

Ken Croswell

DON'T mess with the Milky Way Galaxy. In 1994, astronomers discovered the ruins of a dwarf galaxy that had strayed too close to the shores of our Galaxy and was being ripped apart by its tide. This luckless object was our Galaxy's latest victim, and its stars will soon be devoured by the Milky Way's insatiable appetite.

Our Galaxy's immense power stems from its enormous size and mass. The Milky Way is a giant galaxy, far larger, brighter, and more massive than most other galaxies in the Universe. In addition to the Sun, it harbours hundreds of billions of stars, which orbit the Galactic centre in the same way that the planets of the Solar System circle the Sun. Every star visible to the unaided eye is part of the Milky Way. But, for every star you see, there are 50 million others that also belong to the galaxy we call home.

Astronomers observe the Milky Way not only to learn about our immediate neighbourhood but also to understand other galaxies. Galaxies are the basic building blocks of the Universe, just as atoms are of matter. But only in the Milky Way can

Figure 1: A spiral galaxy: if we viewed our Galaxy from above the disc, we would see a celestial pinwheel. The Orion arm winds through the Sun and contains every bright star in the sky, such as Rigel, Betelgeuse, Antares, and Deneb; the Sagittarius arm lies closer to the Galactic centre; and the Perseus arm lies farther away. All three spiral arms contain clouds of interstellar gas and dust, such as the Orion Nebula (in the Orion arm) and the Lagoon Nebula (in the Sagittarius arm), as well as young star clusters, such as the Pleiades

40. Milky Way

astronomers scrutinise the ages, locations, orbits, and chemical compositions of the faint stars that constitute most of all galaxies, and thereby decipher a galaxy's origin and evolution.

The Milky Way is more than just the galaxy to which we owe our existence; it is a touchstone for the entire Universe.

Stars and gas
Galactic geography

THE MILKY WAY is a **spiral galaxy** that to an external observer would appear as a beautiful pinwheel in space. But, because we live inside it and see stars from different regions superimposed on one another, it took astronomers centuries to discern the Galaxy's spiral shape and, even today, we do not know exactly what the Milky Way looks like. The first spiral galaxy was seen in 1845, when Lord Rosse in Ireland used a large telescope to examine an object named M51. But only in 1951 did the American astronomer William Morgan and his colleagues map out regions of hydrogen gas that are ionised by bright blue stars which populate spiral arms and thus prove that our own Galaxy was also a spiral.

Although the spiral arms outshine the regions in-between, they do not contain more stars. Instead, they harbour clouds of gas and dust that give birth to new stars. A few of these new stars are bright, blue and massive. Massive stars die quickly, before they drift out of the spiral arm that gave birth to them; so nearly all massive stars reside in spiral arms. These stars' brilliance lights the arms and makes them look bright, beautiful and obvious.

The Sun lies in the **Orion arm** of the Milky Way, as do all the bright stars, and most of the faint ones, that you see at night. The Orion arm gets its name because it includes famous objects from that constellation, such as the Orion Nebula and the stars Betelgeuse and Rigel. But stars from the Orion arm appear in every other constellation as well. Of course, there are other arms to our Galaxy. Lying about 6000 light years closer to the Galactic centre than the Orion arm is the **Sagittarius arm**; lying about 6000 light years in the opposite direction is the **Perseus arm**.

The spiral arms belong to the Galaxy's disc, which is shaped like a thin pancake and measures about 130 000 light years across. The Sun is about 27 000 light years from the Galaxy's centre, or about 40 per cent of the way from the centre to the edge. The disc's main component is 2000 light years "thick", and the Sun lies near its midsection, which is called the **Galactic plane**. Stack two CDs atop each other and you have a miniature model of just how thin the Galactic disc is in relation to its diameter.

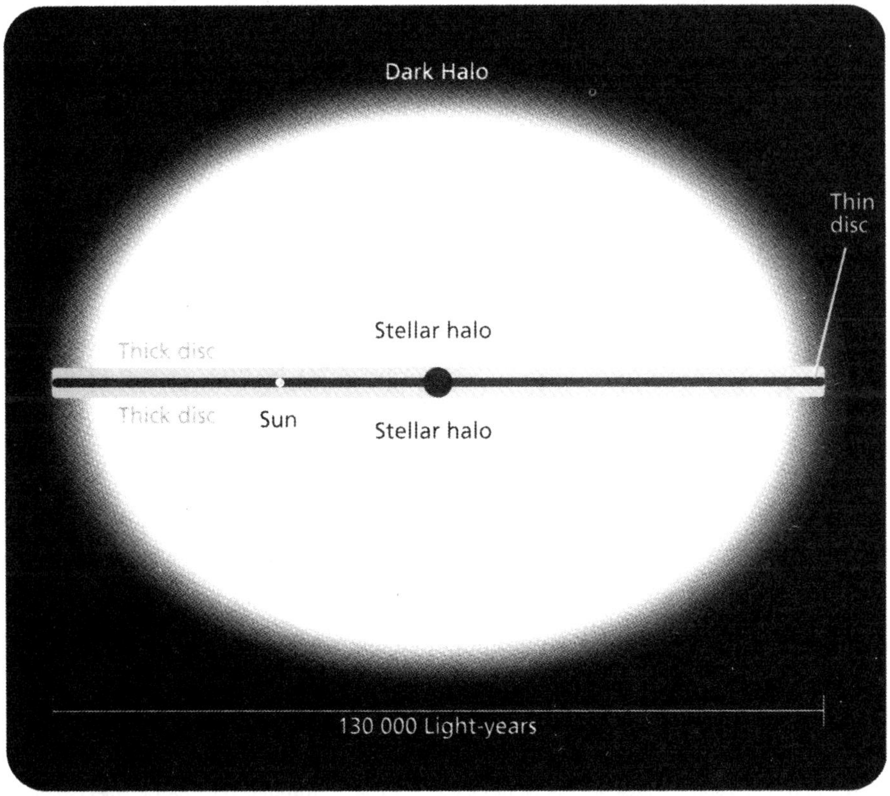

When we look up into this star-filled disc at night, we see a stronger band of white light produced by the combined glow of innumerable stars. This band is sometimes called the "Milky Way" because the Greeks and Romans thought it looked like spilled milk. But every star visible to the naked eye, whether it is in this band or not, is part of the Milky Way Galaxy. It is just like being in a crowd of people. When you look into the crowd, you see lots of people. When you look up or down, you see few or none.

In addition to stars, the disc contains **interstellar gas and dust**. This material is extremely tenuous: on average the disc has only one atom per cubic centimetre. In contrast, terrestrial air has 25 billion billion molecules per cubic centimetre, so the interstellar medium would pass for a "perfect" vacuum on Earth. Yet our Galaxy is so huge that all of the interstellar gas and dust adds up to several billion times more mass than our Sun (see Inside Science, Numbers 45 and 70).

We owe our lives to interstellar matter, because 4·6 billion years ago it gave birth to the Sun and the Earth. Today, clouds of interstellar gas and dust continue to create new stars in the spiral arms. The Milky Way spawns about 10 stars a year, so the youngest stars in the Galaxy are younger than you or me.

All the stars and gas orbit **Sagittarius A*** (pronounced "ay star"), which is probably a black hole at the Galaxy's centre containing a million times more mass than the Sun. Extending for thousands of light years in all directions from the Galactic centre is the **bulge**, which appears in edge-on views of the Milky Way as a bump at the Galaxy's centre jutting above and below the otherwise flattened disc (see Inside Science, Number 70). Recently, astronomers have found that the bulge is somewhat elongated, in a direction roughly towards the Sun, which means the Milky Way may be a **barred spiral** galaxy. If so, it confirms a speculation made by the American astronomer Gérard de Vaucouleurs in 1963.

Clues to the past
Galactic demographics

SURROUNDING the disc is the **stellar halo**, which contains some of the Galaxy's oldest stars, and surrounding that and everything else is the mysterious **dark halo**, which emits little or no light but contains most of the Galaxy's mass. The evidence for the existence of the dark halo comes from the high speed with which stars and gas in the outer disc revolve around the Galaxy, as well as from the motions of the 10 galaxies that orbit ours (see Box). The extent of the dark halo is uncertain, but its diameter is at least a quarter of a million light years, and probably much greater, in which case the nearest galaxies actually reside within it. The dark halo's composition is unknown. It could be made of dark stars, such as brown dwarfs or black holes, or it could be composed of subatomic particles that we have yet to detect.

In 1943, the American astronomer

Figure 2: The Milky Way has four stellar populations: the thin disc, the thick disc, the stellar halo and the bulge (center). As viewed by an external observer, though, only the thin disc and bulge would be bright and prominent

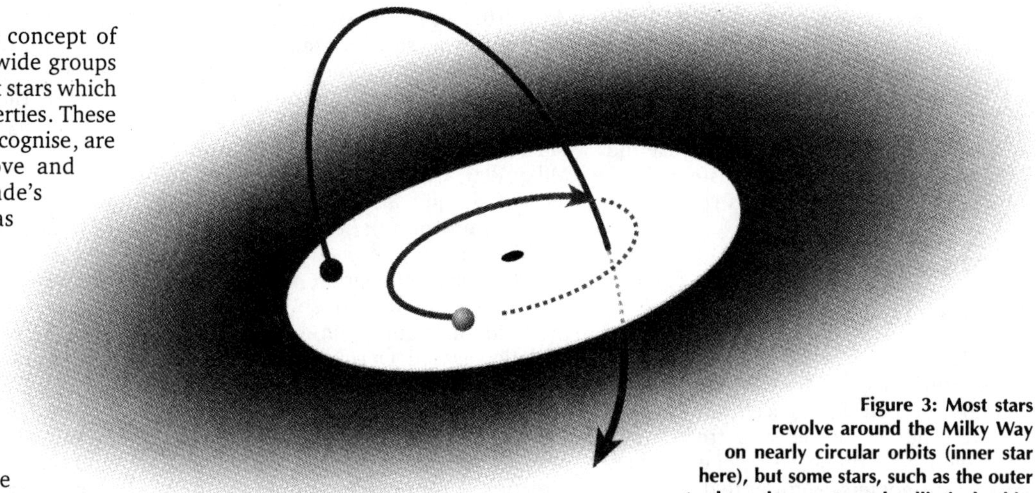

4. THE UNIVERSE: The Milky Way Galaxy

Walter Baade formulated the concept of **stellar populations**—Galaxy-wide groups that contain billions of different stars which nevertheless share similar properties. These properties, astronomers now recognise, are age, location, how stars move and what they are made of. Baade's stellar population concept has allowed astronomers to decode the rich historical record written in our Galaxy's stars in order to paint a vivid portrait of how the Milky Way formed and evolved.

Of the four properties that mark a stellar population, **age** is the most crucial. The difference in age from one stellar population to another allows astronomers to piece together the chronological sequence of the Galaxy's evolution, much as a geologist reconstructs the Earth's past by studying different rock strata.

The second property of a stellar population, **location**, refers primarily to how the stars distribute themselves around the Galactic plane. Stars in some stellar populations cling tightly to the Galactic plane, while stars in others shoot far above and below it.

The third property of a stellar population, **kinematics**, describes how stars move around the Galaxy. Every star follows its own orbit around the Galactic centre. For example, the Sun revolves around the Galaxy once every 230 million years on a fairly circular orbit. But some orbits are so elliptical that stars journey from the remote stellar halo to the Galaxy's central region.

The fourth and final property of a stellar population, **metallicity**, is a star's abundance of elements heavier than hydrogen and helium, which are the two most common elements. This name exists because astronomers consider all heavy elements—even oxygen and neon which normally appear as gases—to be metals. The Sun is metal-rich, with 2 per cent of its mass composed of metals. But metallicity varies greatly from stellar population to stellar population, and some stars have only a fraction of our Sun's metallicity. Metals include such life-giving elements as carbon and oxygen.

Figure 3: Most stars revolve around the Milky Way on nearly circular orbits (inner star here), but some stars, such as the outer star here, have extremely elliptical orbits

Stellar populations
Family features

AS astronomers presently understand the Milky Way, every star falls into one of four stellar populations. The brightest and most prominent is the **thin disc** population, which includes the Sun and 96 per cent of the stars near it, such as Alpha Centauri, Sirius, Vega, Betelgeuse, and Rigel. Stars in the thin disc have a wide range of ages: some are newborn objects, others are middle-aged, like the Sun, and others are older still, with ages of about 10 billion years. As the name implies, thin disc stars lie close to the Galactic plane, usually within 1000 light years of it. Kinematically, the stars have fairly circular orbits, and the metallicity is high—indeed, high enough for life to exist.

The second great stellar population in the Milky Way is the thick disc, which makes up about 4 per cent of the nearby stars. The brightest likely member is the beautiful spring star **Arcturus**. As the name indicates, the **thick disc** is a more distended population, with a typical star lying about 3500 light years from the Galactic plane. Thick disc stars are old—about 10

The vast Galactic empire

THE MILKY WAY is more than just a giant galaxy. It is also the hub of a vast empire that stretches over a million light years and encompasses at least 10 other galaxies. These **satellite galaxies** orbit the Milky Way in the same manner that the Moon circles the Earth.

The two biggest and brightest satellites are the **Large Magellanic Cloud** and the **Small Magellanic Cloud**, which respectively lie 160 000 and 190 000 light years from our Galaxy's centre. Although the Magellanic Clouds are smaller than the Milky Way, they outshine most other galaxies, for they contain billions of stars.

At greater distances lie at least eight **dwarf galaxies**, which bear the names of the constellations in which they lie. The first two dwarf satellites, Sculptor and Fornax, were discovered in 1938. A typical dwarf contains just a few million stars that are quite spread out. The faintest dwarfs emit less light than the single brightest star in the Milky Way. The dimmest known galaxy in the entire Universe is Draco, which is about 240 000 times brighter than the Sun. For comparison, the Milky Way is 15 billion times brighter than the Sun.

The satellite galaxies reveal the Milky Way's mass, because they orbit under the influence of its gravity. The faster the satellites move, the more massive our Galaxy must be, just as the planets of the Solar System would move faster if the Sun had more mass. The motions of the satellite galaxies indicate the Milky Way has roughly a million million times more mass than the Sun.

In part because of this huge mass, the satellite galaxies lead a dangerous life, because our Galaxy's **tide** tries to tear them apart. On Earth, the Moon exerts a tide because one side of the Earth is closer to the Moon than the other and so feels a greater gravitational force. Fortunately, the Earth is strong enough to withstand the tide and is not torn apart. But the satellite galaxies, and especially the dwarf satellites, are much more fragile. The stars of the dwarfs are so spread out that they are only bound weakly to their home galaxy, so it does not take much to pull the galaxies apart.

The Galactic tide depends on distance: halve a satellite's distance from our Galactic centre and the tide strengthens eightfold. Thus the nearest satellites are in the most danger. In 1994, Rodrigo Ibata, Gerard Gilmore and Mike Irwin, while working at the University of Cambridge, discovered the remains of a dwarf galaxy just 60 000 light years from the Galactic centre—closer than any other galaxy. This galaxy had been torn in two by the Milky Way's tide.

Because they are so large, the Magellanic Clouds face an additional danger, called **dynamical friction**. The Magellanic Clouds probably lie within the Milky Way's dark halo. As massive satellites such as these move through the sea of objects constituting the dark halo, their gravity disturbs the dark matter, creating a wake behind whose gravity slows down the satellites. Thus, the satellites lose orbital energy, making them fall closer to the centre of the Milky Way. This causes the Galactic tide on these satellites to strengthen, and they split apart, splattering countless stars into the Milky Way's halo. Within 10 billion years, the Magellanic Clouds will crash on the Galaxy's shore, and the Milky Way will feast on a banquet of billions of new stars that will only augment its already mighty power.

billion years old—and have more elliptical orbits around the Galaxy. Their metallicity is about 25 per cent of the Sun's.

The third stellar population is the **stellar halo**, rare but extremely important stars that give great insight into the Galaxy's origin, for they formed when the Galaxy itself did. Halo stars account for only about one nearby star in a thousand. They often lie at large distances above and below the Galactic plane and have extremely elliptical orbits: in a single orbit, the distance of a halo star from the Galactic centre may vary wildly from a couple of thousand light years to 100 000. In comparison, the Sun's distance varies only from 27 000 to 30 000 light years. Halo stars have low metallicities, usually between 1 and 10 per cent of that of the Sun, meaning that life in the halo is unlikely.

The fourth stellar population is the Galaxy's **bulge**, which surrounds the Milky Way's centre. It is old and generally has a high metallicity. Because of its distance from Earth, it is the least explored stellar population in the Galaxy.

By comparing the three stellar populations nearest to the Sun—the thin disc, the thick disc and the halo—we see that there is an **age-metallicity relation**. The oldest stars—those in the halo—have the lowest metallicity; the somewhat younger, but still old, stars in the thick disc have a higher metallicity; and the youngest stars, in the thin disc, have the greatest metallicity. Since stars usually preserve the metallicity that they were born with, this means that the Milky Way's metallicity has increased over time.

Smooth or messy? Origin of the Galaxy

THIS, in turn, implies that the Galaxy's stars have created the heavy elements by fusing lighter ones together. This creation of new elements is called **nucleosynthesis**. When the stars die, these newly formed heavy elements are ejected into the Galaxy, where they enrich clouds of interstellar gas and dust. These clouds give birth to new stars that consequently inherit higher metallicities than their predecessors. The heavy elements on Earth—the oxygen we breathe, the calcium in our bones, the iron in our blood—were created billions of years ago, by stars that have long since died. In a sense, then, we are the heirs of those ancient stars.

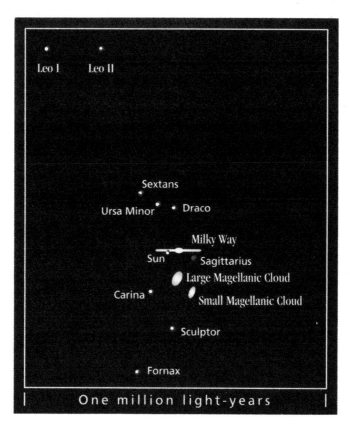

Figure 4: The Galactic empire: the Milky Way rules at least 10 satellite galaxies, and has torn another galaxy—the newly discovered Sagittarius dwarf—apart

Having fathomed the basic geography and stellar content of the Milky Way, astronomers can attempt to piece together the Galaxy's origin and evolution. In 1962, Olin Eggen, Donald Lynden-Bell and Allan Sandage, who were working together in Pasadena, California, proposed that the Galaxy had formed from a single huge ball of low-metallicity gas that was rapidly collapsing. Some stars were born during this collapse and acquired the highly elliptical orbits and low metallicities of the surrounding, infalling gas—these are the ancient stars we see today in the stellar halo. The collapse of the Galaxy was so rapid, said Eggen, Lynden-Bell and Sandage, that all halo stars formed within a very short time of one another; about 200 million years—less than 2 per cent of the Galaxy's total life.

Most of the infalling gas did not immediately condense into stars, however. Instead, it fell into a swirling disc, where collisions among gas clouds had the effect of making their orbits nearly circular. This gas had a higher metallicity, because some of the halo stars had already enriched it with heavy elements. As a result, the stars that formed in the disc had circular orbits and higher metallicities than stars in the halo, thereby explaining the properties of the Galaxy's disc population.

The first serious challenge to the precepts of Eggen, Lynden-Bell and Sandage came in 1978, when Leonard Searle and Robert Zinn, also working in Pasadena, proposed that the Galaxy's origin had been much more chaotic. The Galaxy's outer halo was not born from a single collapsing entity, said Searle and Zinn; instead, it arose from the collision of numerous smaller galaxies, similar to those that orbit the Milky Way today. In fact, the small galaxy mentioned earlier that astronomers discovered in 1994—which the Milky Way is swallowing—shows the Searle and Zinn scenario in action.

During the 1980s, several astronomers who investigated the orbits and metallicities of stars in the halo found evidence favouring the chaotic Searle and Zinn scenario over the more elegant model of Eggen, Lynden-Bell and Sandage.

However, more recent investigations of the kinematics and metallicities of old stars have supported some aspects of the Eggen, Lynden-Bell and Sandage theory, and today many believe that both pictures are necessary to fully understand the Galaxy's origin.

As astronomers continue to study the origin and evolution of the Galaxy, they realise that the Milky Way is a giant that is bustling with hundreds of billions of stars cloaked in an enormous halo of dark matter whose gravitational pull anchors a surrounding empire of at least 10 lesser galaxies and asserts its influence over other galaxies millions of light years away. More than **just our home galaxy, the Milky Way is one of the Universe's most beautiful and spectacular creations.**

FURTHER READING

The Alchemy of the Heavens, by **Ken Croswell** (Oxford University Press, 1996); *Galactic Astronomy*, by **Dimitri Mihalas** and **James Binney** (Freeman, 1981); "Life and Times of a Star", Inside Science, number 76.

Ken Croswell received his PhD from Harvard University for his study of the Milky Way Galaxy.

All graphics are by **Nigel Hawtin**.

THE Dark Side OF THE Galaxy

By the end of this decade, astronomers hope to learn if the dark matter pervading the Milky Way's frontier consists of massive, invisible objects.

by Ken Croswell

We live in a dark galaxy. Although the Milky Way's beautiful disk speckles the night with the glow of countless stars, these same stars mask the Galaxy's dark truth: Most of its mass is invisible, beyond the purview of either human or telescopic eye. Surrounding the luminous disk is an enormous dark halo that stretches for hundreds of thousands of light-years but emits almost no light at all, which means that conventional observations cannot reveal its nature.

"I am completely in the dark about what dark matter is made of," says Princeton astronomer Bohdan Paczynski. "I really have no clue whatsoever, and whatever people claim is just guesswork."

Astronomers know that dark matter surrounds the Milky Way's disk because of the high speed at which stars and gas in the outer Galaxy revolve, as well as from the rapid motions of other galaxies that orbit ours. But the Milky Way's dark halo cannot be made of ordinary stars, which observers could see. Even the faintest such stars — red dwarfs like Proxima Centauri and Barnard's Star — would have been seen by now. And the dark halo cannot be made of gas, because gas would emit detectable radiation. That leaves two broad possibilities for the dark halo's composition: dim astronomical objects and exotic subatomic particles.

If you're an astronomer, you would naturally root for the former, known in the trade as MACHOs, or MAssive Compact Halo Objects, which, though dim, are at least within the range of your experience. A MACHO can have almost any mass, from less than a trillionth of the Sun's up to millions of times the Sun's. From low mass to high mass, possible MACHOs include asteroids; planets, either small (like Earth) or large (like Jupiter); brown dwarfs, stars that have too little mass to ignite their fuel; white dwarfs that have cooled and faded; neutron stars; and black holes ranging from a few times the Sun's mass to millions of times greater.

On the other hand, if you're a particle physicist, you might lean toward some type of exotic subatomic particle, such as WIMPs, or Weakly Interacting Massive Particles. Although WIMPs have yet to be identified, even on Earth, they could exist in such large numbers in the dark halo that they account for all of its mass. But many astronomers don't like WIMPs, because it's hard to relish the thought that most of the Galaxy is made of something that your telescope can't detect.

In a 1986 paper, Paczynski proposed a clever technique that has already given astronomers their first probe of dark matter. If MACHOs populate the dark halo, he said, one would occasionally drift in front of a star in the Large Magellanic Cloud, the Milky Way's brightest satellite galaxy. In so doing, the MACHO's gravity would magnify — or "microlens" — the star's light, bending what would otherwise be diverging light rays together so that, as observed from Earth, the star would appear to brighten, then fade back to its normal brightness. On the other hand, if subatomic particles such as WIMPs make up the dark halo, no microlensing events would be seen.

A MACHO Start

In principle, Paczynski's idea was a clean test that would sepa-

41. Dark Side of the Galaxy

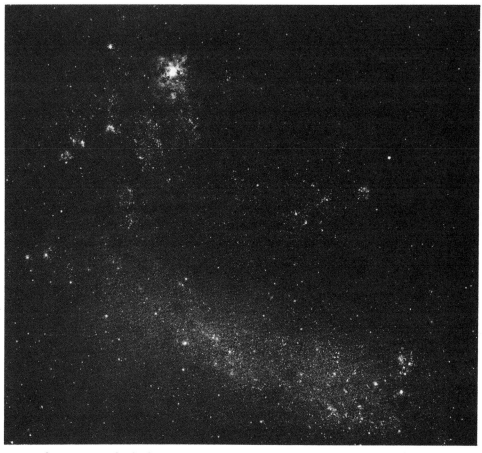

Astronomers patiently observe 10 to 20 million stars in the Large Magellanic Cloud each night for the telltale brightening and fading of MACHO-induced microlensing.

rate the MACHOs from the WIMPs. But it was hardly a plan for the, uh, wimpy, because it called for a huge observational campaign. Even in the best possible case — a dark halo comprised completely of MACHOs — at any time only one star in a million would be getting microlensed. Astronomers would therefore have to monitor millions of stars, night after night, to catch just one instance of MACHO-induced microlensing.

"Before we got started," says Paczynski, "all astronomers I know were extremely skeptical that anything like this could be done. They said we shall be swamped with data, we shall never get it, and there will be all those variable stars." The variations of such stars — Cepheids, RR Lyraes, eclipsing binaries, and more — would far outnumber the rare microlensing events and could even masquerade as those events.

But microlensing events stand out from variable stars in several ways. First, microlensing affects all colors equally. If a MACHO magnifies a star's red light tenfold, then it should do exactly the same to the star's yellow light and its blue light. In contrast, most variable stars vary by different amounts in different colors. Second, the brightness of a microlensed star should not vary again, because the chance that a second MACHO would pass in front of exactly the same star is minuscule.

Four teams of astronomers have now reported microlensing events. The largest group calls itself MACHO and is a collaboration of American, Australian, and British astronomers. Paczynski is affiliated with a second group, based primarily in Poland, called OGLE, short for Optical Gravitational Lensing Experiment. Two other teams are based in France: EROS (Expérience de Recherche d'Objets Sombres) and a newer group, DUO (Disk Unseen Objects).

Members of these teams need more than just cute acronyms; they also need enormous persistence and patience, because huge numbers of stars must be monitored in order to see any microlensing events. For example, the MACHO team routinely measures the brightnesses of between 10 and 20 million stars per night. The experiment would be impossible without advanced technology — in particular, CCDs to acquire the data and powerful computers to analyze it.

The first searches for microlensing began in the early 1990s. The MACHO and EROS groups observe the Large Magellanic Cloud. The OGLE and DUO groups look at the galactic bulge, the central district of the Milky Way, because ordinary stars in the disk should pass in front of bulge

Possible Dark Halo MACHOs

Object	Mass (Solar Masses)	Typical Duration of Microlensing Event
Massive Black Hole	1 million	200 years
Stellar Black Hole	10	8 months
Neutron Star	1.4	3 months
White Dwarf	0.6	2 months
Brown Dwarf	0.02	2 weeks
Jupiterlike Planet	0.001	2 days
Earthlike Planet	0.000003	3 hours
Large Asteroid	10^{-9}	3 minutes
Small Asteroid	10^{-15}	0.2 second

4. THE UNIVERSE: Dark Matter

stars and microlens their light. These events could not reveal the nature of the dark matter in the halo, of course, but they would prove that microlensing can be detected, even if the dark halo contains no MACHOs.

In the fall of 1993, the MACHO team reported its first success, with the detection of a microlensing event in the direction of the dark halo, and the EROS team announced two such events. Meanwhile, OGLE reported a microlensing event toward the bulge. With these announcements, a wave of euphoria swept over astronomers, for they had finally "seen" some of the mysterious dark matter that dominates the Galaxy. Many pundits proclaimed that the heroic MACHOs had won their battle with the evil WIMPs — which astronomers can never hope to detect — and that the dark halo was made entirely of MACHOs.

But such claims were misguided. "Quite a lot of people jump to conclusions before the experimental groups are willing to reach those conclusions themselves," says Charles Alcock, a member of the MACHO team who works at the Lawrence Livermore National Laboratory in California. "In our first paper, we published the discovery of one event, and an enormous amount of speculation was based on that one event, plus two events of lower quality from the EROS group. And it was entirely premature."

Not only were such statements wrong, but they now appear to have been based on faulty data. Two of the three dark-halo MACHOs reported in 1993 are under attack, by no less than Paczynski himself. He believes that both of the brightenings announced by the EROS team were actually the antics of variable stars rather than genuine MACHOs.

"The whole series of announcements in 1993 was really triggered by the EROS claim," says Paczynski. "The EROS people told the MACHO people that they had two microlensing events, and they wanted to go public, so the MACHO team decided to go pub-

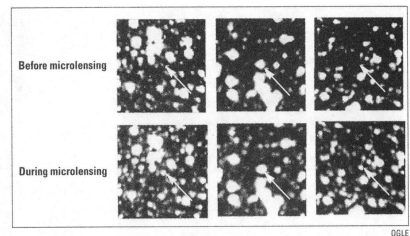

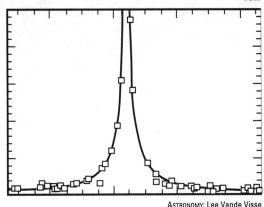

Three stars (top, left to right) in the direction of the galactic bulge brighten as they are microlensed by a foreground object. Microlensing produces a brightening and fading pattern (right), that can be distinguished from variable stars because all star colors are affected equally. The duration of the microlensing event yields information about the mass of the intervening object.

ASTRONOMY: Lee Vande Visse

lic; and as soon as OGLE detected their first event, they went public as well. Now we know that what EROS was announcing was probably not really microlensing. It just shows that people under stress claim things which don't last."

Revenge of the WIMPs

By 1995, the MACHO-versus-WIMP showdown had become incredibly volatile. In the first half of the year, the MACHO team reported that dark-halo MACHOs were few and far between, falling well below the number expected if MACHOs in the mass range to which the experiment was sensitive made up all of the dark halo. This led many people to declare WIMPs the winners. But that, too, was a premature conclusion.

"We hoped the dark halo was made more of MACHOs, because frankly it is more fun to find most of the stuff than some of the stuff," says Alcock. "But what we actually expected is that we'd see either no events or lots of events.

So finding an intermediate number was very surprising."

And confusing. The MACHO team first said that 20 to 25 percent of the dark halo was made of MACHOs having the mass of brown dwarfs. Brown dwarfs range from 1 to 8 percent of the Sun's mass, not enough to spark the nuclear reactions that power ordinary stars. Such stars glow briefly, as they convert gravitational energy into heat, and then fade from view, making them ideal candidates for dark matter.

The 20 to 25 percent estimate was widely quoted, convincing many that WIMPs made up the remainder. "People tend to jump to conclusions," says Paczynski. "I just can't find any serious argument in favor of WIMPs. The interpretation of the results is very ambiguous. All you can say is that most of the dark matter is not red dwarfs and probably not brown dwarfs." He points out that MACHOs can span a wide spectrum of masses — from less than a trillionth of a solar mass up to mil-

41. Dark Side of the Galaxy

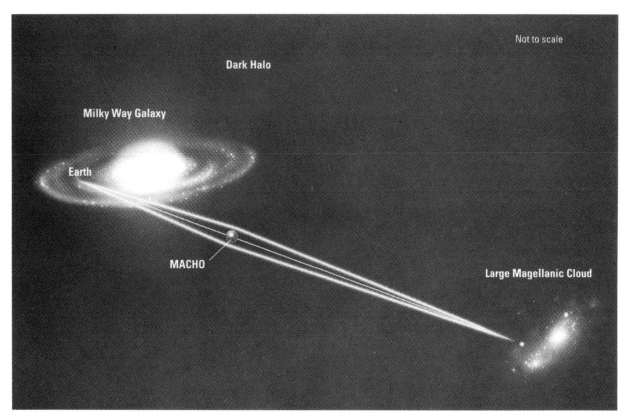

If invisible objects lurk in the Galaxy's dark halo, astronomers can detect them as they pass directly between Earth and stars in the Large Magellanic Cloud. The MACHOs act as a lens, focusing divergent light rays on Earth. This causes the background star to brighten and fade in a characteristic fashion.

lions of solar masses — and that the present experiments are sensitive to only a narrow section of that range. The more massive a MACHO, the longer the microlensing event lasts — which is how Alcock and his colleagues estimate MACHO masses. Low-mass MACHOs cause events lasting only a few minutes and would be missed, whereas massive MACHOs, such as black holes, cause events lasting years and would not yet have been detected.

Paczynski's skepticism proved correct. In late 1995, the tables turned yet again, this time in favor of the MACHOs. When the MACHO team analyzed additional observations, Alcock was stunned: the brown dwarf MACHOs were gone, and in their place were more massive MACHOs, clustering around 0.2 solar mass (see "A MACHO Galaxy?" AstroNews, May 1996). These MACHOs, said the scientists, constituted anywhere from 25 to 100 percent of the Milky Way's dark halo — which means there might be no WIMPs at all.

The exact nature of the MACHOs is uncertain, however. Many red dwarfs have 0.2 solar mass, but Alcock thinks that such stars would already have been seen directly if they really made up most of the dark halo. So he favors instead the idea that the MACHOs are faded white dwarfs. These burned-out remnants of solar-type stars account for 5 percent of the stars in the Galaxy's disk. As they age, white dwarfs fade and cool, so they could populate the dark halo without being seen from Earth. The trouble, though, is that the typical white dwarf — at least in our part of the Galaxy — has 0.6 solar mass, three times the mass of the MACHOs that now seem to constitute at least part of, and perhaps all of, the dark halo.

To further complicate things, there is the chance — which Paczynski thinks quite possible, Alcock very much less so — that most of the microlensing events have not been caused by objects in the Galaxy's dark halo. Instead, ordinary foreground stars in the Large Magellanic Cloud could be responsible, an idea suggested by Kailash Sahu of the Canary Islands' Institute of Astrophysics in 1994. If that is the case, then all of the Milky Way's dark matter could be WIMPs — or else MACHOs having different masses from those the experiment can now detect.

A Bright Future

With the present state of affairs so uncertain, only further observations can clarify the matter. Meanwhile, other types of MACHOs are still viable candidates. In particular, the experiments haven't been running long enough yet to detect black-hole-mass MACHOs.

"The experiment gradually becomes more and more sensitive to more massive objects as we go along," says Alcock. "But so far we have no real sensitivity to MACHOs above 10 solar masses. So if they're there, we wouldn't

4. THE UNIVERSE: Dark Matter

have seen them yet."

Other astronomers are taking aim at the Andromeda Galaxy. In 1992 Arlin Crotts of Columbia University pointed out that both Andromeda's dark halo and our own should microlens some of Andromeda's stars. In addition, Andromeda's greater distance makes it easier to detect lensing objects with as small a mass as the Moon. But Andromeda's stars are very faint due to its large distance, making the observations extremely difficult. In June, Crotts' group reported two possible microlensing events. But he says events are so rare that low-mass MACHOs – brown dwarfs and planets – appear to make up little if any of Andromeda's dark halo.

The future for the dark matter hunt is bright. "What I am optimistic about," says Paczynski, "is that within a few years, we shall cover the whole range of masses from 10^{-8} to 10^6 solar masses with a variety of microlensing experiments. And then if you don't see anything that cannot be accounted for by a few brown dwarfs here and there, I think it will probably point toward WIMPs."

For now, though, not even the scientists most closely involved in the dark matter quest know how it will come out — whether it will be all MACHOs, all WIMPs, or a mixture of both. But when the observations finally run their course, astronomers should at long last know what most of the Galaxy we call home is really made of.

Rather than using the Large Magellanic Cloud as a backdrop, some astronomers are looking toward the Andromeda Galaxy to search for MACHOs.

Ken Croswell is the author of The Alchemy of the Heavens *(Doubleday/Anchor), The book's home page is at: http://www.ccnet.com/~galaxy.*

MACHO Men at the Galactic Bar

Although the microlensing experiments were designed to explore the Milky Way's outermost fringes, the observations have actually told astronomers a lot about its inner district, the galactic bulge. The microlensing experiments provide evidence that the bulge is elongated rather than round, which means that our Galaxy may be a barred spiral, an idea first suggested in the 1960s by Gérard de Vaucouleurs.

From the first, the OGLE team observed bulge stars in order to test the microlensing technique. They knew they would see microlensing as ordinary stars in the disk passed in front of bulge stars. But the astronomers got more than they bargained for: The microlensing rate toward the bulge turned out to be three times higher than they expected, and thirty times the rate toward the Large Magellanic Cloud. The MACHO team, which also observes the bulge, found a high rate of microlensing, too. Because it is so packed with stars, stars in the bulge would microlens other bulge stars at a high rate.

But the actual microlensing rate was so high that the OGLE team had to model the bulge as a bar. This bar points more or less toward us, which increases the chance that a bulge star will microlens another bulge star. This explains the greater-than-expected microlensing rate seen toward the bulge and provides evidence that our Galaxy may be a barred spiral.

Yerkes Observatory

Is the Dark Matter Mystery Solved?

–James Glanz

SAN ANTONIO—To judge from the headlines, astronomers have solved one of their field's greatest mysteries: the identity of the long-sought "dark matter." A widely reported press conference at a meeting of the American Astronomical Society held here last month left the impression that the hitherto undetectable mass whose gravity keeps our own Milky Way and other spiral galaxies from flying apart as they spin on their axes has now been found. But astronomers—including the dark-matter hunters themselves—say reports of the mystery's solution may have been premature.

In their presentation at the meeting, a multi-institutional team of astronomers announced that MACHOs—for massive compact halo objects—most likely make up 50% of the dark matter in our galaxy, more than doubling the team's earlier estimate, which they published less than a year ago in *Physical Review Letters* (*Science*, 5 May 1995, p. 642). Because MACHOs consist of ordinary matter, such as the burnt-out normal stars known as white dwarfs, the revision—if it holds up—could mean that all the galaxy's missing mass consists of ordinary matter rather than the exotic particles some theorists favor. The rest of the mass, some astrophysicists suggested, could be in objects larger than the team has searched for yet, or too far away to detect. "My sentiment is that if half of it is ordinary stuff, then the rest of it is ordinary stuff, too," said astrophysicist John Bahcall of the Institute for Advanced Study in Princeton, New Jersey, who is not a MACHO team member, at the press conference.

And that could pose problems for current astrophysical theories, including some predictions of the big-bang scenario for the origin of the universe and its later evolution. Many theorists believe that during the big bang, a cosmos of purely "ordinary stuff" would have had trouble cooking up observed elemental abundances unless the matter's overall density was extremely low. And later, the primordial soup might not have collapsed into galaxies and the large galaxy clusters known to exist. Some theorists argue, too, that the universe as a whole should have enough mass to slow its expansion toward a standstill. That would require far more mass than could be supplied by ordinary matter and has led some cosmologists to favor swarms of massive neutrinos or other, more exotic particles as candidates for the dark matter, both within galaxies and in the vast reaches in between.

For these reasons, says the University of Chicago's Michael Turner, "if indeed [the MACHO group] has solved the dark-matter problem, they would have started a revolution." But Turner and others, including MACHO team members, note that the data so far are too sketchy to really settle the dark matter issue. "We still only have less than two handfuls of events," concedes team member Kem Cook of Lawrence Livermore National Laboratory.

Even if they're plentiful, after all, MACHOs are hard to find. By definition, these clumps of dark matter can't be seen directly. To detect them, the team, including astronomers from Livermore, the Center for Particle Astrophysics at the University of California, Berkeley, and the Mount Stromlo Observatory in Australia, instead takes advantage of the fact that a MACHO's gravity bends rays of light. When it passes between Earth and a distant star, the MACHO acts as a lens, temporarily increasing the star's apparent brightness.

The group scanned for the objects in the Milky Way's spheroidal halo, where theory says the dark matter should be, using the Large Magellanic Cloud (LMC), a nearby galaxy, as a starry backdrop. The researchers monitored 9 million LMC stars each night for any that showed such transient brightness increases. After 2 years, the team announced, it has picked up seven MACHOs. From the duration of the brightenings, it estimates that their average mass is between 0.1 and 1 solar mass—a size suggesting that they are old, dim, cold white dwarfs.

Extrapolated to the whole halo, the results imply that half its mass consists of MACHOs, double what the group found in their first year's scan. In light of the new results, "it's certainly viable that all the dark matter in our galaxy is made of [ordinary matter]," says team member David Bennett of Livermore.

But uncertainties in the data also open the way to more mundane conclusions. The team already threw out two "really ratty" events in the original first-year data set of three MACHOs, says Bennett, before coming up with the new total of seven over 2 years. And if, as the team now suspects, one of the new lenses is actually in the LMC and not in our galaxy's halo, the most probable fraction dips to 40%. Finally, the unfolding of an object's mass, speed, and relative distance from a single number—the duration of the brightening—depends strongly on the halo model chosen, such as the flatness of the spheroid and whether it rotates or not. Given the doubts, "you could go down to 20% and not be outrageous," says team member Charles Alcock, also at Livermore.

The resolution of the dark matter puzzle will probably have to wait until more MACHOs have been detected, most researchers say. "We are dealing with small-number statistics," says Bohdan Paczyński of Princeton University, whose own team will begin searching the halo for dark matter in 9 months. "I would rather wait 2 or 3 years and see how things look at that time."

Sky high. The stars of the Large Magellanic Cloud provided the backdrop for detecting the brightenings caused by MACHOs, such as the one illustrated by the graph.

The Evolution of Our Galaxy

The seemingly immutable Milky Way is actually a dynamic system whose origins are hotly debated and whose boundaries are unknown.

James Binney

MOST OF THE LUMINOSITY in the universe comes from galaxies similar to the Milky Way. Hence, to ask how the Milky Way formed is to ask how the dominant galaxies in the universe came into being. And because we can observe the inner workings of the Milky Way with a precision unattainable for any other galaxy, we have the opportunity to uncover clues to how these island universes formed and how they evolve.

On the other hand, our view of the Milky Way is uncomfortably local in both space and time. Dust makes it difficult to observe the structure of its disk, and whereas external galaxies can be studied at various epochs in the remote past, we see the Milky Way at essentially one instant. Studies of external galaxies at both high and low redshift (large and small look-back times) have therefore greatly influenced how we interpret observations of the Milky Way.

BOTTOM-UP FORMATION

In the last decade computer simulations of how objects clustered early in the expanding universe have also profoundly affected how we decipher data on the Milky Way. The most influential models are those based on the cold-dark-matter theory, which states that objects much smaller than our galaxy formed first, then merged to form larger entities. The Milky Way, therefore, may have been "assembled" over a significant

James Binney is an astronomer at the University of Oxford. With Scott Tremaine he is author of Galactic Dynamics *(Princeton University Press, 1987), a standard reference for professional astronomers.*

period of time rather than created in a single spectacular event as was once believed.

When entities of comparable size fall together, the event is called a merger. Some tens of billions of years in the future the Milky Way is likely to merge with the Andromeda Galaxy, which is about three times more massive. When a small object falls into a larger one, we speak of cannibalism. The Milky Way is currently ripping apart and eating up the Magellanic Clouds, which together are less massive by about a factor of 10. Astronomers would like to know what mergers and acts of cannibalism lie in the Milky Way's past.

When our galaxy gobbles up a smaller system, it changes in three potentially detectable ways. First, stars ripped from the victim move into orbits through the Milky Way that are similar to one another. The smaller the victim, the narrower the range of orbits over which its stars are distributed, because initially they were more tightly bunched in space and in velocity. Astronomers have identified several groups of stars that move in very similar orbits, many of which possess remarkably similar chemical compositions. This would be expected if the stars formed together in a low-mass system.

A second consequence of the Milky Way's cannibalism is the "thickening" of its disk. A thin galactic disk is nothing more than a very large concentration of stars in a small region of "phase space"—the imaginary six-dimensional space whose points are specified by three spatial coordinates and three corresponding velocity components. But strong concentrations of stars are inherently implausible because any random disturbance of the Milky Way will tend to spread them over a greater volume in phase space. In the language of thermodynamics, the entropy of the system increases. In real space the distur-

43. Evolution of Our Galaxy

ACCRETION OF GAS

The third consequence of the Milky Way's cannibalism is its acquisition of fresh gas. Newly acquired stars go into orbits characteristic of their source. In contrast, the victim's gas tends to settle into the equatorial plane of the Milky Way. Several lines of evidence suggest that the total mass of gas acquired from victims in this way is comparable to the gas currently in the disk. For instance, gas is being locked up in newly formed stars so fast that all of it currently in the disk will be exhausted within the next several billion years. Since the rate of star formation has been nearly constant for a longer period, it seems unlikely that the current gas supply will be so soon exhausted. Instead, the Milky Way may be accreting gas at a rate roughly equal to its star-forming rate, so that its gas content remains roughly constant.

The evolution of star metallicity supports this idea. Early in the life of the disk the proportion of metals in each batch of new stars increased surprisingly rapidly, so that now very few disk stars have less than 10 percent of the Sun's metallicity. But the metallicity of recently formed stars has been rising much more slowly. The lack of extremely metal-poor stars in the solar neighborhood is the G-dwarf problem mentioned in "Meet the Milky Way," the first article in this series on our galaxy (see Sky & Telescope, January 1995 issue, page 26). It asks why, if all the G-type dwarfs ever born in the galaxy are still around, we don't see any low-metallicity examples nearby.

This problem has two possible explanations. First, imagine that initially the galaxy's disk contained only a very small amount of gas and that most of it turned into stars before much

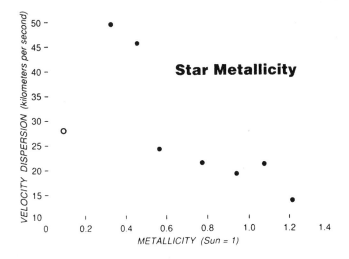

The velocity dispersions for 189 disk stars. A larger dispersion means the stellar orbits are less circular and less constricted to the galactic plane. The metal-richer stars, which are generally younger, seem to have more regular orbits, perhaps because stars gain random motion over time. The velocity dispersion for the lowest-metallicity point is uncertain. Courtesy Bengt Edvardsson, Astronomical Observatory, Uppsala.

bance caused by the gravitational field of a victim that is being cannibalized makes the disk thicker and the random velocities of its stars larger—the disk gets "hotter."

Most astronomers believe that when the overwhelming majority of stars form they have nearly circular orbits in a very thin layer near the disk's midplane, from which they subsequently diffuse out. When we categorize stars by age we find that young stars are indeed confined to this thin, cold disk, while older stars move with larger random velocities through a thicker, hotter disk. However, more impressive data are obtained when astronomers group stars by chemical composition, which is a comparatively easy characteristic to determine spectroscopically for very large samples of stars.

Metallicity is strongly correlated with age because supernovae are constantly pumping metals (elements higher on the periodic table than hydrogen and helium) into the interstellar medium (ISM), which was originally mostly free of them. The metal-poorest stars are distributed almost spherically; stars similar to those in the metal-richest globular clusters form a thick disk; and stars with the Sun's still-higher metal abundance form a thin disk. It is not yet clear, however, whether these three components of the Milky Way are distinct physical entities or the older and younger portions of a continuous, complex structure.

Unfortunately, the thickening of the disk with time does not point unambiguously to cannibalism. Precisely because it is associated with something as basic as entropy increase, the diffusion of stars from the thin to the thick disk will be driven by *any* fluctuations in the gravitational field. Spiral arms and giant molecular clouds also give rise to these fluctuations. However, quantitative analyses of stellar diffusion suggest that they cannot alone be responsible for driving the stars into the thick disk.

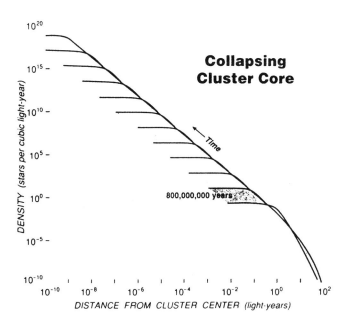

The evolution of an idealized globular cluster undergoing core collapse. The curves represent the density profile at times nearing collapse, with the curves for later times extending to higher central densities. In this simulation the entire collapse takes about a billion years, 80 percent of which occurs between the first two curves. A real globular cluster collapse would be halted by the formation of binary stars long before the highest densities shown here are reached. Courtesy Haldan Cohn, Indiana University.

159

4. THE UNIVERSE: Galaxy Structure and Evolution

more gas could be accreted. After the more massive of these stars exploded as supernovae, the disk consisted of stars of varying metallicity (from metal-poor to metal-rich) and a small supply of metal-rich gas. Now suppose that gas trickled into this low-mass disk at a rate only slightly greater than the rate at which it was locked into stars. Then the masses contained in both stars and gas would have gradually risen to their present values. Furthermore, the metallicities of most of the stars would now be high because they would have formed mostly from the ISM after it had been enriched by the first burst of supernovae.

A different explanation of the G-dwarf problem invokes the existence of metal-poor stars in the halo. These are thought to be the oldest stars in the Milky Way and similar to those in globular clusters, which are certainly considerably older than any disk star. A significant quantity of heavy elements would have been made when the more massive of these halo residents exploded as supernovae. Some astronomers suggest that the gas from which the disk formed was enriched by these heavy elements before a significant number of disk stars could form.

Which of these explanations of the G-dwarf problem is correct? Probably both mechanisms contributed to the paucity of extremely metal-poor disk stars. Astronomers would be aided if they knew how much of the galaxy's total luminosity is attributable to stars much older than the oldest disk stars. This must certainly include the luminosities of the globular clusters and most of the halo stars. And until a few years ago astronomers believed that it would include most of the luminosity from the bulge. Recently, however, doubts have been raised that the bulge is older than the old disk. In fact, there is an intriguing possibility that the bulge formed from the disk.

BARS AND THE BULGE

Since the very earliest days of studying stellar disks by means of numerical N-body simulations (involving many mass points influenced by their mutual gravitational attraction), it has been apparent that self-gravitating disks tend to form bars. For nearly a decade dynamicists fought this prediction of their computer models, believing that at least half of galactic disks, including that of the Milky Way, were unbarred. But recently it has become clear that most disk galaxies, including our own, contain at least a small bar. In fact, we now think that bars probably play a vital role in structuring all stellar disks.

A bar's nonaxisymmetric gravitational field is good at transferring angular momentum from stars near the center of the galaxy to ones farther out. Spiral structure does this too, but less efficiently. Transferring angular momentum outward is the key to increasing the entropy of disks because it enables the inner disk to contract, thus releasing gravitational energy that can be channeled into the random motions of stars. This manifests itself in individual stars moving over a wider radial range and oscillating vertically to greater distances above and below the galaxy's midplane. Computer simulations indicate that the magnitudes of individual radial excursions increase steadily as a bar grows in a disk, but the vertical oscillations are excited abruptly when a previously flat bar briefly buckles when the ratio of the bar's length to its width attains a critical value.

When a disk that has experienced a bar-forming episode is viewed edge on, it often has a characteristic peanut shape. So there was considerable excitement when an infrared image of the inner Milky Way obtained by NASA's Cosmic Background Explorer satellite showed that the bulge does have a slight peanut shape. Was the bulge of the Milky Way formed by the buckling of a bar, and, if so, when did it happen?

Bars change the distribution of gas within a galactic disk even more dramatically than they change the distribution of stars. Gas that lies within the bar streams rather rapidly in toward the nucleus, greatly increasing the galaxy's degree of central concentration. Gas that lies outside the bar is pushed outward and tends to accumulate in a ring whose diameter is rather less than twice the bar's length. The giant molecular clouds in the Milky Way are concentrated into just such a ring 13,000 light-years in radius—about half of the Sun's distance to the galactic center. Moreover, millimeter radio observations suggest that the motion of gas 650 to 3,000 light-years from the center has a number of characteristics expected from gas moving in the gravitational field of a bar that is more than 14,000 light-years long.

Bars probably do not live as long as galaxies because the disorder they generate undermines their structure. They behave a little like yeast, which during fermentation is poisoned by the alcohol it produces. Unfortunately, we cannot accurately date any given bar or say how long we expect it to persist. But it does seem likely that now-extinct bars played an important role in determining the radial density profile of the disk by shifting gas inward and outward as described above. Most of these bars will have been short lived and not have become so long and thin as to first buckle and then become permanently thick.

STAR CLUSTERS

Most stars are born in groups. Some of the assemblages are of very low mass and fly apart as soon as the massive stars become luminous and blow away gas left over from their formation. This ejection of mass disrupts a group's gravitational binding. Unbound groups of stars are called associations. More massive stellar groups survive the loss of residual gas and orbit through the galaxy as star clusters. An open or galactic cluster is one that has less than several thousand members and lives only in the disk; more populous ones are called globular clusters and are found in the disk, bulge, and halo.

Even after a cluster has survived the loss of its original gas, its life is constantly threatened by internal and external effects. It is liable to be shaken apart by fluctuations in the gravitational field through which it moves. For example, a passing giant molecular cloud can briefly squeeze a cluster so that its gravitational binding energy is reduced. Any open cluster will fall apart after a few billion years of this treatment. Globulars are not as affected by molecular clouds, both because they are more compact and because they spend most of their time away from the biggest concentrations of clouds. Nevertheless they are regularly squeezed by the gravitational field of the Milky Way where it changes direction abnormally rapidly—for example, near the disk and the inner bulge. Also, encounters between globular clusters and any massive black holes in the Milky Way's dark halo would be very destructive.

Besides these external threats to the lives of clusters, there are internal threats, too. One is the "evaporation" of stars. As a star moves through a cluster it exchanges energy with other stars by gravitationally interacting with its neighbors. Sometimes it picks up a little energy, perhaps gaining enough to speed right out of the cluster. This process is called evaporation because it is very similar to what happens when a drop of water evaporates through fast molecules speeding off into the air.

The loss of a cluster's most energetic stars leaves the remainder more tightly bound and thus more compact. This increases the rate at which stars interact and thus the rate of evaporation. It also speeds up the transfer of energy from the center to the edge of the cluster. This transfer is closely analogous to thermal conduction in a metal object. On average the stars at the center move faster than those at the edge, just as electrons move faster in hot metal than in cold. Consequently, encounters between stars that spend most of their time at the center and ones that live mostly at the edge tend to remove energy from the cluster's central stars. This causes the core of the cluster to become more gravitationally bound; it shrinks. Paradoxically, this shrinkage increases the mean kinetic energy of the central stars, which in turn accelerates the rate at which energy is conducted from the center to the edge. So a runaway develops in which the cluster core shrinks faster and faster and becomes ever more compact. The outer envelope swells, increasing the rate at which stars are lost by evaporation or are ripped off by fluctuations in the Milky Way's gravitational field.

The shrinkage of a cluster core is eventually halted by binary stars. As the mean star density increases, more and more binaries are penetrated by an intruder that happens to be passing through. When a third star enters a binary, all three stars usually dance about each other for a time in highly complex and unstable orbits. This dance is eventually resolved by one star (not necessarily the intruder) being flung out at high speed and the remaining stars forming a new binary. This pair is usually closer than the original one. The kinetic energy of the ejected star enables the cluster core to supply energy to the outer edges by conduction without further contraction.

We see then that there are several processes that cause clusters to lose mass, and the rate of loss increases as the cluster becomes less massive. Hence clusters can be totally dissolved, especially the least massive ones. It has been suggested that all halo stars were once in globulars, but 99 percent of them are now free because they were in low-mass clusters that have dissolved.

At the opposite end of the scale, very massive clusters are dragged toward the center of the galaxy (in the same way that cannibalized galaxies are) and are ultimately torn up by the strong gravitational field of the nucleus. Observed globular clusters have roughly the range of masses that escape being either dragged to the center or dissolved by the combined actions of conduction, evaporation, and squeezing.

How globular clusters were formed is an especially interesting problem since both their low metallicities (the metal-poorest have less than 1 percent of the Sun's metal abundance) and estimates of their ages indicate they are as old as any known objects—in fact, their ages exceed some estimates of the age of the universe!

There are two schools of thought as to how globulars came about. One is that they formed due to thermal and gravitational instabilities in a great cloud of gas that would go on to become the entire Milky Way. The attraction of this theory is that the characteristic mass of a globular cluster emerges fairly naturally from the physical parameters, such as temperature and density, that one would expect in a protogalactic cloud. The theory remains highly speculative, however, because the physics involved is complex and impossible to work through in detail. Also, simulations of galaxy formation have discouraged the idea that the Milky Way formed through the rapid collapse of a single, huge protogalactic cloud within which simple instabilities might occur.

An alternative school of thought starts from the observation that the disk of the Large Magellanic Cloud (LMC) contains young clusters that look set to evolve into low-mass globulars. Thus, while no globular clusters have formed in the Milky Way in a very long time, they seem still to be forming in the metal-poor disk of the LMC. Young globular-clusterlike objects have also recently been discovered by the Hubble Space Telescope in the giant active galaxy NGC 1275. The disk of the LMC is disturbed by both the Milky Way and the Small Magellanic Cloud, while NGC 1275 seems to be a superposition of two galaxies colliding at high speed. These observations suggest that globular-cluster formation may be facilitated by, or entirely confined to, periods during which the disk is strongly disturbed by interaction with another galaxy.

Two further observations support the conjecture that globular clusters formed in now-destroyed disks. First, correlations have been found between the orbits of globular clusters and the metallicity of their stars. Such correlations might arise because clusters on similar orbits formed in a common disk. Second, different galaxies have widely varying numbers of globular clusters per unit galactic luminosity, and there are reasons to believe that the galaxies with the most clusters are those that have been the most voracious cannibals. Much more work will be needed before the origin of our galaxy's globulars is known with any surety.

THE OUTER MILKY WAY

Where does the Milky Way end? This is one of the hardest and most important questions to answer concerning our galaxy's dynamics. It is now generally accepted that galaxies are

4. THE UNIVERSE: Galaxy Structure and Evolution

enveloped in halos of dark matter, whose densities fall off roughly as the inverse square of distance from the center (*S&T:* January 1994, page 20). The dark halo starts to dominate the Milky Way's mass budget at roughly the Sun's distance from the galaxy's center. The halo probably continues at least to the Magellanic Clouds, six times farther out than the Sun. Very little is known about the mass distribution beyond, though a couple of arguments suggest that the dark halo ultimately touches that of the Andromeda Galaxy.

Over the last two decades it has become clear that even the visible parts of galaxies show signs of not being in a steady state. The farther one goes from the center, the longer it takes a typical star to orbit once around the galaxy, and the less likely the system is to be in equilibrium. Much of the dark halo is likely to be very unsteady too. Does this have an observable manifestation?

Cold gas clouds nearer to the galaxy's center than we are lie in a thin layer that is remarkably flat over more than a factor of 10 in radius. A little beyond the Sun the layer warps away from the plane that it defines farther in, above the plane in one direction and below it in the opposite direction. A similar phenomenon is observed in a large fraction of other spirals. Gas in the warped part of the disk is spinning about a different axis than gas farther in. A possible explanation of this warp is that the symmetry axis of the dark halo at large radii differs from that of the inner Milky Way. The time scale upon which this misalignment is ironed out may be very long, and the warp may be a long-lived but transient manifestation of this disordered state of affairs.

LONG WAY TO GO

The wide variety of stellar types and orbits found in the Milky Way makes it a very complex dynamical system. After more than 60 years of work, we still have no satisfactory dynamical model of the Milky Way as a whole. Such a model is necessary to delineate the Milky Way's gravitational potential, since the majority of its mass is believed to take the form of dark matter. In all probability this dark matter can only be traced through its gravitational influence, which requires careful dynamical modeling of stellar motions. Also, we should be able to read much of the formation history of the Milky Way from its current dynamical state.

At the moment the known dynamics of the Milky Way are analogous to an incompletely translated fragment of a stone tablet. An efficacious dynamical model for the Milky Way would be the Rosetta stone that would reveal the whole fascinating story.

Seeing how galaxies form

Craig J. Hogan

Craig J. Hogan is in the Department of Astronomy and Physics, University of Washington, Seattle, Washington 98195, USA.

Gas in the early Big Bang was nearly uniformly distributed, as we know from the abundances of light elements and the isotrophy of the cosmic microwave background radiation. In broad terms, we also know that this gas formed into galaxies because of gravity: a slightly overdense region of the Universe expands more slowly than the average gas and gradually falls farther and farther behind, until it stops expanding altogether and collapses. Within each galaxy, some of the gas collapses to high density and forms stars, whose production of energy eventually suppresses further star formation and thereby leads to a regulated steady state, in which a galaxy's remaining reservoir of gas is steadily converted into stars over a long period of time.

Until recently, the details of how and when these processes and others moulded the properties of galaxies were obscure, as they had to be pieced together from fossil evidence in the contents and structure of galaxies in the present-day Universe. With new astronomical techniques, however, it is now possible to see so far away that we can effectively look back to the period in the history of the Universe when galaxy formation was occurring. On page 603 of [*Nature*, **377**, 19 October 1995[1]], Cowie, Hu and Songaila describe the latest step in this programme, perhaps a watershed: evidence that observations have now reached that critical formation epoch of most of the stars of the main galaxy populations.

Cowie *et al.* apply the light-collecting power of the Keck telescope to collect better spectra of faint galaxies than was previously possible. For many of them—galaxies with strong emission lines—this had led to redshifts being measured, giving a much more precise idea of their distance and of their physical properties, such as the overall brightness, the continuum spectrum in the rest frame and the strength of emission lines. The new survey includes the most distant galaxy ever found by its starlight (as opposed to the light from an active nucleus such as a quasar) which, with a redshift $z = 1.61$, was emitted when the Universe was 2.61 times smaller than it is today.

The most interesting thing about these galaxies is not their record-breaking distance, but the rate at which they are forming fresh stars out of gas. The star formation rate can be estimated from the supply of ultraviolet ionizing radiation in a galaxy, which can in turn be estimated from the amount of light emerging in fluorescent emission lines. Ionizing radiation is a good way to estimate star formation rate because it can be produced only by hot massive stars; these burn out so quickly that if they are found at all in a galaxy we know that they must be constantly replenished by new ones forming out of interstellar gas, at a rate directly proportional to the ultraviolet luminosity. Although there are variations and uncertainties in the conversion factor from line luminosity to star formation rate, the quantitative connection has been thoroughly documented in detail in nearby galaxies[2].

Cowie *et al.* have found in the new deep spectra a population of galaxies with high star formation rates, high enough that, if the same rate were to continue for the age of the Universe then (about a quarter of what it is now, or a few billion years), each galaxy would have formed all the stars in a typical bright galaxy today. Such high star formation rates are very rare in the Universe today, but appear commonly in the sample taken by Cowie and co-workers.

The observed number of these bright emission line systems is significant: not only are the individual galaxies forming stars fast, but the total production of stars is prodigious. The total volume density of stars formed in just the observed galaxies is 4–20 per cent of all the stars in galaxies today. Adding galaxies whose spectra have not yet been measured (but whose brightnesses and colour are similar) seems likely to double this number; there are also likely to be many individually fainter systems at the same redshift not yet in the spectroscopic sample. (The fraction could also go up if the Universe is open or has a cosmological constant.) It is therefore plausible that we are viewing a substantial portion of the population of modern giant galaxies caught at the time when they are forming the bulk of their stars out of gas.

4. THE UNIVERSE: Galaxy Structure and Evolution

This agrees with the picture we get from directly viewing the gas at high redshift rather than the stars. The cosmic inventory of neutral hydrogen gas, measured from damped Lyman-alpha absorption in quasar spectra, includes about the same amount of material as present-day stars at $z = 3$, but a much smaller amount at the present epoch[3]. This makes sense if the damped absorbers are protogalazies which form most of their stars between about $z = 1$ and 2. This period should then correspond to most of the cosmic production of heavy elements, also roughly in accord with what is seen in quasar absorbers[4,5].

Not only do we now know the distance and the physical conditions in these galaxies, we know what they look like. The Hubble Space Telescope now images them with enough clarity to reveal their detailed morphology. Instead of appearing as generic faint blue blobs (as they do in ground-based images), they show up as having a rich and varied structure strikingly different from most galaxies today. Instead of finding the symmetric disks, well organized spirals, and smooth triaxial shapes most common in bright galaxies close to dynamical equilibrium, we find[6] irregular patterns, multiple centres of concentration, chains and one-sided arms that characterize galaxies in a period of rapid dynamical evolution, before they have settled into their final quasi-stable regulated study state.

It now seems likely that many of the galaxies in these images are not faint blue dwarfs at moderate redshifts of the order of 0.5 which were already revealed in deep redshift surveys[7], but are progenitors of large galaxies like the Milky Way at somewhat earlier times. We are beginning to observe directly how galaxies got the way they are, by viewing in several different ways—spectroscopy, absorption, and direct imaging—the evolution of the structure, chemical enrichment, gas content, star formation history, clustering and other properties of the main population of galaxies in the Universe.

Notes

1. Cowie, L.L., Hu, E.M. & Songaila, A. *Nature* **377,** 603–605 (1995).
2. Kennicutt, R. C. *Astrophys. J.* **388,** 310–327 (1992).
3. Wolfe, A. M. *Ann. N. Y. Acad. Sci.* **688,** 281–296 (1993).
4. Steidel, C. C. & Sargent, W. L. W. *Astrophys. J.* **80,** 1–108 (1992).
5. Pettini, M., Smith, L. J., Hunstead, R. W. & King, D. L. *Astrophys. J.* **426,** 79–96 (1994).
6. Driver, S. P. *et al. Astrophys. J.* (in the press).
7. Hammer, F., Le-Fevre, O., Tresse, L., Lilly, S. & Crampton, D. *Astrophys. Lett. Comm.* **31,** 49–54 (1995).

Before Galaxies Were GALAXIES

Astronomers are closing in on one of the big questions:
How did galaxies come to be?

by William Keel

GALAXIES CAUSE ASTRONOMERS problems. They're lumpy. And yet the earliest light we can see in the universe is smooth. So how did the universe go from smooth to lumpy, and how did galaxies form?

It's one of the great questions of modern astrophysics and cosmology. For many astronomers, searching for the answer has been like the Quest for the Holy Grail, with potential answers constantly receding from view just as they have approached what they thought was the end. Finally, in the last couple of years, new developments may be showing us how galaxies, and the universe as we recognize it today, took form. We can finally look at galaxies — and perhaps their building blocks — within a few billion years of the Big Bang, and dimly glimpse what it all looked like. Fortunately, the picture is richer and more complex than most of the theories predicted.

The Fossil Record

To appreciate the origins of today's ideas, we need to travel back to the early 1960s. Many astronomers grew up with a view of galaxy formation based on an epochal 1962 paper by Olin Eggen, Donald Lynden-Bell, and Allan Sandage, who brought together information on the motions and chemical makeup of stars in our own corner of the Milky Way. They showed that the older stars, poorer in the heavy elements that have been created in the cores of later generations of stars, form a thicker, rounder distribution than the familiar, younger stars such as the sun.

Because the form and extent of a star's orbit is essentially frozen once the star has formed from interstellar gas, the paper argued, the oldest stars orbited in near circles around the galaxy's center while younger stars formed later from the gas clouds that settled into the galaxy's thin, rotating disk. This occurred because the galaxy was a sphere in its infancy, when the oldest stars formed, before the galaxy flattened out and formed the disk. Astronomers see so many old stars in giant, circular orbits that they must have formed relatively quickly. The galaxies must form with a single, smooth collapse, they reasoned.

This so-called collapse picture suggests why we see different kinds of galaxies today. Consider the Hubble sequence for a moment. This classification scheme of elliptical, spiral, barred

4. THE UNIVERSE: Galaxy Structure and Evolution

spiral, and irregular is still so widely used because it does more than Hubble originally intended, which was to describe galaxies on photographs. He used subclasses to indicate the degree of winding of spiral arms or the apparent shape of ellipticals. A galaxy's Hubble type correlates nicely with the kinds of stars it contains, the amount and type of interstellar material it has, and even how its mass is concentrated toward the center. Elliptical galaxies are composed almost solely of old stars, lacking the cool gas needed to make more. Spirals have central bulges of old stars and disks where star formation continues. Irregulars seem to have been making stars at about the same pace for their whole histories. In the collapse picture, ellipticals are the galaxies that were most efficient, leaving almost no gas after their first spree of starbirth, while irregulars are the lazy relatives that made a few stars at the outset and left almost all the gas for later. Because the bulges of spiral and elliptical galaxies are all more or less the same age, there must have been major fireworks in the early universe as all the galaxies contracted and made their first, brilliant generations of stars. It would have been a spectacular view, with galaxies larger, brighter, and much closer together than today, and studded with brilliant starforming regions and young star clusters — a time we might fondly wish for as we peer at the dim blurs of galaxies in our eyepieces.

By the late 1960s astronomers were in a position to look for signs of how galaxies have evolved. Beyond this, some were thinking ahead to how primeval galaxies, or protogalaxies, might appear. The hunt for primeval galaxies, loosely defined as galaxies forming their first important generation of stars, has been a long-running quest of Arthurian proportions — and, some argue, just as embellished with myth. If most large galaxies really formed during a single short period, they would have been loaded with brilliant, short-lived blue stars, making the galaxies correspondingly powerful. If this were so, the sky would be peppered with primeval galaxies at about 20th magnitude. But these galaxies are nowhere to be found.

To further confuse things, Leonard Searle and Robert Zinn showed in a 1978 study that the chemistry of star clusters in the Milky Way's halo is too consistent to allow a single, smooth collapse like that envisioned by Eggen, Lynden-Bell, and Sandage. A single, smooth collapse would concentrate the elements cooked in massive stars toward the developing galaxy's center as the gas dropped inward, while a lumpy collapse would blend the different materials widely. Because Searle

---

"... there must have been major fireworks in the early universe as all the galaxies contracted and made their first, brilliant generation of stars."

---

and Zinn found a uniform distribution of heavy elements, they proposed that the galaxy formed not from a single, smoothly collapsing gas cloud, but from a very lumpy set of clouds already forming stars before they fell together. These lumps might look somewhat like dwarf galaxies.

If lumps continued accreting after they started making stars, the process might last a good deal longer than originally believed. In fact, in a sense, galaxy formation by this process is still going on. In our neighborhood, astronomers still see examples of galaxies merging. When one of the galaxies is much smaller than the other, we tend not to dignify it as a merger, but as cannibalism — the galactic equivalent of a hostile takeover. If Searle and Zinn are correct, the early universe would have looked quite different, with many small galaxies swarming together at various rates instead of huge, full-grown galaxies all coming together at once. This also makes primeval galaxies fainter than we would have expected. They would have fewer of the brightest young stars at any one time, although the process would last longer.

Telescopes as Time Machines

The question for primeval galaxy hunters has become, "Where do you look?" Virtually all astronomers agree that primeval galaxies must exist at high redshifts, extreme distances near the beginning of the observable universe. The light we see from them, then, has traveled since the universe's earliest days. So the work to find primeval galaxies means finding galaxies with the highest redshifts and learning about them by comparing their spectra and structures to those of nearby, fa-

45. Before Galaxies Were Galaxies

miliar galaxies. The problem is that not all extremely faint galaxies are extremely distant. Many intrinsically faint galaxies lie closer to home. How can astronomers tell just which 24th-magnitude galaxy to spend a whole night researching?

Various groups of astronomers have taken different approaches to finding galaxies at high redshifts. One choice is to look at quasars, because they have long held the records for the most distant objects. This thing in our corner of the universe. These are so odd that astronomers must wonder what they can conclude about ordinary galaxies from them. Are they representative of later, normal galaxies, or are they such rare and extreme objects that they tell us only about themselves?

Galaxies by Radio

Several groups have tried to circumvent this problem by looking at weak radio galax-

---

"... quasars are intimately related to galaxies. They probably represent the energetic, unstable centers of infant galaxies."

---

year's record has a redshift, z, of 4.95, meaning it lies more than 8 billion light-years distant. Clearly, quasars are intimately related to galaxies. They probably represent the energetic, unstable centers of infant galaxies. But at the highest redshifts, even the Hubble telescope fails to show galaxies surrounding quasars. More creatively, some astronomers have looked at objects directly in front of quasars, partly blocking their light at particular redshifts and introducing "dips" into the quasars' spectrum. Many of the intervening objects appear to be immense hydrogen clouds. Some of these contain carbon, silicon, and magnesium atoms and appear to have faint galaxies associated with them. While yielding only a few objects directly, this technique allows the selection of galaxies based on their gas content, not on how bright they are.

Many radio galaxies are nearly as distant as quasars. In our neighborhood, spectacular double radio sources are associated almost exclusively with very bright galaxies. Thus, if astronomers see a tiny double radio source that matches a very faint object in visible light, the odds are good that the galaxy is in fact very distant and luminous. Radio galaxies from the 1960-vintage Third Cambridge (3C) catalog have redshifts as large as $z = 2.5$ (7.4 billion light-years), and later surveys which go beyond the bright radio sources have turned up galaxies with redshifts as large as $z = 4.25$ (8 billion light-years). Going after radio sources provides a good way of finding galaxies distant enough that they must be very young. And some of these have bizarre forms and powerfully glowing clouds of gas which indicate they are unlike any-

ies — those thousands of times less powerful than the high-redshift galaxies found by the 3C survey. These objects can be observed by the Very Large Array radio telescope in New Mexico. The unusual radio emission flags these distant galaxies, making them identifiable, and because they are so numerous the hope is they are likely to have become today's run-of-the-mill galaxies, perhaps after a loud phase of adolescent rebellion. As astronomers had hoped, it proved possible to find many of these galaxies at redshifts above 1, so they could start using them to examine galaxy evolution and, eventually, formation.

An especially interesting target was the galaxy 53W002, which led to spectacular new findings with the Hubble Space Telescope backed by a host of ground-based observations. The radio source, number 2 in the 53rd catalog from the Westerbork Synthesis Radio Telescope in the Netherlands, was found by Rogier Windhorst. At 23rd magnitude, it's faint enough that several years elapsed before Windhorst could get a spectrum showing measurable features, at a redshift $z = 2.39$ (7.3 billion light-years).

Far and away the highest redshift found for any of his radio sources, the radio source was weak enough to offer a chance to see the properties of the stars in the surrounding galaxy clearly. Windhorst used the 4.4-meter Multiple Mirror Telescope in Arizona and the 5-meter Palomar telescope in California, which suggested that the stars in this galaxy are much younger than we see for nearby radio galaxies. But the structure of the galaxy eluded this work, since all the action takes place in a region not much more than an arcsecond across.

4. THE UNIVERSE: Galaxy Structure and Evolution

I got to know this galaxy through a chance encounter over a coffee table at the Baltimore meeting of the International Astronomical Union in 1987. A group discussion began by asking how we could compare the spectrum of 53W002, redshifted so strongly that Windhorst and his colleagues had measured only the deep ultraviolet part, with nearby radio galaxies, so we could see how they compared. At the time, this was possible only by using the International Ultraviolet Explorer satellite, which carried an 18-inch telescope and ultraviolet spectrometers in geosynchronous orbit. Our group observed the handful of radio galaxies close enough for measurements with IUE over the next three years, and found that indeed their stellar populations were normal for bright elliptical galaxies with or without radio sources. However, something rather different popped up in 53W002 — evidence of light from large numbers of hot, young stars. Was this the trail of galaxy formation, of the initial widespread epoch of starbirth?

> *"Even more crucial, and difficult to pin down, is the role of dark matter."*

Building Blocks at Last?

Clearly, more work needed to be done. Did 53W002 look like today's elliptical galaxies, or did it have some other strange form? The Hubble Space Telescope would help us find out. Our group was able to show that 53W002 is very close to the form of an elliptical galaxy, with its light falling off around the edges in just the way it does for nearby ellipticals, and not like spirals or irregular galaxies. The comparison was obvious. We had an elliptical galaxy of the same size and form as nearby ones such as M87 in Virgo, but loaded with much younger stars. Our group speculated at this point that we might be seeing part of a single collapse as hypothesized by Eggen, Lynden-Bell, and Sandage — stars had been forming, were still forming, in the telltale elliptical framework, but the spectrum, along with millimeter-wave measurements, showed that lots of gas was still present.

More detailed images could naturally tell us more. Because orbital periods increase with distance from a galaxy's center, the distance at which the stars have settled into a smooth, symmetric distribution can act as a crude clock. Images taken with several filters could show whether the stars started forming on the inside, outside, in patches, or everywhere at once. Our group obtained Hubble images that extended our earlier conclusions, showing a blue cloud associated with the radio source, probably caused by light from the nucleus reflecting off dust. It also suggested that the youngest, bluest stars were in the center, indicating a smooth collapse.

The most interesting result from these data, though, hinged on serendipity. Windhorst's continued observations with the Multiple Mirror Telescope included measuring spectra of brighter galaxies in the field for statistical purposes. Sam Pascarelle, a doctoral student, did this one night and followed the field closer to the horizon than is desirable. At this low elevation, atmospheric refraction moved another faint object into one of the several small spectrograph apertures, and Pascarelle immediately saw that it showed the same redshift as 53W002. Subsequent Hubble observations allowed us to capture 17 objects at the same distance. Are these the long-sought protogalaxies?

If they are, we're seeing a piece-by-piece buildup of several galaxies. These objects have the right structures to be ellipticals, but are much too small, taxing the ability of even Hubble's camera to trace them. They have typical radii of only about 3,000 light-years, smaller than we would expect the progenitors of galaxies in clusters to be. If this group represents normal galaxies at this stage in the universe, these objects must somehow grow — a lot — to become today's bigger galaxies.

A State of the Galaxies Report

A new study by Lennox Cowie of the University of Hawaii and colleagues, using the Keck telescope, provides food for thought that may resolve some otherwise puzzling facts. This group finds evidence that the time needed to form galaxies (and most of their stars) corresponds to the brightness of a galaxy. The most massive galaxies may have finished the process by the time the universe was half its current age, and progressively smaller galaxies are continuing the process.

But astronomers don't yet know the ultimate answers. Beyond the cluster our group found, similar searches are turning up more possible primeval galaxies. Caltech's Charles Steidel and his coworkers note that any substantial amount of interstellar hydrogen in a galaxy will chop out virtually all its light at wavelengths immediately short of 912 angstroms, deep in the ultraviolet. Thus, a galaxy which is easy to see at red wavelengths but disappears going into the blue region is likely to be very distant, so that its deep-ultraviolet light shows up in the visible range. Steidel and coworkers have found galaxies at redshifts of up to $z = 3.4$ (7.8 billion light-years) in this way. Two more groups have searched the Hubble Deep Field data for even higher-redshift galaxies, with cutoffs redshifted into the near-infrared. Several such objects were found, but they are so faint that further spectroscopic confirmation is beyond present capabilities.

Such searches may answer some of the big questions about how galaxies formed: Theories have predicted that galaxies should form big to small or small to big, depending on what gravitational and radiation processes were most important in the early universe.

Which is Correct?

Even more crucial, and difficult to pin down, is the role of dark matter. Ample evidence exists that galaxies contain an enormous amount of unseen material, detected only by gravitational influence, and there must be so much dark matter that its gravity completely dominated galaxy formation — all that we can see went along for the ride. The precise nature of dark matter also controls just when galaxies started forming, and what the seeds were. The dark matter holds the key, and much of the mystery, then, to galaxy formation.

How can we penetrate the desert between the microwave background, when the universe was less than a million years old, and the first quasars and galaxies astronomers can see, at ages of several billion years? The answer will probably lie in the infrared and submillimeter regions of the spectrum, if for no other reason than the redshift from the expanding universe. What astronomers want to see here is light now shifted so far to the red that there is basically nothing to see in visible light.

A major step should come with the installation of Hubble's first true infrared instrument, NICMOS. In principle, NICMOS will detect galaxies at redshifts almost 2.5 times

"Seeing galaxies form and evolve is most important, . . . as the ultimate signal that ours is indeed an evolving universe."

greater than the current Hubble camera. Furthermore, it will see in unparalleled detail what started out as visible light before it was redshifted, so astronomers can compare near and distant objects.

Impatient astronomers are already dreaming of the next generation of telescopes. One advisory panel to NASA has called for an orbiting telescope with aperture about 4 meters optimized for work deep into the infrared, and cooled for greater sensitivity. New technologies will be required to produce a system that stays cold enough to avoid seeing its own thermal infrared radiation, and can be operated reliably far enough away from Earth to avoid the planet's own infrared glow. Conveniently, this instrument represents the next step in studying the formation and evolution of both galaxies and planetary systems.

Seeing galaxies form and evolve is most important, perhaps, not for the details, but as the ultimate signal that ours is indeed an evolving universe. Few can be unmoved by the spectacle of the deep universe as unveiled in these deep images. Many more of them will be forthcoming from the electronic eyes of Hubble. The goal, though, is to do intellectually what these pictures lead us to aesthetically — to appreciate the span of space and time that led us here.

University of Alabama astronomer William Keel researched his dissertation on the nuclei of galaxies at Lick Observatory, followed by postdoctoral work at Kitt Peak National Observatory and Leiden Observatory. Check his extragalactic activities at http://www.astr.ua.edu/keel.

What Makes Galaxies Change?

Conflicting evidence suggests galaxies are driven either by their own character or the environment that surrounds them.

by Marcia Bartusiak

The Big Bang gets all the press, but a much larger story remains to be told. How did we get to here (the Milky Way) from there (a vast primordial explosion)? How did those vast collections of stars we call galaxies evolve into the universe we currently behold? Why did some galaxies become spiraling disks of gas, while others turned into bulbous ellipticals? And how did the populations of each galactic type change over the eons?

Edwin Hubble first pondered these questions more than 60 years ago, soon after he confirmed that the Milky Way was but one of billions of other galaxies roaming the vast gulfs of space. A few hardy souls, the most patient and persistent observers in extragalactic astronomy, eventually joined the pursuit, including Augustus Oemler, who has been wrestling with these mysteries for more than two decades. "Very few things in astronomy, from discovery to understanding, happen quickly," notes Oemler, who is director of the Observatories of the Carnegie Institution in Pasadena, California.

Answers have remained elusive for good reason: The deeper astronomers peer into extragalactic space (and consequently back into time), the more they must strain their telescopic "eyes." Even with the largest telescopes, galaxies that reside more than a couple billion light-years away appear dim, small, and indistinct. A fundamental quest in astronomy, determining how galaxies originated and evolved, seemed an impossible endeavor.

About 10 percent of large galaxies are neither spirals nor ellipticals, but are known as S0 galaxies, or lenticular galaxies, which have characteristics of both types. This lenticular is NGC 5102.

46. What Makes Galaxies Change?

Elliptical galaxies like M87 in Virgo comprise about one quarter of the cosmos' large galaxies. Their dominant population of old stars gives them their characteristic reddish color.

Oemler, have been independently classifying galaxies in the Hubble images. As a cross-check for accuracy, they separately assessed each galaxy's size, type, and magnitude — many thousands of objects in all. "Hubble finally allows us to sort these far galaxies into categories," explains Ellis. "We can now tell the spirals from the ellipticals. Previously, from the ground, we could only say that something was red or blue, big or small."

Another group member, Ian Smail of the University of Durham in England, is attempting to measure the total mass of each distant cluster. Rich clusters of galaxies, with their huge gravitational fields, can act as lenses. Like a stream of water that comes upon a rock

---

"Very few things in astronomy, from discovery to understanding, happen quickly."
— *Augustus Oemler*

---

and gets diverted to either side of the stone, the light from a distant galaxy caught behind a cluster can be split up. This gravitational lensing — an effect rooted in Einstein's general theory of relativity — also magnifies the light. Smail observes how massive clusters bend the beams of light passing by from more distant galaxies. That will give him a handle on the clusters' assorted environments.

Each of these researchers has been pursuing the same goal; they want to know whether a galaxy is shaped primarily by "nature" — the celestial "genes" it inherits at birth, such as its size, mass, and rotation rate — or by "nurture" — the galaxy's later encounters with its environment, such as neighboring galaxies or intergalactic gas.

Such questions were not being discussed when Oemler first started out in astronomy. "If you asked the average astronomer in the 1950s or 1960s, 'Do galaxies change?' they would have said no," recalls Oemler. They were being swayed by the theories of the day; then-current models of our galaxy's construction suggested that it collapsed rather quickly out of a spherical blob of gas. Most astronomers assumed this happened for all types of galaxies, each system forming during a brief, galaxy-forming epoch shortly after the Big Bang. These galaxies, as the story went, then coasted along in the ensuing years, remaining serene and immutable.

The largest number, about two-thirds, emerged as spiral galaxies, with bright round or barred cores surrounded by lush pinwheels of dust and gas. Others became ellipticals: egg-shaped conglomerations that are populated with old, red stars and largely exhausted of the gaseous resources needed to make stars. About 10 percent ended up in a state known as S0s, or lenticular galaxies, which have characteristics of both spirals and ellipticals. Lenticulars consist of a

But now there is a palpable excitement in the air. Oemler and a loose confederation — five other astronomers hailing from England, the Netherlands, Australia, and California — have joined forces in recent years to interpret a blizzard of data from the Hubble Space Telescope. Based on the group's proposal, the space telescope had intermittently focused on 10 rich clusters of galaxies scattered about the celestial hemisphere, each cluster situated from 4 to 5 billion light-years away, nearly halfway back to the moment of creation. Distant galaxies that were mere smudges could at last be viewed with impressive clarity. And what the Hubble search revealed was a universe far different from the one observed today.

Five billion years ago, an age when our solar system was just starting to form, there were simply more spiral galaxies around. Oemler and his fellow cosmic evolutionists want to know how the universe got from that vibrant state several billion years ago to the more settled, tidy version we now see around us. More than that, they'd like to find out what happened to those extra galaxies.

Formerly competitors, these extragalactic astronomers teamed up to use the Hubble Space Telescope more efficiently as an international resource. Richard Ellis of Cambridge University, Alan Dressler of the Observatories of the Carnegie Institution, Warrick Couch of the University of New South Wales in Australia, and Harvey Butcher of the Netherlands Foundation for Research in Astronomy, as well as

4. THE UNIVERSE: Galaxy Structure and Evolution

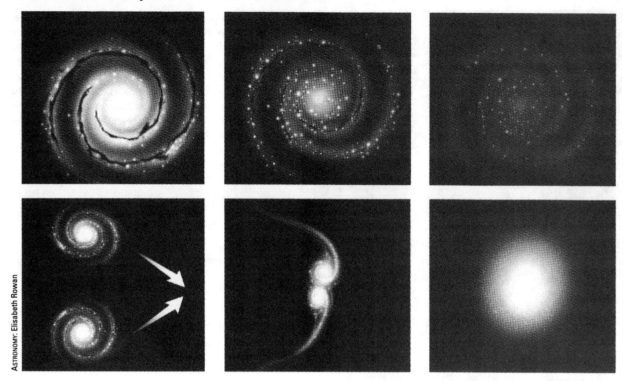

Is galaxy evolution due to nature or nurture? Do galaxies change because of internal processes, or do they change due to outside influences? It's probably a little of both. On the top, we see an example of nature. A spiral galaxy gradually converts its gas into stars. Over time, the stars fade and die. Today, the galaxy is a dim shadow of its former self. On the bottom, we see the forces of nurture at work. Two spiral galaxies in a cluster collide, forming long tidal tails. Eventually the pair merge to form an elliptical galaxy.

sizable bulge enwreathed with completely smooth disks of stars devoid of spiral arms.

Yet clues were scattered throughout the heavens that a galaxy's life is not that simple — that galaxies have not drifted through space in tranquillity since the Big Bang. Even Hubble noticed that gas-depleted ellipticals and lenticulars were predominant in the universe's "urban areas," the centers of rich, packed clusters, where hundreds to thousands of individual galaxies huddle together. Spirals are more commonly found in the "field," the more sparsely populated or rural regions of space.

Did this segregation happen from the start, based on the initial conditions of galaxy formation? Perhaps ellipticals and lenticulars were selectively born where the primordial gas was densest. Dressler long championed that idea. Or possibly those ellipticals were created later on, when an initial population of gas-rich spirals fell into a cluster and merged. Oemler favored that scenario. "I used to insist that everything was due

Elliptical Galaxies Through Time

12 billion years ago

9 billion years ago

46. What Makes Galaxies Change?

to evolution in clusters," says Oemler. "Alan and I argued these points in journal papers for more than a decade." Today, they are coming to believe it's probably a little of both.

The roots of this debate can be traced back to 1978, the year that astronomers obtained the first direct evidence that galaxies were indeed evolving over time. Using the 84-inch telescope on Kitt Peak in southern Arizona, Oemler and his longtime collaborator Harvey Butcher analyzed the colors emanating from two galaxy clusters situated some 5 billion light-years away (hence 5 billion years back in time). And what they saw was somewhat unexpected: Both clusters' galaxies were radiating more blue light than the clusters seen near us today, which tend to be

---

"Galaxies just don't look like that today, I hardly know how to classify them."
— Alan Dressler

---

red. This bluish characteristic of populations of galaxies in distant clusters was soon dubbed the Butcher-Oemler effect.

Even as Butcher and Oemler were gathering their data, Yale theorist Beatrice Tinsley was modeling how entire populations of stars age within a galaxy. She concluded that galaxies can undergo substantial internal evolution through time, more than previously assumed. A young galaxy can start out bright and blue, when the gaseous resources needed to form stars are at their peak, and then gently redden with age and fade away over the eons as the stars age and die.

But others wondered whether the Butcher-Oemler effect was a sign that great upheavals were taking place in that far-off time: Could galaxies be dimmed, brightened, or altered owing to influences from outside the galaxy? Perhaps the galaxies were colliding, merging, and consequently "starbursting," vigorously forming stars at hyper-accelerated rates. Theory was on their side. At MIT in the 1970s, mathematician Alar Toomre was beginning to conduct his groundbreaking computer simulations that showed how two spiral galaxies can merge to form an elliptical galaxy.

Something was obviously happening in these clusters. Following up on Butcher and Oemler's finding, first Dressler in California and later Couch in Australia obtained spectra of the distant galaxies and saw evidence of heightened star formation. "A spectrum can tell you a lot," explains Couch. "We found emission lines, which means that star formation is definitely going on. But we also found lots of hydrogen absorption lines from A-type stars, which hang around for only a billion years or so. So, we were seeing galaxies within a billion years of a starburst."

Obtaining additional information over the next 15 years, though, was a struggle. Astronomers switched from photographic plates to CCDs — digital imaging — which offered a better look at the clusters. But not good enough. The blue objects in Oemler and Butcher's far clusters were still indistinguishable, no more than faint blobs.

It was the amazing resolution of the Hubble Space Telescope that provided a quantum leap in this endeavor. "It really hits you when you look at these Hubble images, just how beautiful the galaxies are," says Couch. "Exquisite detail." Astronomers could at last see directly what was only guessed at or suspected. The bluer objects in the far clusters were definitely spiral galaxies and, more important, they existed in numbers greater than in clusters today. "That's a big step forward," says Oemler. "It could have been that the blue things we saw had nothing to do with normal galaxies. We always suspected they were spirals, but we couldn't prove that until we had the Hubble Space Telescope."

The appearance of these distant galaxies, though, is making astronomers redefine the term "normal."

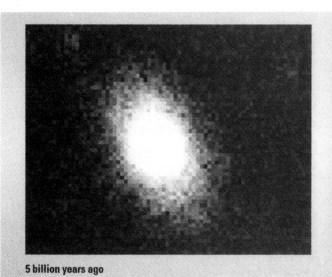

5 billion years ago

A few million years ago

4. THE UNIVERSE: Galaxy Structure and Evolution

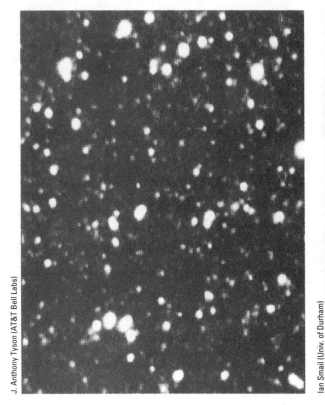

Ground-based telescopes show distant blue galaxies as smudges of light devoid of shape. To understand how galaxies change with time, astronomers needed the Hubble Space Telescope.

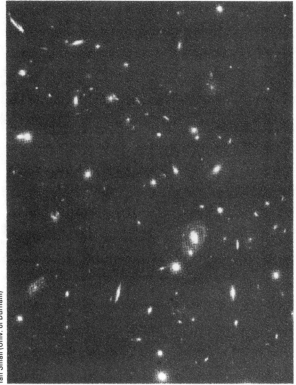

Hubble reveals distant galaxies in all their glory. By comparing the shapes of distant galaxies to their nearby counterparts, astronomers can piece together the story of galactic evolution.

Harvey Butcher and his colleagues have been independently classifying galaxies in the Hubble images.

They are spirals, but very strange-looking spirals indeed. Oemler brings up the image on a computer workstation of one particular cluster under study.

Scanning over the picture, he points to many galaxies in the cluster that are ragged and asymmetric. "You also see all these rings," he notes. "Galaxies just don't look like that today." Dressler adds, "It's full of twisted, funny spiral galaxies. I hardly know how to classify them." One of the group's tasks is gauging the degree of "blobbiness" in each galaxy — in other words, how much the galaxy is distorted compared to a more normal disk galaxy. In this way, they hope to glean clues on the origins of those distortions.

At first, the group was tempted to conclude that

Spiral Galaxies Through Time

12 billion years ago

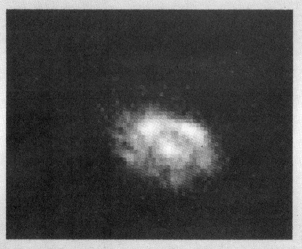

9 billion years ago

46. What Makes Galaxies Change?

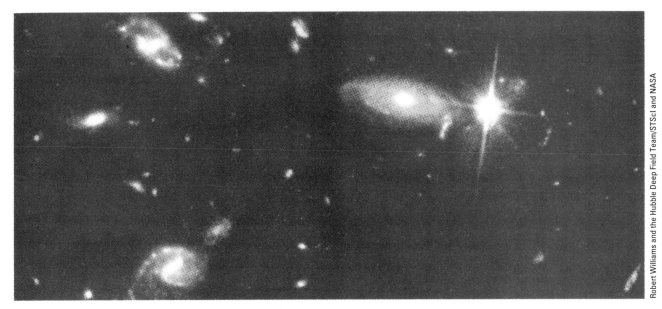

Galaxies through time: Last year the Hubble Space Telescope took a 10-day exposure of a tiny patch of sky, revealing galaxies in the local, intermediate, and distant universe.

mergers explained it all — that these unusual spirals, types no longer seen today, simply paired up to become something else, most likely elliptical galaxies. Interactions are certainly occurring; some of the galaxies are elongated and have tidal bridges between them (see "Galaxies Colliding in the Night," November 1996). But that simple scenario no longer explains all that they are seeing.

For one, there are problems with mergers within a cluster: Galaxy velocities are so fast within a cluster, up to 1,000 kilometers per second, and stars are spaced so far apart, that the stars would merely pass by one another with hardly a fare-thee-well. Even if the spirals did somehow merge, the newly formed ellipticals would be rich in blue, star-forming light because the colliding clouds of gas and dust would trigger a new round of star formation. But all the big elliptical galaxies in these clusters appear red and long settled. Ellis is checking to see if any of the ellipticals display a smidgen of residual blue light, a sign of prior merging and star formation. "But so far," he says, "the bulk of the luminous ellipticals in these clusters look very old." Their births seemed to have

> *"It really hits you when you look at the Hubble images, just how beautiful the galaxies are."*
> — *Warrick Couch*

occurred much further in the past. This seems to bear out Dressler's long-cherished belief that ellipticals are a product of "nature" more than "nurture."

But the spirals may be another matter entirely. The group is now considering whether the bizarre-looking spirals were, instead, altered in some way. A spiral's gas and dust could have been be stripped away as it

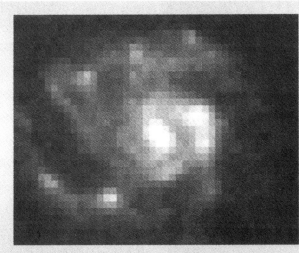

5 billion years ago

A few million years ago

4. THE UNIVERSE: Galaxy Structure and Evolution

zoomed through the relatively dense gases hovering throughout the cluster. Maybe in that way, suggests Couch, the spirals turned into lenticulars. "The fraction of lenticulars in nearby clusters is very high," he notes, "but it is low in these younger clusters."

Or the distant galaxies could have been "harassed" (as astronomers like to call it these days), whittled down by tidal forces as they gravitationally encountered other passing galaxies. This encounter would compress their gases, and new stars would flare up,

> *"Hubble finally allows us to sort these far galaxies into categories."*
> — Richard Ellis

speedily depleting the galaxy's gaseous resources. Today, too dim to be photographed, these spiral galaxies seem to have vanished from the universe. Astronomers have recently begun finding hordes of dim galaxies that contain very few stars (see "Ghost Galaxies of the Cosmos," June 1996). Maybe these are the ghostly remains of these distant blue spirals that once burned brightly many eons ago. But that's only speculation at this point. "I'm not yet convinced that we've found enough dim galaxies nearby to explain what's going on," says Ellis.

"We need to find the smoking gun," adds Oemler.

What would help would be to compare what's happening in the urban areas, that is the central area of the clusters, with what's happening in the field. Fortunately, Hubble is also carrying out the Medium Deep Survey, a set of long-term exposures of distant galaxies in the field. And the view there is just as bizarre as in the clusters. Astronomers are seeing many faint, blue, irregular galaxies in deep space, far more than seen today (see "Our Strange, Scrappy Ancestors," December 1995). Either they've stopped forming stars and so have now faded away, or they've come together to form more normal-looking galaxies. Whatever the reason, the change provides further evidence that the Universe has indeed evolved over time.

If galaxies both in the field and in the clusters are following the same course, then it would point to nature having the upper hand. Where a galaxy is located doesn't matter; it's just following its own internal life cycle. "In other words, those distant spiral galaxies could be ragged and scruffy simply because that's what galaxies looked like back then," points out Dressler. But if the evolution of a galaxy differs in those disparate environments—field and urban center—then a strong

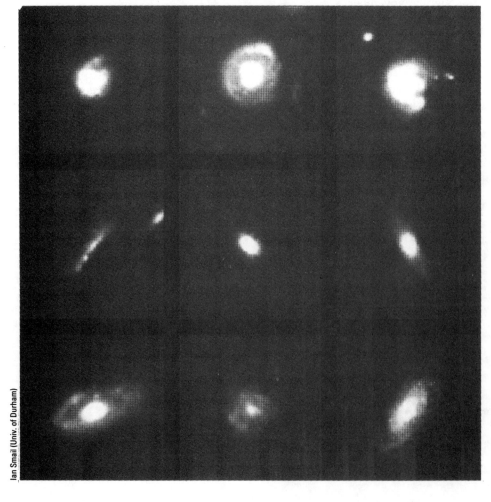

Nine spiral galaxies as they existed halfway back to the moment of creation. Many spirals in that distant epoch had distorted shapes. These spirals have no counterparts in the modern-day universe.

finger is pointed at nurture doing the job. The change a galaxy makes would depend on its surroundings, its encounters with other galaxies over time.

Ian Smail observes how massive clusters bend the beams of light passing by from more distant galaxies.

Answers will arrive with additional observations, especially spectral measurements from the ground. Only there are the mirrors big enough (particularly on the new 8- and 10-meter-class telescopes) to capture sufficient light from those distant regions to dissect the light into its spectral components. That will help astronomers determine what is causing the bursts of star formation in those far galaxies. Hubble's relatively small 2.4-meter mirror, while superb for imaging because of its position above the atmosphere, cannot gather enough light to do spectroscopy on particularly dim objects.

"We've got a series of snapshots," says Oemler. But they need the see the entire movie, including scenes closer in as well as farther out. A new wider-field camera on Hubble, scheduled for installation in 1999, will help astronomers view nearby clusters, which extend over a broader swath of sky. But getting deeper images will eventually require new instrumentation because Hubble is not optimized for viewing the primeval universe.

Waves of light from distant space stretch out with the universe's expansion, so going deeper into space means peering at longer and longer wavelengths, beyond the visible spectrum — into the infrared. Dressler recently chaired a committee that recommended the construction of a future 4-meter infrared space telescope, nearly twice the size of Hubble, to view the era of galaxy formation some 10 billion light-years distant or more. "In the current competition for financial resources, we're probably talking about something 15 to 20 years from now," says Dressler. But NASA has jumped on the Dressler committee suggestion, and is now looking at the possibility of building an even larger space telescope with a 6- to 8-meter mirror and launching it as early as one decade from now. "The power of such a telescope to look back to the era of galaxy birth would be awesome," adds Dressler.

In the meantime, Smail is working on a clever technique to discern some information on galaxies farther out. As pointed out earlier, rich galaxy clusters can act as lenses. Smail is testing a method of estimating the distance of those far galaxies by the amount their images are distorted — elongated — by the lensing effect.

But, for now, the group is simply relishing the view from Hubble. "That's the most fun of all," notes Oemler, "just staring at these pictures." Dressler muses: "For as long as all of us have been trying to figure out whether nature or nurture is the key to galaxy evolution, it's hard to believe that Hubble and the giant new telescopes like Keck are finally giving us some answers — it's some of both. It's funny how many things turn out that way."

Marcia Bartusiak is a Massachusetts-based science writer. Her previous article for ASTRONOMY, *"The New Dark Age of Astronomy," appeared in the October 1996 issue.*

Galactic Engines

By Neil de Grasse Tyson

Galaxies are phenomenal objects in every way. They are the basic organization of visible matter in the universe, which contains as many as a hundred billion of them. They each commonly pack hundreds of billions of stars. Many are found solo in space; others are found in gravitationally linked pairs, familial groups, and clusters. They can be spiral, elliptical, or irregular in shape. Most are photogenic. Their morphological diversity has prompted all manner of classification schemes, supplying a conversational vocabulary for astrophysicists. One variety, the "active" galaxy, emits an unusual amount of energy in one or more bands of light. This excess energy usually comes from the galaxy's center. The center is where you will find a galactic engine. The center is where you will find a supermassive black hole.

The names for active galaxies read like a manifest for a carnival grab bag: BL Lacertae galaxies, Seyfert galaxies (type I and II), blazars, N-galaxies, LINERs, radio galaxies, and of course, the royalty of active galaxies—quasars. What they have in common is that their extraordinary luminosities appear to be driven by mysterious activity from a very small region buried deep within their nuclei.

Quasars, discovered in the early 1960s, are up to a thousand times as luminous as

The extraordinary luminosity of quasars and other active galaxies is powered by activity in a very small region of their nuclei.

our own Milky Way galaxy, yet their energy hails from regions that would fit comfortably within the planetary orbits of our solar system. Curiously, none are nearby. The closest one is about 1.5 billion light-years away—its light takes 1.5 billion years to reach us. Most are over 10 billion light-years away, yet we can see even farther out into the universe to a time before quasars turned on. Possessed of small size and extreme distance, they are almost indistinguishable on photographs from the pointlike images left by stars in the Milky Way. Quasars were first found with radio telescopes. And because stars are not known for emitting copious amounts of radio waves, it was clear that a new class of object had been discovered. In the we-call-them-as-we-see-them tradition of astrophysicists, these objects were dubbed Quasi-Stellar Radio Sources or, more affectionately, "quasars."

What manner of beast are they?

One's ability to describe and understand a new phenomenon is always limited by the contents of the prevailing scientific and technological toolbox. An eighteenth-century person who was briefly thrust into the twentieth century would return to describe a car as a horse-drawn carriage without a horse, and a light bulb as a candle without a flame. With no knowledge of internal combustion engines or electricity, the possibility of true understanding would be remote indeed. With that as a disclaimer, allow me to declare that we think we understand the basic principles of what drives a quasar. In what has come to be known as the standard model, black holes have been implicated as the engines of quasars and of all active galaxies.

Within a black hole's invisible boundary—its "event horizon"—the concentration of matter is so great that space curves back on itself and not even light can escape. When you fall into a black hole, you fall in for good, even if you are made of light.

How, you might ask, can something that emits no light be a power source for something that emits more light than anything else in the universe? When many of the exotic properties of black holes were explored in the late 1960s and 1970s, people swiftly realized that they provided a remarkable addition to the theorist's toolbox. According to some well-known laws of gravitational physics, gaseous matter funneling toward a black hole must heat up and radiate profusely

47. Galactic Engines

before it descends through the event horizon. The energy comes from the efficient conversion of gravity's potential energy into heat.

While it is not a household notion, most people have, at some point, had a run-in with converted gravitational potential energy. If you have ever dropped something to the floor and broken it, or if you have ever set a pie to cool on a windowsill and watched it fall and splatter on the ground below, then you understand the power of gravitational potential energy. As objects fall, their potential energy is continuously being converted to energy of motion. And if something stops the fall, all the energy the object has gained swiftly converts into the kind of energy that breaks or splatters things. This is the reason that you are more likely to die if you jump off a tall building instead of a short one.

But if something prevents the object from gaining speed, and the object continues to fall, then the converted potential energy reveals itself in some other way—usually in the form of heat. Good examples include space vehicles and meteors that heat up as they plunge through Earth's lower atmosphere—they want to speed up, but air resistance prevents it. In a famous experiment, the nineteenth-century English physicist James Joule created a device—rotating paddles powered by falling weights—that stirred a jar of water. The potential energy of the weights was transferred into the water and successfully raised its temperature. Joule describes his effort:

The paddle moved with great resistance in the can of water, so that the weights (each of four pounds) descended at the slow rate of about one foot per second. The height of the pulleys from the ground was twelve yards, and consequently, when the weights had descended through that distance, they had to be wound up again in order to renew the motion of the paddle. After this operation had been repeated sixteen times, the increase of the temperature of the water was ascertained by means of a very sensible and accurate thermometer....

I may therefore conclude that the existence of an equivalent relation between heat and the ordinary forms of mechanical power is proved.... If my views are correct, the temperature of the river Niagara will be raised about one fifth of a degree by its fall of 160 feet.

Joule's concluding thought experiment refers, of course, to the great Niagara Falls. But had he known of black holes, he might have said instead, "If my views are correct, the temperature of gas funneled toward a black hole will be raised by about a million degrees by its fall of a billion miles."

Joule's paddle experiment highlights one of the most profound applications of physical laws to the cosmos. If a star gets ripped apart in its descent to the black hole's event horizon, then some of the star's original gravitational potential energy will go to heating the gas, while the rest will be radiated back into space.

One can easily imagine that a black hole would have a prodigious appetite for stars that wander too close. The secret to powering the galactic engine lies in a black hole's ability to reliably and ruthlessly rip apart stars before they cross the event horizon. Like a cosmic paper shredder, the black hole's tidal forces of gravity elongate the otherwise spherical stars in much the same way that the Moon's tidal forces elongate Earth's oceans to create high and low tides. Fortunately, the Moon's tidal forces are not strong enough to shred Earth, but black holes are not so gentle. Once pulled apart, gas that was

For a quasar to stay healthy, its central black hole must devour about ten stars every year.

formerly part of stars (and possibly ordinary gas clouds) cannot simply gain speed and fall. The gas of previously shredded stars impedes wanton free fall. The result? A star's gravitational potential energy is converted to prodigious amounts of heat and radiation.

Faced with the proliferation of classes of unusual galaxies, the late astrophysicist Gerard de Vaucouleurs, a consummate morphologist, was quick to remind the astronomical community that a car that has been wrecked does not suddenly become a different kind of car. This car-wreck philosophy has led to a standard model that unifies the zoo of active galaxies. The model is endowed with enough tweakable parts to explain most of the basic, observed features. For example, the funneling gas often forms an opaque, rotating disk before it descends through the event horizon. If the outward flow of radiation cannot penetrate the disk of accreted gas, then radiation will fly out from above and below the disk to create titanic jets of matter and energy. The observed properties of the galaxy depend on whether the galaxy's jet happens to be pointing toward you or at some other angle—and on whether the ejected material moves slowly or at speeds close to the speed of light. The thickness of the disk will also influence the galaxy's appearance, as will the rate at which stars are consumed. In a healthy quasar, the black hole eats as many as ten stars per year. Other, less active galaxies require many fewer shredded stars per year to maintain their luminosity.

The luminosity of many quasars varies over days and even hours. Allow me to impress you with how extraordinary this is. If the active part of a quasar were the size of the Milky Way, 100,000 light-years across, and if it all brightened at once, then you would first learn about it from the side of the galaxy that was closest to you, and 100,000 years later, the last of the galaxy's light would reach you. In other words, you would have to observe a quasar for 100,000 years to see it brighten fully. For a quasar to brighten within hours means that the dimensions of the engine cannot be greater than light-hours across. How big is that? About the size of the solar system.

With a careful analysis of the light fluctuations in all bands, a crude, but informative three-dimensional structure can be deduced for the black hole's surrounding material. For example, the luminosity in X-rays might vary over a time scale of

4. THE UNIVERSE: Quasars

hours, but the red light might vary over weeks. Comparing the two allows you to conclude that the red-light-emitting part of the active galaxy is much larger than the X-ray-emitting part. This exercise can be performed for many bands of light to derive a remarkably complete picture of the system.

If most of this action took place long ago in distant quasars, then why isn't it still happening? Why are there no local quasars? Good explanations are available. The most obvious one is that the nuclear regions of local galaxies have run out of stars to feed their engines. In other words, all stars whose orbits came too close to the black hole have already been vacuumed up. No more food, no more prodigious regurgitations.

A more interesting shut-off mechanism comes from what happens to the tidal forces as the black hole's mass (and thus its event horizon) grows and grows. Since tidal forces have nothing to do with the total gravity felt by an object—just the difference in gravity across it, which increases dramatically as you near an object's center—large, massive black holes actually exert lower tidal forces than small, low-mass black holes. No mystery here. The Sun's pull on Earth dwarfs that of the Moon, yet the proximity of the Moon enables it to exert a considerably higher tidal force.

What all this means is that there is a point where a black hole has eaten so much, and its event horizon has grown so large, that its tidal forces are no longer sufficient to shred a star before it crosses the event horizon. When this happens, all of the star's gravitational potential energy is converted into speed (like that of the freshly baked pie as it fell out of the window but before it hit the ground) and the stars are eaten whole as they plunge through the event horizon. No more conversion to heat and radiation. This shut-off valve kicks in for a black hole when it becomes about a billion times the mass of the Sun.

The unified picture would account for a scenario in which quasars and other active galaxies are just early chapters in the life of ordinary galaxy nuclei. If this is true, specially exposed images of quasars should reveal the surrounding fuzz of a host galaxy. The observational challenge is similar to that faced by planet hunters who try to detect planets hidden in the glare of a host star. The quasar is so much brighter than the surrounding galaxy that masking techniques must be used to detect anything other than the quasar. Sure enough, nearly all high-resolution images of quasars have revealed surrounding galaxy fuzz. The several exceptions—uncloaked quasars—continue to confound expectations based on the standard model. Or are the host galaxies simply too faint to be detected? The data remain controversial.

The unified picture also accounts for quasars eventually shutting themselves off. Actually, the absence of nearby quasars requires this. But it also means that black holes in galactic nuclei should be common, whether or not the galaxy has an active nucleus. Indeed, the list of nearby galaxies with dormant supermassive black holes in their nuclei has grown to two dozen—and includes the Milky Way. The smoking gun in each galaxy is the unexpectedly astronomical speed that stars achieve as they orbit close (but not too close) to the central black hole.

Such scientific models are always seductive, but one should occasionally ask whether the model actually captures some deep truths about the universe, or whether it was constructed with so many tunable variables that you can use it to explain anything at all. Similarly, have we been sufficiently clever today, or are we missing a tool that will be invented or discovered tomorrow? The English physicist Dennis Sciama, of the University of Oxford, was asking himself a similar question when he penned:

Since we find it difficult to make a suitable model of a certain type, Nature must find it difficult too. This argument neglects the possibility that Nature may be cleverer than we are. It even neglects the possibility that we may be cleverer tomorrow than we are today.

Neil de Grasse Tyson is the Frederick P. Rose Director of New York City's Hayden Planetarium. He also teaches astrophysics at Princeton University.

Beyond the Soapsuds Universe

IN 1986, MARGARET GELLER DISCOVERED THAT GALAXIES ARE ARRAYED ACROSS THE COSMOS LIKE A VAST CONGREGATION OF BUBBLES. NOW A PAIR OF ROBOTS ARE HELPING HER LOOK BACK IN TIME TO FIND OUT JUST HOW THE STARS GOT THAT WAY.

GARY TAUBES

MARGARET GELLER FIRST MET THE stickman in the fall of 1986. While the exact date has faded from her recollection, she remembers the time as midafternoon and her reaction as a kind of euphoria. No one had ever seen the stickman before—at least, not really. Valérie de Lapparent, who was Geller's graduate student, noticed it but says she was too inexperienced to understand its implication. John Huchra, who was Geller's collaborator at the Harvard-Smithsonian Center for Astrophysics (CFA), says he took one look at the stickman and assumed he had botched his observations. It took Geller's eye to recognize the stickman as something real and important.

Geller, Huchra, and De Lapparent had mapped the nearby universe, taking several months to carefully measure the distance to 1,000 galaxies, some as near as 30 million light-years away, others as far as 650 million. De Lapparent had fed the distance and positions of those galaxies into a computer program that printed out a two-dimensional representation of their three-dimensional distribution in the universe. On the printout was this slice of the northern sky, sprinkled with 1,000 distant galaxies, and smack in the middle, says Geller, was "this remarkable stickman figure." The distribution of galaxies looked like a child's drawing of a somewhat bowlegged person. It's a whimsical name for a grand figure: the stickman extended 500 million light-years across the universe. Its torso was composed of hundreds of galaxies, a massive congregation known to astronomers as the Coma cluster. Its arms were two more sheets of galaxies streaming across the night sky.

The stickman was grand not just in dimension but in destiny. You might even say it changed our understanding of the universe. Until the stickman, the universe appeared to be a smooth and homogeneous place. Astronomers believed that galaxies were distributed at random, although they might occasionally form clusters like Coma containing as many as a thousand or so galaxies like the Milky Way. There was even some evidence that the universe contained at least one enormous void, in the constellation Boötes, which seemed to extend for some 200 million light-years—and other suggestions that galaxies could be found strung out on long filaments. But in 1985 most astronomers assumed these structures were products not of the universe itself but of the methods used to survey it.

Then Geller saw the stickman, which constituted compelling evidence that galaxies were congregating on two-dimensional structures, as though they had condensed out of the cosmic nothingness on the surfaces of invisible bubbles. Indeed, when Geller later wrote up the results of the CFA galaxy survey, she described the distribution of galaxies in the universe as looking like a slice through suds in the kitchen sink. Her metaphor implied that astronomers were mightily confused about how the universe had formed.

The very early universe, around the time of the Big Bang, was a smooth place. We know that because the Big Bang left an imprint: the cosmic background radiation, which is a radiation 3 degrees above absolute zero that pervades the entire universe. That background radiation is considerably smoother than a baby's behind, and it means the universe, when it was a couple of hundred thousand years old (and maybe even younger), was equally smooth. Now it's not. It's full of these enormous two-dimensional structures. Perhaps the most awe-inspiring is one that Geller and Huchra discovered in 1989, known as the Great Wall: a sheet of galaxies extending for at least 500 million light-years, stretching across the entire northern sky. It may indeed be bigger than 500 million light-years, but no one can yet tell.

The confusion comes about because astronomers can see the huge structures at the very limits of their vision, which means when the universe was considerably younger than it is today. When we look out into space, we're looking back in time; the light from a galaxy a billion light-years away, for instance, will take a billion years to reach us. "It's an amazing thing," says Geller. "The history is there for us to see. It's not mushed up like the geologic record of Earth. You can just see it exactly as it was."

SO WHAT HAPPENED? THE universe is full of these prodigious two-dimensional structures as far out as we can see, and thus was full of them as far back as we can see. In the 10 or 15 billion years the universe has taken to grow up, it has evolved from something unimaginably smooth into this sink suds of a structure, and no one yet knows how or why.

Geller is at work trying to answer this question, as are at least a hundred other astronomers around the world. Mapping

4. THE UNIVERSE: Quasars

the structure of the universe has become a cottage industry in astronomy; since the stickman made his appearance, astronomers have initiated more than a dozen surveys to chart the distribution of galaxies. Geller's, with CFA astrophysicist Dan Fabricant, may be one of the deepest. The two are working on a survey that should begin probing the universe in late 1998 at the rate of thousands of galaxies a night. By the time they're done, they will have surveyed more than 50,000 galaxies and mapped strips of the universe out to a distance of 5 billion light-years. This might just be far enough out—that is, far enough back in time—to understand why the universe we see appears so profoundly different from the universe of the Big Bang. "Now that we know something about what the nearby universe looks like," says Geller, "the issue everybody wants to understand is how it got that way. And the race is on to find that out."

The discovery of the stickman may have changed Margaret Geller as much as it did our conception of the universe. It launched her into the stratosphere of science and linked her name with the idea of mapping the universe and with the structure of the universe itself. The morning after Geller first displayed the stickman to her fellow astronomers at a meeting in Houston, she made an appearance on the *Today* show, complete with the stickman and the news that the universe was a considerably more perplexing place than previously imagined. She went on to win a MacArthur Fellowship and to become a star of the Smithsonian Center for Astrophysics and the Harvard Observatory. All those achievements were not enough, however, to make Geller feel at home in astronomy.

There is a theory that creativity arises when individuals are out of sync with their environment. To put it simply, people who fit in with their communities have insufficient motivation to risk their psyches in creating something truly new, while those who are out of sync are driven by the constant need to prove their worth. They have less to lose and more to gain. The theory of asynchronicity might help explain Geller, who has been struggling to fit in since she began studying astronomy. Geller's first love was acting, but her father, a physical chemist who worked on crystal structures, did what he could to encourage her in science. He would take her to his lab at Bell Laboratories in New Jersey, then in its heyday, where she would play with his state-of-the-art, hand-cranked calculator. "The biggest challenge was to make it make a big racket for as long as possible," says Geller. By the time she was ten, she was working out simple calculations. Because she didn't like going to school, her parents let her teach herself. Her mother, from whom Geller inherited her fascination for language, would take her to the library and help her choose books, then supervise her study at home. Geller would show up at school to take tests and little more.

As an undergraduate at the University of California at Berkeley, Geller first took up math but moved on to physics. "I didn't know what kinds of questions to ask in mathematics," she says. "In physics, I could see there were things that were known and things that weren't." She went back east to Princeton for graduate school, where she studied astrophysics, learned about galaxy surveys, and pored over galaxy catalogs. She also lost her confidence, falling out of sync with her environment. Until then, she says, "it never occurred to me that there might be something I wasn't able to do."

She hated Princeton. Only one woman—Glennys Farrar, now at Rutgers University—had successfully obtained a doctorate from the Princeton physics department before Geller arrived in 1970. Princeton had just admitted its first women as undergraduates that year, and the atmosphere in the physics department, according to Farrar, was "terrible." Geller says she wasn't mature enough to deal with it: "Students would ask me what I was doing in physics at Princeton when men couldn't get jobs in physics, or they would say, 'Only one woman has passed her generals in the department and three have been admitted since, and they all failed. So there's a 75 percent chance you'll fail.'"

Until she got to Princeton, Geller says, she simply never realized how few women there were in science. "Somehow it hadn't registered. But I never had a single female professor throughout my whole education, from the beginning of university to the end. Even all the books were about men; I never really liked reading books about the history of science, and I never really understood why." At Berkeley, it hadn't bothered her, perhaps because the physicists were used to dealing with women as undergraduates. "I had a great deal of confidence when I graduated from Berkeley," says Geller. "I had almost none when I was at Princeton. After a while when people tell you you can't do something because you're a woman, you begin to believe maybe they're right. It's amazing, because even though you know these things are totally irrational, they stick with you for many, many years."

Geller often thought of quitting Princeton, but her parents talked her out of it. "They said I shouldn't quit, because I'd never failed at anything, and if I left I'd feel that I failed and that would haunt me my whole life. Get the degree and *then* quit." She got the degree but didn't quit. The experience toughened her. The Nobel laureate Steven Weinberg, who was at Harvard in the mid-1970s when Geller arrived from Princeton, says this is what stood out about her. "I admired her," he says. "Not only her intelligence, but she had a certain toughness, a strength of purpose in pursuing her science, in making sure what she thought should be done would be done."

Nonetheless, as Geller admits, her first years at Harvard could be called her lost years. She had yet to recover from the Princeton experience, and she now had mixed feelings about astrophysics: "I felt I could do it, but did I want to do it?" Finally, in 1979, she spent a year at Cambridge University in England and thought things over.

Her self-appointed task at Cambridge was to assess the state of knowledge of the universe's structure. "I realized almost nothing was known about anything outside our galaxy," she says. Galactic surveys had been done, but they were small and useless for drawing firm conclusions about how galaxies were distributed in the universe. To gauge galactic distances, the surveys studied redshifts, which are measures of how much the wavelengths of light from distant objects stretch toward the red end of the spectrum the farther away those objects happen to be. Huchra and Harvard astronomer Marc Davis, now at Berkeley, had measured redshifts of 2,400 galaxies going out barely 300 million light-years, our own backyard by the standards of the universe. That survey was too shallow to see the remarkable patterns that were uncovered later.

Bob Kirshner, who was then at Michigan and is now at Harvard, had led a deeper survey of a few hundred galaxies but had measured them all in a few tiny regions of the sky. Geller compares that technique to "sticking a few needles in a haystack" to map the internal structure

48. Beyond the Soapsuds Universe

> By 1990 astronomers had mapped a paltry 10,000 galaxies. Interpreting the structure of the universe from the positions of just 10,000 galaxies, Geller says, is like trying to understand the surface of Earth from a map of Rhode Island.

of the stack. The survey had detected the great void in the constellation Boötes, but few astronomers believed the void was real. "Everybody was skeptical," says Geller. "I thought there was something wrong with the survey, because the void was much larger than any structure anybody thought existed."

Geller wrote no papers during her year in England. "What I did was much more valuable," she says. "I figured out the kinds of problems I thought were interesting." Back in Cambridge, Massachusetts, she began studying galaxy clusters with John Huchra, which led her to think about the distribution of galaxies on a larger scale. Somebody, she decided, should do a survey reaching deep into the universe, one that would be able to see very large patterns like the Boötes void, if it was real. She had no trouble convincing Huchra, who had worked on the redshift survey with Davis and knew that the field needed a deeper one. Geller would design the project and crunch the data, and Huchra and De Lapparent would spend the nights at the telescope making the observations. They just had to decide what part of the sky to map.

That's where they got lucky, as they all admit today. Geller wanted to measure the redshifts of galaxies that could be found in a continuous strip across the sky. She argued that by examining long continuous strips rather than isolated patches, they could find large structures and understand their geometry. Huchra agreed because strips are easy to measure; you just let the sky roll over your telescope, and it does the hard work for you. "The place where luck was involved was the width of the strip," says Geller. "We were lucky it was wide enough to see the stickman."

Geller thinks visually, seeing problems as patterns and geometries, so the appearance of the stickman represented a convergence of some dominant themes in her life. The choice of strips was a geometric choice, and the stickman itself was a pattern that emerged from the background clutter of the universe. It had been waiting, in a sense, for Geller to interpret its meaning. In the months after the discovery, she spent considerable time creating a video of the stickman, one that would convey to her colleagues the beauty of the pattern, and with it her sense of wonder and revelation. Using what she calls "incredibly primitive graphics," she and her CFA colleague Michael Kurtz made a display of their slice of the universe. "We would sit there absolutely mesmerized by this one slice moving around," says Geller. "We would stare at this thing over and over and over again. It was as if we were high on something."

With the discovery of the stickman, Geller had arrived. She had established herself, in Weinberg's words, as an "adornment to the astronomy establishment at Harvard." And when she and Huchra completed four more slices of the redshift survey, she wrote a paper for *Science* on the Great Wall and the art of universe mapping. She then won the MacArthur Fellowship, which gave her the cachet to raise $200,000 to make a documentary film about her work with Huchra and other colleagues. The film rejuvenated her, taught her new skills, and gave her new ideas.

With all the accolades, however, Geller was still only a senior researcher at the Smithsonian Astrophysical Observatory, and although she has a professorial appointment at Harvard, she has never been given tenure. "I drive myself crazy with this periodically," she says. "I keep wondering, 'Why me, why don't I have this?' I've had periods where I find it very difficult to work because of this. Then I think, 'The hell with it, I've

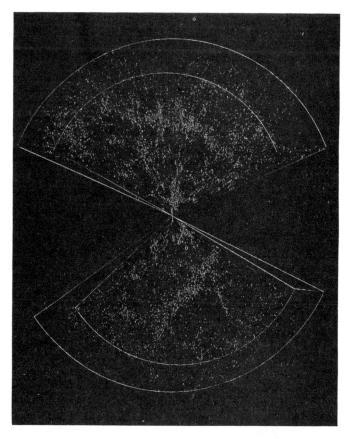

The Great Wall (top, horizontal) and the Southern Wall (bottom, diagonal) flank enormous galaxy-free voids.

shown I can do it. Why don't I go do something else?'"

That question is borderline rhetorical. If it needs an answer (beyond her passion for her pursuit), one might be the new survey and the instrument Dan Fabricant is constructing. The program Geller set in motion at the CFA is about to pay off, and she wants to be there to see it. While Fabricant's reputation puts him in the top handful of astronomical-instrument builders, Geller gets some small credit for his accomplishments. She met Fabricant, who's now 45, while he was still in graduate school. Later, when the Harvard-Smithsonian optical astronomy community was ignoring his ambitions, Geller was cultivating them.

Fabricant started his career building instruments to do X-ray astronomy from rockets and later from satellites. But when the space shuttle *Challenger* exploded in 1986, the launch of any scientific satellite, X-ray or not, began to look like a debatable proposition. Fabricant switched his focus to optical astronomy, which was about to undergo its third major revolution in 300 years. The technology that would make it happen is known as multiplexing—not to be confused with having a choice of 17 movies every time you go to the theater, although the concepts are similar.

The first revolution began when Galileo invented the telescope. For the next 300 years, the science of astronomy advanced because astronomers made bigger and bigger telescopes, which allowed them to collect more and more light and see ever fainter objects. By the 1950s, with the advent of the 200-inch (5.1-meter) mirror at Mount Palomar near San Diego, telescopes had gotten about as big as they were going to get. In the following 30 years, the advances came from new technology to collect the light hitting the equipment. This was revolution number two. Astronomers stopped using photographic plates, which might capture one-half of 1 percent of the incoming light, and turned to electronic detectors, which can capture more than 90 percent.

In the mid-1980s, astronomy was embarking on revolution number three. New technologies were emerging that allowed astronomers once again to build bigger telescopes—ones like the 10-meter Keck in Hawaii, which dwarfed Mount Palomar's. Meanwhile, researchers had been developing instruments that could look at multiple objects simultaneously (hence the term *multiplexing*). A telescope that could measure the redshift from 100 galaxies simultaneously, for instance, would be a 100-fold improvement over a telescope that could look at just one. "We don't get factors of 100 very often for astronomy," says Fabricant.

When the *Challenger* disaster made Fabricant turn to optical astronomy, the Smithsonian and the CFA were trying to decide what to do with an optical telescope at Mount Hopkins near Tucson, Arizona, known as the Multiple Mirror Telescope, or MMT. The telescope, designed in the 1970s, was composed of six identical mirrors that worked together as though they were one telescope six times as large. It was a remarkable idea because building six modest mirrors was considerably cheaper than building one huge one. But the MMT focused on one object at a time, which pretty much condemned it to obsolescence from the day it opened its shutters to what astronomers call first light.

At the suggestion of colleagues at the University of Arizona, the CFA astronomers thought about rebuilding the MMT as one mirror that could focus clearly on a patch of sky four times the size of the full moon, instead of on a single star or galaxy. No one was sure just what they would do with this new MMT (which they planned to rename the Magnum Mirror Telescope so the initials could remain the same). Fabricant, who was only a young X-ray astronomer, attended meetings, but no one took his ideas very seriously. "They were kind of discouraging," says Fabricant. "Their attitude was, 'We don't really need anyone for this, and if we did, it wouldn't be you.'"

Geller was the exception. The stickman had convinced her that a massive sky survey was needed, as deep into the universe as technology would allow. By 1990 astronomers had mapped a paltry 10,000 galaxies; Geller liked to say that interpreting the structure of the universe from the positions of 10,000 galaxies was like trying to understand the surface of Earth from a map of Rhode Island. The deep redshift survey Geller wanted would be possible only with some new multiplexing instrument, because the deeper into the universe you look, the more galaxies there are in each patch of sky. When Geller and Huchra mapped the nearby universe a few hundred million light-years out and uncovered the stickman, they were lucky to see a single galaxy in every square degree of sky. Look 5 billion light-years out into the universe, however, and you'll see more than a thousand galaxies in the same amount of sky. "That just gives you an idea how big the universe is," says Geller. With the right instrument, a sky survey could map a good number of those thousand galaxies simultaneously, and Geller believed Fabricant, X-ray astronomer or not, had the talent to help build it.

Geller also felt that she now had enough prestige to throw her weight around. Over the next six or seven years, she managed to get Fabricant the funding and permission to build two new instruments for redshift surveys. The first—called the Decaspec because it could look at ten galaxies simultaneously—went on a 2.4-meter telescope on Kitt Peak near Tucson. It worked the very first night, unlike most instruments in the history of astronomy. The second, which went on a smaller telescope at Mount Hopkins, not only worked perfectly the first night but also was the most efficient instrument of its type ever built. Fabricant and Geller used that instrument to perform a galaxy survey twice as deep as the one that revealed the stickman and the Great Wall. Their new survey showed the same inexplicable two-dimensional structures, with still no sign of how and when they formed.

Now Fabricant is building the ultimate redshift collector, the one they will use to map 50,000 galaxies. It will be mounted on the tail end of the converted MMT, which is expected to observe first light in late 1998 with a single, immense, 6.5-meter mirror. Fabricant's instrument, called the Hectospec, will collect light hitting the mirror from 300 galaxies at a time. (*Hecto* means 100 in Greek, which is the right order of magnitude and less of a mouthful than "Trihectospec.") It will then robotically redistribute its 300 light-catching fibers, one at a time, so that five minutes later it can begin to observe redshifts from another 300 galaxies. And so on through the night, every night Fabricant and Geller can get time on the telescope.

FABRICANT AND HIS colleagues are finishing up the Hectospec design in a laboratory in Cambridge across the street from the Harvard-Smithsonian. To understand what Fabricant is building and how it will work, first imagine the telescope itself. The light from the heavens comes down and

48. Beyond the Soapsuds Universe

bounces off the huge mirror and then bounces back off a secondary mirror 6 meters above the main one. The doubly reflected light then comes back down and lands on a surface called the focal plane, which is sometimes covered with photographic film or electronic detecting sensors. In this case, it will be covered with the Hectospec.

The Hectospec's light collectors are 300 tiny glass fibers, each of which ends in an equally tiny prism that sits in a metal button sticking magnetically to the focal plane. If the button is placed correctly, light from a galaxy will fall on the prism, which will direct it down the fiber, which will run, along with the 299 other such fibers, into a spectrograph—the instrument that breaks the light into its component colors and makes it possible to measure the redshift.

That's the relatively simple part. The hard part is positioning those 300 buttons: figuring out how to pick them up, one at a time, and put them down again exactly where the next galaxy's light will fall, without tangling the glass fibers. To make matters even tougher, all this repositioning has to happen on the telescope, which might be pointed off toward the heavens at who knows what angle and at temperatures ranging from a comfortable 70 degrees Fahrenheit to a very chilly 20, depending on the time of year. Speed counts, too. The faster the fibers are repositioned, the more galaxies can be surveyed, which is crucial because, as already noted, it's a big universe out there.

The job of moving the buttons goes to a pair of robots, which Geller describes as "pretty massive pieces of machinery." If you met them on the street, she adds, you wouldn't recognize them as robots. Like refugees from some mechanized, futuristic drafting table, they move on a pair of perpendicular rails that allow them to cover the entire focal plane. On the bottom of each robot is a clamp that can close around the button on the end of the fiber. Once that clamp locks on, the robot will reposition the fiber one foot per second, which is about the speed of a violinist's bow playing andante.

The guiding philosophy behind the Hectospec, says Fabricant, was to build the most ambitious project they could pull off, measuring redshifts from the greatest number of galaxies while still functioning the first day on the telescope. "You can run the risk of being not ambitious enough and having something no one finds competitive by the time it's finished," he says. "But if the instrument is temperamental during your assigned nights, you're out of luck. You have to be awfully sure the thing is going to work." Because of the tricky nature of the job and the Hectospec's idiosyncratic requirements, Fabricant and his friends couldn't entice any industrial builder to take on the challenge, so they're building the instrument themselves. They expect to have it done by late 1997, which gives them plenty of time to make sure it works before the MMT opens its single enormous eye one year later.

At that point, Geller and Fabricant will start measuring their galaxies—300 at a time, as many as 3,000 a night, and tens of thousands a year. While they're doing it, they'll be competing with a half-dozen other redshift surveys. Some will map fewer galaxies but get on-line sooner. Some will map more galaxies but won't go as deep. They'll all be looking for signs of the beginning of structure, for the time when the universe started to form what Geller calls these beautiful patterns. And this may be why Geller is still in the business. "It's the grandeur and the aesthetics of the problem," she says. "I've seen these beautiful patterns that the universe makes, and I'd like to know how it does it."

Ideas, Hypotheses, and Theories

Space and Time (Article 49)
Extraterrestrial Life (Articles 50–52)
The Big Bang Theory (Articles 53–56)
The Age Paradox (Articles 57–60)

Unit 5 deals with the interpretations of the data astronomers have acquired about space and time, extraterrestrial life, the Big Bang theory, and the age paradox of the universe.

Steven Hawking and Roger Penrose, arguably two of the most brilliant physicists alive today, begin this unit by comparing notes on the nature of time and space in their essay, "The Nature of Space and Time."

The next three articles address the search for extraterrestrial life. Rather than UFOs visiting our planet, the search for extraterrestrial life lies along rather more mundane pursuits: possible organic structures in a Martian meteorite or possible organic material on Europa. The search for extraterrestrial intelligent life is currently being conducted by listening to millions of radio frequencies for signals from other civilizations.

The Big Bang theory is the "big" theory in cosmology. The two articles from *Nature* ("The Best Cosmology There Is" and "Holes in the Big Bang") explain clearly the evidence for and the weaknesses of the Big Bang theory. Both essays are characterized by their honesty and clarity. Critics of science often misunderstand the role that problems, puzzles, and anomalies play in science. Rather than simply being excuses to abandon the theory in question, they serve as avenues for further research and data gathering. In "In Defense of the Big Bang," Neil de Grasse Tyson sets forth the reasons why the theory still remains the most compelling explanation of the origin of the universe. Then, Richard Talcott in "Everything You Wanted to Know about the Big Bang" answers 10 of the most common questions students have about the theory.

The four articles discussing the age paradox that were chosen show how scientists deal with anomalies. The age paradox is this: calculating the ages of the oldest stars in our galaxy produces an age greater than that obtained from standard cosmological models. Since the stars in the universe cannot be older than the universe, something is amiss. "Conflict Over the Age of the Universe," which may be the most difficult article to read in the collection, describes the paradox in detail. Then, the three articles that follow all deal with the two major issues of the paradox—the ages of stars and the Hubble parameter.

Looking Ahead: Challenge Questions

In what ways do Steve Hawking and Roger Penrose (see "The Nature of Space and Time") differ in their views on space and time?

Describe the methods scientists are using to try to discover intelligent extraterrestrial life.

What evidence supports the Big Bang?

What problems or anomalies does the Big Bang theory have?

What is the age paradox? What recent discoveries have helped in solving the paradox?

UNIT 5

Article 49

The Nature of Space and Time

Two relativists present their distinctive views on the universe, its evolution and the impact of quantum theory

Stephen W. Hawking and Roger Penrose

In 1994 Stephen W. Hawking and Roger Penrose gave a series of public lectures on general relativity at the Isaac Newton Institute for Mathematical Sciences at the University of Cambridge. From these lectures, published this year by Princeton University Press as The Nature of Space and Time, SCIENTIFIC AMERICAN has culled excerpts that serve to compare and contrast the perspectives of the two scientists. Although they share a common heritage in physics—Penrose served on Hawking's Ph.D. thesis committee at Cambridge—the lecturers differ in their vision of quantum mechanics and its impact on the evolution of the universe. In particular, Hawking and Penrose disagree on what happens to the information stored in a black hole and on why the beginning of the universe differs from the end.

One of Hawking's major discoveries, made in 1973, was that quantum effects will cause black holes to emit particles. The black hole will evaporate in the process, so that ultimately perhaps nothing of the original mass will be left. But during their formation, black holes swallow a lot of data—the types, properties and configurations of the particles that fall in. Although quantum theory requires that such information must be conserved, what finally happens to it remains a topic of contentious debate. Hawking and Penrose both believe that when a black hole radiates, it loses the information it held. But Hawking insists that the loss is irretrievable, whereas Penrose argues that the loss is balanced by spontaneous measurements of quantum states that introduce information back into the system.

Both scientists agree that a future quantum theory of gravity is needed to describe nature. But they differ in their view of some aspects of this theory. Penrose thinks that even though the fundamental forces of particle physics are symmetric in time—unchanged if time is reversed—quantum gravity will violate time symmetry. The time asymmetry will then explain why in the beginning the universe was so uniform, as evinced by the microwave background radiation left over from the big bang, whereas the end of the universe must be messy.

Penrose attempts to encapsulate this time asymmetry in his Weyl curvature hypothesis. Space-time, as Albert Einstein discovered, is curved by the presence of matter. But space-time can also have some intrinsic bending, a quantity designated by the Weyl curvature. Gravitational waves and black holes, for example, allow space-time to curve even in regions that are empty. In the early universe the Weyl curvature was probably zero, but in a dying universe the large number of black holes, Penrose argues, will give rise to a high Weyl curvature. This property will distinguish the end of the universe from the beginning.

Hawking agrees that the big bang and the final "big crunch" will be different, but he does not subscribe to a time asymmetry in the laws of nature. The underlying reason for the difference, he thinks, is the way in which the universe's evolution is programmed. He postulates a kind of democracy, stating that no point in the universe can be special; therefore, the universe cannot have a boundary. This no-boundary proposal, Hawking claims, explains the uniformity in the microwave background radiation.

The physicists diverge, ultimately, in their interpretation of quantum mechanics. Hawking believes that all a theory has to do is provide predictions that agree with data. Penrose thinks that simply comparing predictions with experiments is not enough to explain reality. He points out that quantum theory requires wave functions to be "superposed," a concept that can lead to absurdities. The scientists thus pick up the threads of the famous debates between Einstein and Niels Bohr on the bizarre implications of quantum theory.
—The Editors

Stephen Hawking on quantum black holes:

The quantum theory of black holes... seems to lead to a new level of unpredictability in physics over and above the usual uncertainty associated with quantum mechanics. This is because black holes appear to have intrinsic entropy and to lose information from our region of the universe. I should say that these claims are controversial: many people working on quantum gravity, including almost all those who entered it from particle physics, would instinctively reject the idea that information about the quantum state of a system could be lost. However, they have had very little success in showing how information can get out of a black hole. Eventually I believe they will be forced to accept my suggestion that it is lost, just as they were forced to agree that black holes radiate, which went against all their preconceptions....

The fact that gravity is attractive means that it will tend to draw the matter in the universe together to form objects like stars and galaxies. These can support themselves for a time against further contraction by thermal pressure, in the case of stars, or by rotation and internal motions, in the case of galaxies. However, eventually the heat or the angular momentum will be carried away and the object will begin to shrink. If the mass is less than about one and a half times that of the Sun, the contraction can be stopped by the *degeneracy pressure* of electrons or neutrons. The object will settle down to be a white dwarf or a neutron star, respectively. However, if the mass is greater than this limit there is nothing that can hold it up

---

DEGENERACY PRESSURE

No two electrons or neutrons can occupy the same quantum state. Thus, when any collection of these particles is squeezed into a small volume, those in the highest quantum states become very energetic. The system then resists further compression, exerting an outward push called degeneracy pressure.

49. Nature of Space and Time

and stop it continuing to contract. Once it has shrunk to a certain critical size the gravitational field at its surface will be so strong that the *light cones* will be bent inward.... You can see that even the outgoing light rays are bent toward each other and so are converging rather than diverging. This means that there is a closed trapped surface....

Thus there must be a region of space-time from which it is not possible to escape to infinity. This region is said to be a black hole. Its boundary is called the event horizon and is a *null surface* formed by the light rays that just fail to get away to infinity....

[A] large amount of information is lost when a body collapses to form a black hole. The collapsing body is described by a very large number of parameters. There are the types of matter and the *multipole moments* of the mass distribution. Yet the black hole that forms is completely independent of the type of matter and rapidly loses all the multipole moments except the first two: the monopole moment, which is the mass, and the dipole moment, which is the angular momentum.

This loss of information didn't really matter in the classical theory. One could say that all the information about the collapsing body was still inside the black hole. It would be very difficult for an observer outside the black hole to determine what the collapsing body was like. However, in the classical theory it was still possible in principle. The observer would never actually lose sight of the collapsing body. Instead it would appear to slow down and get very dim as it approached the event horizon. But the observer could still see what it was made of and how the mass was distributed.

However, quantum theory changed all this. First, the collapsing body would send out only a limited number of photons before it crossed the event horizon. They would be quite insufficient to carry all the information about the collapsing body. This means that in quantum theory there's no way an outside observer can measure the state of the collapsed body. One might not think that this mattered too much, because the information would still be inside the black hole even if one couldn't measure it from the outside. But this is where the second effect of quantum theory on black holes comes in....

[Quantum] theory will cause black holes to radiate and lose mass. It seems that they will eventually disappear completely, taking with them the information inside them. I will give arguments that this information really is lost and doesn't come back in some form. As I will show, this loss of information would introduce a new level of uncertainty into physics over and above the usual uncertainty associated with quantum theory. Unfortunately, unlike Heisenberg's uncertainty principle, this extra level will be rather difficult to confirm experimentally in the case of black holes.

Roger Penrose on quantum theory and space-time:

The great physical theories of the 20th century have been quantum theory, special relativity, general relativity and quantum field theory. These theories are not independent of each other: general relativity was built on special relativity, and quantum field theory has special relativity and quantum theory as inputs.

It has been said that quantum field theory is the most accurate physical theory ever, being accurate to about one part in about 10^{11}. However, I would like to point out that general relativity has, in a certain clear sense, now been tested to be correct to one part in 10^{14} (and this accuracy has apparently been limited merely by the accuracy of clocks on Earth). I am speaking of the Hulse-Taylor binary *pulsar* PSR 1913 + 16, a pair of neutron stars orbiting each other, one of which is a pulsar. General relativity predicts that this orbit will slowly decay (and the period

---

LIGHT CONES
To depict space-time, physicists routinely plot time on a vertical axis and space on a horizontal. In this scheme, light rays emanating from any point in space fan out along the surface of a vertical cone. Because no physical signal can cover more distance in a given time than light can, any signals originating at that point are confined within the volume of the light cone.

NULL SURFACE
A surface in space along which light travels is known as a null surface. The null surface surrounding a black hole, called an event horizon, has the shape of a spherical shell. Nothing that falls inside the event horizon can come back out.

MULTIPOLE MOMENTS
The dynamics of an object can be summarized by determining its multipole moments. Each moment is calculated by dividing an object into tiny elements, multiplying the mass of each element by its distance from the center zero, one or more times, then adding these terms for all the elements. A sphere, for example, has a monopole moment, whereas a dumbbell has a dipole moment, which allows it to acquire angular momentum easily.

PULSARS
Some dying suns collapse into neutron stars, massive objects made entirely of densely packed neutrons. Rapidly rotating neutron stars become pulsars, so called because they emit pulses of electromagnetic radiation at astonishingly regular millisecond-to-second intervals. A pulsar sometimes orbits another neutron star, forming a binary pair.

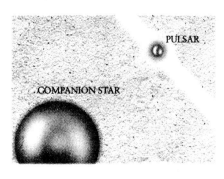

5. IDEAS, HYPOTHESES, AND THEORIES: Space and Time

shorten) because energy is lost through the emission of gravitational waves. This has indeed been observed, and the entire description of the motion...agrees with general relativity (which I am taking to include Newtonian theory) to the remarkable accuracy, noted above, over an accumulated period of 20 years. The discoverers of this system have now rightly been awarded Nobel Prizes for their work. The quantum theorists have always claimed that because of the accuracy of their theory, it should be general relativity that is changed to fit their mold, but I think now that it is quantum field theory that has some catching up to do.

Although these four theories have been remarkably successful, they are not without their problems.... General relativity predicts the existence of space-time *singularities*. In quantum theory there is the "measurement problem"—I shall describe this later. It may be taken that the solution to the various problems of these theories lies in the fact that they are incomplete on their own. For example, it is anticipated by many that quantum field theory might "smear" out the singularities of general relativity in some way....

I should now like to talk about information loss in black holes, which I claim is relevant to this last issue. I agree with nearly all that Stephen had to say on this. But while Stephen regards the information loss due to black holes as an extra uncertainty in physics, above and beyond the uncertainty from quantum theory, I regard it as a "complementary" uncertainty.... It is possible that a little bit of information escapes at the moment of the black hole evaporation...but this tiny information gain will be much smaller than the information loss in the collapse (in what I regard as any reasonable picture of the hole's final disappearance).

If we enclose the system in a vast box, as a thought experiment, we can consider the phase-space evolution of matter inside the box. In the region of *phase space* corresponding to situations in which a black hole is present, trajectories of physical evolution will converge and volumes following these trajectories will shrink. This is due to the information lost into the singularity in the black hole. This shrinking is in direct contradiction to the theorem in classical mechanics, called Liouville's Theorem, which says that volumes in phase space remain constant.... Thus a black hole space-time violates this conservation. However, in my picture, this loss of phase-space volume is balanced by a process of "spontaneous" quantum measurement in which information is gained and phase-space volumes increase. This is why I regard the uncertainty due to information loss in black holes as being "complementary" to the uncertainty in quantum theory: one is the other side of the coin to the other....

[Let] us consider the *Schrödinger's cat* thought experiment. It describes the plight of a cat in a box, where (let us say) a photon is emitted which encounters a half-silvered mirror, and the transmitted part of the photon's wave function encounters a detector which, if it detects the photon, automatically fires a gun, killing the cat. If it fails to detect the photon, then the cat is alive and well. (I know Stephen does not approve of mistreating cats, even in a thought experiment!) The wave function of the system is a superposition of these two possibilities.... But why does our perception not allow us to perceive macroscopic superpositions, of states such as these, and not just the macroscopic alternatives "cat is dead" and "cat is alive"?...

I am suggesting that something goes wrong with superpositions of the alternative space-time geometries that would occur when general relativity begins to become involved. Perhaps a superposition of two different geometries is unstable and decays into one of the two alternatives. For example, the geometries

SINGULARITIES
According to general relativity, under certain extreme conditions some regions of space-time develop infinitely large curvatures, thus becoming singularities where the normal laws of physics break down. Black holes, for example, should contain singularities hidden inside the event horizon.

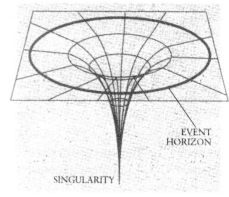

PHASE SPACE
A phase-space diagram is a mathematical volume of many dimensions formed when coordinate axes are assigned to each of the distance and momentum values of each particle. The motion of a group of particles can then be represented by a moving element of volume in phase space.

SCHRÖDINGER'S CAT
Penrose invokes a thought experiment originally invented by Einstein and used by Erwin Schrödinger to study the conceptual knots tied by wave functions. Prior to a measurement, a system is assumed to be in a "superposition" of quantum states or waves, so that the value of, say, the momentum is uncertain. After a measurement, the value of a quantity becomes known, and the system suddenly assumes the one state that corresponds to the result. The significance of the original superposition and the process by which the system "collapses" into one state are highlighted by Schrödinger's cat paradox.

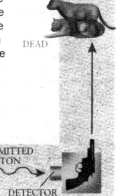
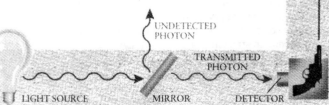

49. Nature of Space and Time

PLANCK SCALE

The Planck scale is an unattainably small distance—related, by quantum mechanics, to an impossibly small time span and high energy—that emerges when the fundamental constants for gravitational attraction, the velocity of light and quantum mechanics are appropriately combined. The scale represents the distance or energy at which current concepts of space, time and matter break down, and a future theory, quantum gravity, presumably takes over.

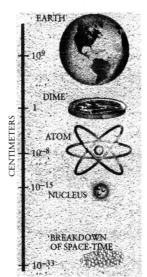

CPT (CHARGE-PARITY-TIME) INVARIANCE

This powerful principle requires that theories describing particles must remain true even when the charge, parity (or handedness) and time simultaneously reverse. In other words, the behavior of a negatively charged electron with clockwise spin moving forward in time must be identical to that of a positively charged positron with anticlockwise spin moving backward in time.

WEYL TENSOR

The curvature of space-time has two components. One derives from the presence of matter in space-time; the other, recognized by the Ger-

man mathematician Hermann Weyl, occurs even in the absence of matter. The mathematical quantity that describes this curvature is called the Weyl tensor.

NO-BOUNDARY PROPOSAL

Hawking suggests that the evolution of the universe is explained by the no-boundary proposal, put forth in 1983 by him and James B. Hartle of the University of California at Santa Barbara. The idea that the universe has no boundary places constraints on how the equations of cosmology are solved. Hawking believes these conditions will lead to the ends of the universe being different, thereby determining the direction of time's arrow.

might be the space-times of a live cat, or a dead one. I call this decay into one or the other alternative objective reduction, which I like as a name because it has an appropriately nice acronym (OR). How does the Planck length 10^{-33} centimeter relate to this? Nature's criterion for determining when two geometries are significantly different would depend upon the *Planck scale,* and this fixes the timescale in which the reduction into different alternatives occurs.

Hawking on quantum cosmology:

I will end this lecture on a topic on which Roger and I have very different views—the arrow of time. There is a very clear distinction between the forward and the backward directions of time in our region of the universe. One only has to watch a film being run backward to see the difference. Instead of cups falling off tables and getting broken, they would mend themselves and jump back on the table. If only real life were like that.

The local laws that physical fields obey are time symmetric, or more precisely, *CPT (charge-parity-time) invariant.* Thus, the observed difference between the past and the future must come from the boundary conditions of the universe. Let us take it that the universe is spatially closed and that it expands to a maximum size and collapses again. As Roger has emphasized, the universe will be very different at the two ends of this history. At what we call the beginning of the universe, it seems to have been very smooth and regular. However, when it collapses again, we expect it to be very disordered and irregular. Because there are so many more disordered configurations than ordered ones, this means that the initial conditions would have had to be chosen incredibly precisely.

It seems, therefore, that there must be different boundary conditions at the two ends of time. Roger's proposal is that the *Weyl tensor* should vanish at one end of time but not the other. The Weyl tensor is that part of the curvature of space-time that is not locally determined by the matter through the Einstein equations. It would have been small in the smooth, ordered early stages but large in the collapsing universe. Thus, this proposal would distinguish the two ends of time and so might explain the arrow of time.

I think Roger's proposal is Weyl in more than one sense of the word. First, it is not CPT invariant. Roger sees this as a virtue, but I feel one should hang on to symmetries unless there are compelling reasons to give them up. Second, if the Weyl tensor had been exactly zero in the early universe, it would have been exactly homogeneous and isotropic and would have remained so for all time. Roger's Weyl hypothesis could not explain the fluctuations in the background nor the perturbations that give rise to galaxies and bodies like ourselves.

Despite all this, I think Roger has put his finger on an important difference between the two ends of time. But the fact that the Weyl tensor was small at one end should not be imposed as an ad hoc boundary condition but should be deduced from a more fundamental principle, the *no-boundary proposal....*

How can the two ends of time be different? Why should perturbations be small at one end but not the other? The reason is there are two possible complex solutions of the field equations.... Obviously, one solution corresponds to one end of time and the other to the other.... At one end, the universe was very smooth and the Weyl tensor was very small. It could not, however, be exactly zero, for that would have been a violation of the uncertainty principle. Instead there would have been small fluctuations that later grew into galaxies and bodies like us. By contrast, the universe would have been very irregular and chaotic at the other end of time with a Weyl tensor that was typically large. This would explain the observed arrow of time and why cups fall off tables and break rather than mend themselves and jump back on.

5. IDEAS, HYPOTHESES, AND THEORIES: Space and Time

Penrose on quantum cosmology:

From what I understand of Stephen's position, I don't think that our disagreement is very great on this point [the *Weyl curvature hypothesis*]. For an initial singularity the Weyl curvature is approximately zero.... Stephen argued that there must be small quantum fluctuations in the initial state and thus pointed out that the hypothesis that the initial Weyl curvature is zero at the initial singularity is classical, and there is certainly some flexibility as to the precise statement of the hypothesis. Small perturbations are acceptable from my point of view, certainly in the quantum regime. We just need something to constrain it very near to zero....

Maybe the no-boundary proposal of [James B.] Hartle and Hawking is a good candidate for the structure of the *initial* state. However, it seems to me that we need something very different to cope with the *final* state. In particular, a theory that explains the structure of singularities would have to violate [CPT and other symmetries] in order that something of the nature of the Weyl curvature hypothesis can arise. This failure of time-symmetry might be quite subtle; it would have to be implicit in the rules of that theory which goes beyond quantum mechanics.

Hawking on physics and reality:

These lectures have shown very clearly the difference between Roger and me. He's a Platonist and I'm a positivist. He's worried that Schrödinger's cat is in a quantum state, where it is half alive and half dead. He feels that can't correspond to reality. But that doesn't bother me. I don't demand that a theory correspond to reality because I don't know what it is. Reality is not a quality you can test with litmus paper. All I'm concerned with is that the theory should predict the results of measurements. Quantum theory does this very successfully....

Roger feels that...the collapse of the wave function introduces CPT violation into physics. He sees such violations at work in at least two situations: cosmology and black holes. I agree that we may introduce time asymmetry in the way we ask questions about observations. But I totally reject the idea that there is some physical process that corresponds to the reduction of the wave function or that this has anything to do with quantum gravity or consciousness. That sounds like magic to me, not science.

Penrose on physics and reality:

Quantum mechanics has only been around for 75 years. This is not very long if one compares it, for example, with Newton's theory of gravity. Therefore it wouldn't surprise me if quantum mechanics will have to be modified for very macroscopic objects.

At the beginning of this debate, Stephen said that he thinks that he is a positivist, whereas I am a Platonist. I am happy with him being a positivist, but I think that the crucial point here is, rather, that I am a realist. Also, if one compares this debate with the famous debate of *Bohr and Einstein,* some 70 years ago, I should think that Stephen plays the role of Bohr, whereas I play Einstein's role! For Einstein argued that there should exist something like a real world, not necessarily represented by a wave function, whereas Bohr stressed that the wave function doesn't describe a "real" microworld but only "knowledge" useful for making predictions.

Bohr was perceived to have won the argument. In fact, according to the recent biography of Einstein by [Abraham] Pais, Einstein might as well have gone fishing from 1925 onward. Indeed, it is true that he didn't make many big advances, even though his penetrating criticisms were very useful. I believe that the reason why Einstein didn't continue to make big advances in quantum theory was that a crucial ingredient was missing from quantum theory. This missing ingredient was Stephen's discovery, 50 years later, of black hole radiation. It is this information loss, connected with black hole radiation, which provides the new twist.

WEYL CURVATURE HYPOTHESIS
The universe just after the big bang has a small Weyl curvature, whereas near the end of time it has a large Weyl curvature. Penrose suggests that this curvature, therefore, accounts for the direction in which the arrow of time points.

NIELS BOHR

ALBERT EINSTEIN

The Editors acknowledge the assistance of Gary T. Horowitz of the University of California at Santa Barbara.

Searching for Alien Earth

Finding another planet just like Earth orbiting some distant sun may not be quite as hard as is sounds. *John Davies* looks at the missions on the drawing board.

IS the Earth alone in the Universe or are there vast numbers of planets orbiting other stars, occupied perhaps by alien civilisations? Earth-bound astronomers have never succeeded in detecting signals broadcast from outer space, and the "flying saucer" image of these searches makes them unpopular with budget-conscious politicians, who recently cancelled NASA's search for extraterrestrial intelligence (SETI) programme. But another approach may prove more fruitful—looking for Earthlike planets rather than aliens. Next week, astronomers and both European and American space agency officials are gathering in Boulder, Colorado, to thrash out their ideas for how to put this into practice.

'Is the Earth alone in the Universe or are there vast numbers of planets orbiting other stars?'

NASA has been interested in the search for Earthlike planets for some time, and recently initiated a programme called ASEPS (Astronomical Studies of Extrasolar Planetary Systems). [In April 1995], ASEPS hosted a workshop ... at the Jet Propulsion Laboratory in California to work out the best strategy for detecting planets around other stars. Meanwhile in Europe, ESA's Horizon 2000+ consultation process has identified planet detection as a priority and supporters of two rival European missions will be at Boulder to present their ideas.

Beyond the obvious political problems of who works with whom on what, and who picks up the tab, there are significant technical challenges ahead. One of the biggest is that even large planets are very faint compared with the star they orbit. From about 13 light years away—next door by astronomical standards—the Earth would appear as a point of light only 0.25 seconds of arc from the Sun and almost 10 billion times fainter than [the Sun]. Even detecting a planet the size of Jupiter, the largest planet in our Solar System with a mass more than three hundred times that of Earth, would be difficult because tiny imperfections in telescope mirrors scatter starlight into a halo that can swamp the feeble light from nearby planets.

Even a perfect telescope cannot concentrate light from a distant star into a single point. Instead, diffraction spreads the light into a series of bright and dark rings and the amount of light in these rings, called the diffraction pattern, can easily drown out the image of a nearby planet. Worse still, distortions caused by the Earth's atmosphere usually limit the resolution of star images to about 0.5 arc seconds. Until recently this has precluded attempts to image extrasolar planets from Earth, but according to Roger Angel, an astronomer from the University of Arizona, adaptive optics ("New eyes for an ageing star", New Scientist, 14 January) could soon change this.

Angel believes that fitting adaptive optics to large telescopes, with mirrors at least 6 metres across, will make it possible to detect Jupiter-sized planets orbiting around nearby stars. If the planet was at its widest separation from the star, as seen from the Earth, he says it might take only a few hours of observing time to pick it out.

In the meantime, several different groups have been trying more indirect approaches, looking at the behaviour of the central star for hints of a lurking planet. One such approach, called astrometry, relies on the fact that very large planets—even bigger than Jupiter—should exert a detectable gravitational pull on their star. Jupiter is a cold gaseous world unable to support life, and other Jupiter-sized planets would presumably have formed by similar processes and have similar characteristics. But the presence of a Jupiterlike planet around another star may also indicate the existence of smaller, more hospitable planets in the same system.

Wobbly stars

The gravitational effect of a Jupiter-sized planet would be detectable because planets do not actually orbit around a star. Instead, the whole system revolves around the combined centre of mass, or barycentre. If the planets are small, the barycentre will more or less coincide with the centre of the star. But with one (or more) massive planets, the barycentre will be some distance from the star's own centre. As the star circles around the barycentre, there is a telltale "wobble" in its position, with a period equal to the "year" of the massive planet—12 years in the case of our own Solar System which is dominated by the Sun and Jupiter.

The only astrometric search today is being carried out by a small team led by George Gatewood of the Allegheny Observatory of the University of Pittsburgh. Almost every clear night, Gatewood and his colleagues use an instrument called a Multichannel Astrometric Photometer (MAP) to measure the position of about twenty nearby stars. This instrument moves a grating of clear and opaque lines along the focal plane of the telescope. The stars in the viewing field blink on and off as the lines from the grating pass by, and the time sequence of the blinking depends on the relative positions of the stars in the field. If one of the stars has moved since the last observation, the blinking pattern will have changed slightly. Since the blinking, and hence the star positions, can be measured very accurately, it is possible to detect small movements of the star and to search for the wobbles caused by large planets.

Even though Gatewood has been observing for seven years he has seen no evidence for any planets. The problem is not the quality of the data—by now, he should have detected any planets with half the mass of Jupiter around nearby Barnard's star or objects twice Jupiter's size around a number of other stars. "It seems that very large planets are rare," he says. "It was once thought that Jupiter was a run-of-the-mill giant planet, but perhaps Jupiter is a giant giant planet and not typical at all." Gatewood is now trying to develop a more advanced version of the MAP with ten times better resolution; this should be able to detect planets the size of Uranus and Neptune, both around fifteen times the mass of the Earth, which he hopes are more common than "Jupiters".

Meanwhile, other astronomers are trying a different approach—looking for the motion of a star along rather than across the line of sight. If a giant planet is orbiting in a plane that is more or less edge on as seen from the Earth, the movement of the central star around the barycentre will cause the star to move towards and away from the observer over a period of years. This produces small but measurable changes in the frequency of the star's light.

The effect is tiny: Jupiter causes our own Sun to move backwards and forwards with a radial velocity of only 12 metres per second, introducing a shift in the frequency of light of less than 3 parts in 10 million. But although most ways of studying the radial velocities of stars are only accurate up to a few hundred metres per second, various researchers have produced more accurate systems to search for planets.

Bruce Campbell of the University of Victoria in Canada and his colleagues recently completed a 12-year search using the 3.6 metre Canada-France-Hawaii Telescope on Mauna Kea, Hawaii. They placed a cell of hydrogen fluoride gas in the telescope beam to calibrate the spectrum from the star. The idea was that instrumental variations, which would otherwise limit the accuracy, affected both the starlight and the reference beam and cancelled out, producing an impressive accuracy of 13 metres per second.

Similar techniques have been used by other teams, including Bill Cochran at the University of Texas McDonald Observatory, who uses an iodine cell as a calibrator. These other searches have been in progress for about seven years but so far have not detected any planets, even though they should by now have been able to find any really massive objects, the "Super-Jupiters", around the stars observed.

Spectroscopic techniques are improving all the time. Geoff Marcy of San Francisco State University has equipment that has recently been upgraded, and is beginning to deliver accuracies of between 2 and 4 metres per second. This would certainly be good enough to detect a planet the size of Jupiter, and perhaps even slightly smaller.

The lack of success of any of these programmes to date makes it clear that finding even a Jupiter-sized planet is no easy task. Finding an Earth-sized one will be even harder. According to Angel, even with adaptive optics, detecting Earth-sized planets is unlikely to be possible from the ground and so a move into space is probably necessary. Not that it would be exactly easy there: even above the distorting effects of the Earth's atmosphere, taking a direct image of any Earthlike planet would need a space telescope with a mirror about twice the size and ten times smoother than the Hubble Space Telescope. This would be very expensive, and is unlikely to be built for many years, if ever.

Planets in transit

A much cheaper alternative is a mission called FRESIP (Frequency of Earth-Sized Planets), put forward late last year by Bill Borucki of NASA's Ames Research Center in California as part of NASA's "Discovery" programme. FRESIP would monitor the brightness of Sunlike stars to search for changes caused by planets as they pass across them. The effect is small—the Sun loses only 0.01 per cent of its brightness as the Earth passes in front of it—but the diminution stays the same for several hours, lasts for a fixed time and happens once every year, which should make it possible to pick it out against a background of random variations in the star's brightness caused by flares or starspots. Borucki says that three such repeatable events will be enough to establish the existence of an Earthlike planet and should allow researchers to predict the next transit, which would prove the issue.

Transits can be detected only if the observer is roughly level with the orbital plane of the star's planets and from geometric considerations the likelihood of this is only about 1 per cent. Since there is no way to know in advance which stars to start with, the best plan is to observe a large number of stars in the hope of picking some at least that have planets in the right plane, and catching the planets in the act of passing in front of their star. These sorts of observations require round-the-clock monitoring and are best done from space, where there is no interference from daylight, or bad weather.

Borucki's FRESIP would use a 1-metre wide-angle telescope with a field of view of about 10° and a massive array of very sensitive detectors. It would be placed in orbit around one of the Earth-Sun Lagrangian points, regions in which the gravitational influences of the Sun and Earth virtually cancel each other out and a satellite can stay in position using very little fuel. From here, FRESIP could stare at the same area of sky all the year round, pointing in a direction chosen to include about 5000 stars like our own Sun. If planetary systems are very common, FRESIP should detect about 50 transits per year.

While FRESIP was described by the 100-strong peer-review committee assessing it earlier this year as the only feasible technique to detect Earthlike planets, there were some doubts about whether it could be built within the Discovery programme budgetary guidelines of less than $150 million per mission. Two independent panels are now reviewing the estimates and if FRESIP is approved as part of the Discovery programme next year, it would be launched around the turn of the century.

Because of their much smaller mass and hence smaller influence on their parent star, detecting Earthlike planets by astrometry would require instruments that are precise to about a tenth of a microarc second. This could be achieved using interferometry, where the beams of light from two telescopes some distance apart are combined to mimic the resolving power of a single telescope with a mirror as wide as the distance between the two telescopes.

At NASA's Jet Propulsion Laboratory in Pasadena, Michael Shao and his colleagues are building an infrared interferometer that can be used for very high precision astrometry. They want to detect extrasolar planets about the size of Uranus and Neptune around stars up to 40 light years away. Shao's interferometer—two 40 centimetre telescopes 100 metres apart—is being funded by NASA to demonstrate the technology.

NASA is also making arrangements to use the twin 10-metre diameter Keck telescopes on Mauna Kea, Hawaii, for planet searches. One approach is to build a number of small "outrigger" telescopes near to the main Keck telescopes and run in parallel with one of the main Kecks as a multi-element interferometer. Shao believes that with such a set-up accuracies of a few micro arc seconds and detection of planets somewhere between the sizes of Earth and Neptune should be possible.

Pulsar planets

THE first extrasolar planets have already been found, but not around a Sunlike star. In 1992, Alexander Wolszczan of Pennsylvania State University reported in *Nature* that at least two planets were orbiting the millisecond pulsar PSR B1257+12 (*New Scientist*, "Puzzle of the pulsar planets", 18 July 1992). Wolszczan suspected the existence of the planets when very accurate timing of the pulses from this collapsed star revealed systematic inconsistencies, which could not be explained by changes in the pulsar itself. Instead, he reasoned that they were caused by minute gravitational tugs from small planets orbiting the pulsar.

These planets are most unlikely to be habitable and were probably formed from the wreckage of a companion to PSR B1257+12, destroyed many millions of years ago.

'It will still leave open the question of whether the planets are habitable, or indeed inhabited'

The final step, detection of an Earth-sized planet by astrometry, will be very challenging. Subtle complicating factors, like the effects of starspots on the light from the target star, may introduce errors that are difficult or impossible to remove. Even if an astrometric programme eventually succeeds, it will still leave open the question of whether the planets are habitable, or whether they are inhabited. DARWIN, a major mission proposed to ESA late in 1993 by a team led by Alain Léger of the University of Paris, would look for signs of life using a space-based interferometer made up of two or more infrared telescopes between 1 and 2 metres in diameter and about 10–30 metres apart. A mission of this general type is also being considered as part of the ASEPS deliberations.

Sharper contrast

DARWIN's equipment would observe in the infrared region of the spectrum for a number of reasons. First, the contrast between a star and a planet is greater for infrared than for visible light. This is because Sunlike stars, with temperatures of about 5000 K, emit most of their light in the visible region of the spectrum whereas planets, with temperatures of only a few 100 K or less, have

5. IDEAS, HYPOTHESES, AND THEORIES: Extraterrestrial Life

their peak emission in the infrared. Although the greater size and higher temperature of a star makes it much brighter than a planet at all wavelengths, in the infrared end of the spectrum, the difference in brightness is less, so the planet is easier to spot. At a wavelength of about 10 micrometres the Earth is the brightest planet in the solar system, although it is still about 10 million times fainter than the Sun. The DARWIN interferometer would be arranged so that starlight in the various beams would interfere destructively and largely cancel each other out, making it easier to detect the faint signal from an Earth-like planet.

> **'GAIA should detect planets, but the technically more challenging DARWIN could detect life'**

Since oxygen has no easily detectable spectral lines in the infrared, the planet's signal would be examined for a strong absorption band from ozone at a wavelength of 9.6 microns, right in the wavelength range in which an Earth-like planet should be relatively bright compared with its central star. As with the Earth, the presence of an ozone layer would indicate considerable amounts of oxygen in the atmosphere below. Since oxygen is so reactive that it is normally removed quickly from the atmosphere, its presence would suggest that the oxygen was being replaced by biological activity akin to photosynthesis, and that the planet supported life.

DARWIN is one of two planet-detecting projects being considered by ESA. Between now and the year 2000 DARWIN, and its astrometric rival GAIA, which would cost about the same as DARWIN, will be studied in more detail and eventually one will be adopted as a cornerstone of the ESA's Horizon 2000+ long-range space programme. It will be a difficult choice. GAIA should be able to detect planets, but the technically more challenging DARWIN could detect life as well.

Actually seeing what a distant planet looked like would be the most difficult task of all—even the projects that detect the planet directly would see it only as a point of light. It would require space-based interferometers with telescopes separated by vast distances to mimic huge telescope mirrors. This could not be done with a single spacecraft, but Shao believes that it will soon be possible to use different spacecraft as components of a giant interferometer.

He envisages three spacecraft in solar orbit, placed at the corners of an equilateral triangle with sides 1000 kilometres long. Two would be telescopes, the third a platform at which the beams would be combined and analysed. Lasers would be used to make extremely precise measurements of the distance between the three elements so that the beams could be combined correctly. Such a project would form the final stage of ASEPS and would deliver the first detailed image of a world beyond the Sun.

It will be a picture well worth waiting for.

John Davies *is an astronomer working on the UK Infrared Telescope (UKIRT) in Hawaii.*

Is Anybody (Like Us) Out There?

By Neil de Grasse Tyson

Neil de Grasse Tyson, an astrophysicist, is the Frederick P. Rose Director of New York City's Hayden Planetarium and a research scientist at Princeton University.

The recent discovery of about half a dozen planets around stars other than the Sun has triggered tremendous public interest. Attention was generated not so much by the discovery of extrasolar planets but by the prospect of their hosting intelligent life. In any case, the media frenzy that followed was somewhat out of proportion to the events.

Why? Because planets cannot be all that rare in the universe if the Sun happens to have nine of them. Also, the newly discovered planets are all oversized, gaseous giants that resemble Jupiter, which means they have no convenient surface upon which life as we know it could exist. And even if the planets were teeming with buoyant aliens, the odds against these life forms being intelligent are astronomical.

Ordinarily, there is no riskier step that a scientist (or anyone) can take than to make a sweeping generalization from just one example. At the moment, life on Earth is the only known life in the universe, but compelling arguments suggest that we are not alone. Indeed, most astrophysicists accept the probability of life elsewhere. The reasoning is easy: if our solar system is not unusual, then the number of planets in the universe would, for example, outnumber the sum of all sounds and words ever uttered by every human who has ever lived. To declare that Earth must be the only planet in the universe with life would be inexcusably bigheaded of us.

Many generations of thinkers, both religious and scientific, have been led astray by anthropocentric assumptions and simple ignorance. In the absence of dogma and data, it is safer to be guided by the notion that we are not special, which is generally known as the Copernican principle. It was the Polish astronomer Nicholas Copernicus, who, in the mid-1500s, put the Sun back in the middle of our solar system where it belongs. In spite of a third century B.C. account of a Sun-centered universe (proposed by the Greek philosopher Aristarchus), the Earth-centered universe has been by far the most popular view for most of the past 2,000 years. In the West, it was codified by the teachings of Aristotle and Ptolemy and the preachings of the Roman Catholic Church, and the geocentric theory was generally accepted. That Earth was the center of all motion was self-evident: it not only looked that way, but God surely made it so.

The Copernican principle comes with no guarantees that it will guide us correctly for all scientific discoveries to come. But it has revealed itself in our humble realizations that Earth is not in the center of the solar system, the solar system is not in the center of the Milky Way galaxy, and the Milky Way galaxy is not in the center of the universe. And in case you are one of those people who think that the edge may be a special place, we are not at the edge of anything either.

A wise contemporary posture would be to assume that life on Earth is not immune to the Copernican principle. How then can the appearance or the chemistry of life on Earth provide clues to what life might be like elsewhere in the universe?

I do not know whether biologists walk around every day awestruck by the diversity of life. I certainly do. On our planet, there coexist (among countless other life forms) algae, beetles, sponges, jellyfish, snakes, condors, and giant sequoias. Imagine these seven living organisms lined up next to one another in size-place. If you didn't know better, you would be hard pressed to believe that they all came from the same universe, much less the same planet.

Given the diversity of life on Earth, one might expect diversity among Hol-

5. IDEAS, HYPOTHESES, AND THEORIES: Extraterrestrial Life

lywood aliens. But I am consistently amazed by the film industry's lack of creativity. With a few notable exceptions—such as life forms in *The Blob* (1958) and in *2001: A Space Odyssey* (1968)—Hollywood's aliens look remarkably humanoid. No matter how ugly (or cute) they are, nearly all of them have two eyes, a nose, a mouth, two ears, a neck, shoulders, arms, hands, fingers, a torso, two legs, two feet—and they can walk. From an anatomical view, these creatures are practically indistinguishable from humans, yet they are supposed to have come from another planet. If anything is certain, it is that life elsewhere in the universe, intelligent or otherwise, will look at least as exotic as some of Earth's own life forms.

The chemical composition of Earth-based life is primarily derived from a select few ingredients. The elements hydrogen, oxygen, and carbon account for more than 95 percent of the atoms in the human body and all other known life. Of the three, carbon has the chemical structure that allows it to bond readily and strongly with itself and with many other elements in many different ways—which is why we are considered to be carbon-based life, and why the study of molecules that contain carbon is generally known as "organic" chemistry. Curiously, the study of life elsewhere in the universe is known as exobiology, one of the few disciplines that attempt to function in the complete absence of firsthand data.

Is life chemically special? The Copernican principle suggests that it probably isn't. Aliens need not look like us to resemble us in more fundamental ways. Consider that the four most common elements in the universe are hydrogen, helium, carbon, and oxygen. Helium is inert. So the three most abundant, chemically active ingredients in the cosmos are also the top three ingredients in life on Earth. For this reason, you can bet that if life is found on another planet, it will be made of a similar mix of elements. Conversely, if life on Earth were composed primarily of molybdenum, bismuth, and plutonium, then we would have excellent reason to suspect that we were something special in the universe.

Appealing once again to the Copernican principle, we can assume that an alien organism is not likely to be ridiculously large compared with life as we know it. There are cogent structural reasons why you would not expect to find a life form the size of the Empire State Building strutting around a planet. Even if we ignore these engineering limitations of biological matter, we approach another, more fundamental limit. If we assume that an alien has control of its own appendages, or more generally, if we assume the organism functions coherently as a system, then its size would ultimately be constrained by its ability to send signals within itself at the speed of light—the fastest allowable speed in the universe. For an admittedly extreme example, if an organism were as big as the entire solar system (about ten light-hours across), and if it wanted to scratch its head, then this simple act would take no less than ten hours to accomplish. Subslothlike behavior such as this would be evolutionarily self-limiting, because the time since the beginning of the universe may be insufficient for the creature to have evolved from smaller forms.

How about intelligence? When Hollywood aliens manage to visit Earth, one might expect them to be remarkably smart. But I know of some that should have been embarrassed by their stupidity. During a four-hour car trip from Boston to New York City, I surfed the FM dial, and I came upon a radio play in progress that, as best as I could determine, was about evil aliens that were terrorizing earthlings. Apparently, they needed hydrogen atoms to survive so they kept swooping down to Earth to suck up its oceans and extract the hydrogen from all the H_2O molecules. Now those were some dumb aliens. They must not have been looking at other planets en route to Earth because Jupiter, for example, contains more than 200 times the entire mass of Earth in pure hydrogen. I guess nobody told them that more than 90 percent of all atoms in the universe are hydrogen.

Then there were the aliens in the 1977 film *Close Encounters of the Third Kind,* who, in advance of their arrival, beamed to Earth a mysterious sequence of numbers that were eventually decoded by the earthlings to be the latitude and longitude of their upcoming landing site. But Earth's longitude has a completely arbitrary starting point—the prime meridian—which passes through Greenwich, England, by international agreement. And both longitude and latitude are measured in unnatural units we call degrees, 360 of which are in a circle. Armed with this much knowledge of human culture, it seems to me that the aliens could have just learned English and beamed the message, "We're going to land a little bit to the side of Devil's Tower National Monument in Wyoming. And because we're arriving in a flying saucer, we won't need runway lights."

And don't get me started on this summer's blockbuster, *Independence Day.* Actually, I find nothing particularly offensive about evil aliens. There would be no science-fiction film industry

> To declare that Earth is the only planet in the universe with life would be inexcusably bigheaded of us.

> Two excellent places to search for life are in Mars's dried riverbeds and under the ice on Jupiter's moon Europa.

51. Is Anybody (Like Us) Out There?

without them. The aliens in *Independence Day* are definitely evil. They look like a genetic cross between a Portuguese man-of-war, a hammerhead shark, and a human being. While more creatively conceived than most Hollywood aliens, why are their flying saucers equipped with upholstered, high-back chairs with armrests?

I'm glad that, in the end, the humans win. We conquer the *Independence Day* aliens by having a Macintosh laptop computer upload a software virus to the mothership (which happens to be one-fifth the mass of the Moon), thus disarming its protective force field. I don't know about you, but I have trouble just uploading files to other computers within my own department, especially when the operating systems are different. There is only one solution: the entire defense system for the alien mothership must have been powered by the same release of Apple Computer's system software as the laptop computer that delivered the virus.

Let us assume, for the sake of argument, that humans are the only species on Earth to have evolved high-level intelligence. (I mean no disrespect to other big-brained mammals. While most of them cannot do astrophysics, my conclusions are not substantially altered if you wish to include them.) If life on Earth offers any measure of life elsewhere in the universe, then intelligence must be rare. By some estimates, there have been more than ten billion species in the history of life on Earth. It follows that, among all extraterrestrial life forms, we might expect no better than about one in ten billion to be as intelligent as we are—not to mention the odds against the intelligent life having an advanced technology *and* a desire to communicate through the vast distances of interstellar space.

On the chance that such a civilization exists, radio waves would be the communication band of choice because of their ability to traverse the galaxy unimpeded by interstellar gas and dust clouds. But we humans have understood the electromagnetic spectrum for less than a century. More depressingly put, had aliens been trying to send radio signals to earthlings for most of human history, we would have been incapable of receiving them. For all we know, the aliens may have tried to get in touch and have unwittingly concluded that there is no intelligent life on Earth. They would now be looking elsewhere. A more humbling possibility is that aliens did become aware of the technologically proficient species that now inhabits Earth, and drew the same conclusion.

Our Copernican perspective for life-on-Earth, intelligent or otherwise, requires us to presume that liquid water is a prerequisite to life elsewhere. To support life, a planet cannot orbit its host star too closely, or else the temperature would be too high and the planet's water content would vaporize. Also, the orbit should not be too far away, or else the temperature would be too low and the planet's water content would freeze. In other words, conditions on the planet must allow the temperature to stay within the 180°F range of liquid water. As in the three-bowls-of-food scene in *Goldilocks and the Three Bears,* the temperature has to be just right. When I was interviewed about this subject recently on a syndicated radio talk show, the host commented, "Clearly, what you should be looking for is a planet made of porridge!"

While distance from the host planet is an important factor for the existence of life as we know it, a planet's ability to trap stellar radiation matters too. Venus is a textbook example of this "greenhouse" phenomenon. Any visible sunlight that manages to pass through its thick atmosphere of carbon dioxide gets absorbed by Venus's surface and then reradiated in the infrared part of the spectrum. The infrared, in turn, gets trapped by the atmosphere. The unpleasant consequence is an air temperature that hovers at about 900°F, which is much hotter than we would expect, given Venus's distance from the Sun. At this temperature, lead would swiftly become molten.

The discovery of simple, unintelligent life forms elsewhere in the universe (or evidence that they once existed) would be far more likely and, for me, only slightly less exciting than the discovery of intelligent life. Two excellent, nearby places to look are the dried riverbeds of Mars (where there may be fossil evidence of life that thrived when waters flowed) and the subsurface oceans that are theorized to exist under the frozen ice layers of Jupiter's moon Europa. Once again, the promise of liquid water leads our search.

Other commonly invoked prerequisites for the evolution of life in the uni-

> For all we know, aliens may have concluded that there is no intelligent life on Earth.

verse involve a planet in a stable, nearly circular orbit around a single star. With binary and multiple star systems, which make up over half of all stars in the galaxy, orbits tend to be strongly elongated and chaotic, which induces extreme temperature swings that would undermine the evolution of stable life forms. We also require sufficient time for evolution to run its course. High-mass stars are so short-lived (a few million years) that life on an Earthlike planet in orbit around one of them would never have a chance to evolve.

The set of conditions to support life as we know it is loosely quantified though what is known as the Drake equation, named for the American astronomer Frank Drake. The Drake equation is more accurately viewed as a fertile idea rather than as a rigorous statement of how the physical universe works. It separates the overall probability of finding life in the galaxy into a set of simpler probabilities that correspond

5. IDEAS, HYPOTHESES, AND THEORIES: Extraterrestrial Life

to our preconceived notions of suitable cosmic conditions. In the end, after you argue with your colleagues about the value of each probability term in the equation, you are left with an estimate for the total number of intelligent, technologically proficient civilizations in the galaxy. Depending on your bias-level—and your knowledge of biology, chemistry, celestial mechanics, and astrophysics—you may use it to estimate from at least one (ours) up to millions of civilizations in the Milky Way alone.

If we consider the possibility that we may rank as primitive among the universe's technologically competent life forms—however rare they may be—then the best we can do is keep alert for signals sent by others, because it is far more expensive to send rather than receive them. Presumably, an advanced civilization would have easy access to an abundant source of energy, such as its host star. These are the civilizations that would be more likely to send rather than receive.

The search for extraterrestrial intelligence (affectionately known by its acronym, SETI) has taken many forms. The most advanced efforts today use a cleverly designed electronic detector that monitors, in its latest version, billions of radio channels in search of a signal that might rise above the cosmic noise. The discovery of extraterrestrial intelligence, if and when it happens, will impart a change in human self-perception that may be impossible to anticipate. My only hope is that every other civilization isn't doing exactly what we are doing because then everybody would be listening, nobody would be receiving, and we would collectively conclude that there is no other intelligent life in the universe.

When E.T. Calls Us

A message from an extraterrestrial civilization could truly shake the world.

by Seth Shostak

WITHOUT WARNING, A DRAMA of endless consequence could begin in the cramped, cluttered control room of a radio telescope in West Virginia. Here, where a hundred tons of steel face off against the pinpoint glints of the night sky, a back-burner experiment could change the world.

It would begin slowly, perhaps with a lone astronomer fighting to stay awake during the wee hours of the night. Seated in front of a bank of computer displays, she scribbles routine notes in a logbook. On the screens, blocks of text slowly update the status of the observations and monitor the electronics that are sifting through the cosmic static collected by the telescope. Occasionally, the astronomer looks up to read the displays' laconic reports. There are no high-tech sound effects and no theatrical music, only the constant drone of muffin fans in the electronic racks and the faint, distant grind of the telescope's tracking motor.

This is Project Phoenix, the most comprehensive search ever undertaken for intelligent company among the stars. Run by the SETI Institute of Mountain View, California, it is the privately funded descendant of a former NASA program. Here, at the National Radio Astronomy Observatory's 140-foot telescope in Green Bank, Project Phoenix scientists are systematically scrutinizing a thousand nearby sun-like stars for the faint signal that would betray intelligent habitation. So far, they have found nothing — not a single, extraterrestrial peep.

But tonight could be different. The observation of one star system has turned up an interesting signal. On the screen, a single line of text tells the tale, a string of numbers giving the signal strength and exact frequency, terminated by the cryptic words "confirmed by FUDD." At first blush, there is nothing special in this: With a receiver that simultaneously monitors 28 million channels plugged into one of the world's largest antennas,

5. IDEAS, HYPOTHESES, AND THEORIES: Extraterrestrial Life

such signals are a routine event. In fact, the growing horde of telecommunications satellites that whirl above our heads have saturated the airwaves with signals that for all the world appear to be alien transmissions.

Project Phoenix has tried to inoculate itself against this noisome contamination by using a second antenna, the 30-meter Georgia Tech telescope located 50 miles south of Atlanta. The telescopes in West Virginia and Georgia check out promising signals with an automatic reobservation scheme orchestrated by a piece of digital hardware known as a Follow-Up Detection Device, or FUDD.

Now the FUDD has concluded that one channel out of 28 million bears the hallmarks of extraterrestrial origin. The solitary astronomer, while taking note, does not get excited. After all, the system finds such candidate signals five or six times a week. Without prompting, the observing software swings the radio telescope two degrees away from the targeted star system. Several minutes go by, and the FUDD reports that the signal has disappeared, as would be expected if it came from the star system itself. The astronomer is paying close attention now, but her blood pressure remains unchanged. Satellites parading across the sky routinely move in and out of the antenna's sidelobes and can easily, if briefly, mimic extraterrestrial signals. Or perhaps the signal comes from a military jet whose pilot, by chance, has just switched off his radar.

The 140-foot scope slews back to the source. Ten minutes later, the FUDD reports that the signal has returned. The astronomer, her attention now riveted and her skin starting to tingle, waits as the telescope points to sky coordinates halfway between the on- and off-source positions. In another ten minutes she witnesses the next step in a staggering sequence of events, a sequence that has never before occurred. The insistent signal is still present, but at reduced intensity. Several hours later, the target star sets and the signal winks out on cue. Within a few days, and after enlisting the help of astronomers at another observatory, she will have observational proof that other thinking beings populate the galaxy.

Spreading the Word

How would the world react to such a discovery? Since 1989, a document has laid out what should be done in case of a SETI success — the "Declaration of Principles Concerning Activities Following the Detection of Extraterrestrial Intelligence." This short protocol with the long title has been adopted voluntarily by all SETI research groups. Thoroughly uncontroversial, the declaration describes procedures for verifying that the signal in question is truly extraterrestrial. These include, if possible, a confirming observation by a distant radio telescope (which minimizes the chance of a false alarm caused by fiendish interfer-

Software developer Sally Page holds one of the dozens of circuit boards that make up the receiver used by Project Phoenix scientists to look for possible E.T. signals.

ence, a bug in the system, or a sophisticated college prank). Once the discoverers are confident that the signal is truly not of this Earth, they are directed by the declaration to make their find public; to announce the startling result to the astronomical community, the authorities, and the world at large.

Fine. But recent polls show that the majority of Americans not only believe that extraterrestrials exist, but that they make house calls to our planet. According to the popular view, the government has hidden the evidence of aliens, and maybe even the aliens themselves. Extending these conspiratorial suspicions to SETI, many among the lay public expect that government agencies would swoop down on the radio observatories to protect the citizenry from the possibly dangerous news that extraterrestrials have been found.

This is essentially impossible. To begin with, there is no policy of secrecy and no precautions to

The Impact of a Detection

There's little doubt that a SETI success would precipitate intense scientific inquiry. But would it also provoke a dramatic response from society at large? Would there be panic? Disbelief? A sudden eruption of international goodwill?

The immediate reaction can probably be judged by the response that followed last summer's announcement of possible microfossils in a martian meteorite. There were two days of headlines, after which the discovery dropped off the public's radar screen. The story then became the province of long-timescale media such as books, documentary television, and magazines like *Astronomy*. There was no panic, and philosophical introspection was largely limited to newspaper op-ed columnists. Of course, the impact of the announcement of fossilized martian life was weakened by the uncertainty of the evidence. But the impact of a SETI success would also be compromised by the widespread belief that the aliens have long ago been sighted or overheard.

No matter how anemic the short-term reaction, most pundits assume that the long-term consequences of uncovering extraterrestrials will be profound. To get a handle on this, sociologists have often invoked historical analogs. The Copernican reorganization of astronomy and Darwin's revision of biology caused a "paradigm shift" in the perception of how humans fit in to the big picture of existence. We lost our central role in the physical and biological realms. Demonstrable proof by SETI that we have intelligent, cosmic company would surely deliver a roundhouse punch to any remaining hubris, such as the belief that we are intellectually, culturally, or morally superior. This is especially true because detecting one other technological society would presumably lead to the rapid discovery of more. We would find ourselves surrounded by advanced civilizations.

This could lead to a situation in which there's both good news and bad. The good news is the possibility that we can benefit from the knowledge and presumed wisdom of advanced beings. The bad news is that we might revert to a childlike mental state in which we assume that only our superiors can do anything that's really significant. Would an alchemist of the 12th century continue to toil in his laboratory if faced with the chemical knowledge of the 20th?

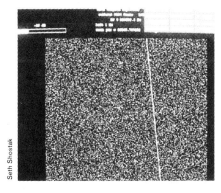

It may not look like something to write home about, but the straight line on the display screen heralds an intelligent signal. In this case, however, it is just a test signal of earthly origin.

Aside from science, there's the matter of beliefs. What effect would a SETI success have on religion? The answer is still vague. Michael Ashkenazi, of Israel's Ben Gurion University, made a study of a small number of religious authorities associated with the so-called "Adamist" religions — Christianity, Judaism, and Islam. Adamist religions share a story of man's creation, but also posit a special relationship between God and man. They hold that we are special. Consequently, extraterrestrial life might be expected to threaten that special relationship.

However, when Ashkenazi presented religious authorities with a scenario in which extraterrestrials have been discovered, "The most common response was amusement." The majority believes in E.T.'s existence, but none felt that this belief caused a problem for their religion. They said that it would be interesting to find the aliens, but their existence, and even their possible atheism, wouldn't affect earthly religions.

Given the historically severe reactions to both the Copernican and Darwinian revolutions, this restrained response might seem a bit *too* restrained. In fact, it may only reflect a failure to give the matter much thought. A hundred years of fictional aliens have misled us into thinking that we will understand E.T. The extraterrestrials in our stories are, for dramatic reasons, similar to us in motivation and intelligence (even if they often choose to emulate our destructive tendencies). They are not so "alien" that they can't be understood in the context of human experience. Consequently, the yet-to-be-discovered extraterrestrials that SETI seeks are seen as no more threat or challenge than someone from an obscure part of our own planet.

But as the English biologist J. B. S. Haldane said, "The universe is not only queerer than we suppose, it is queerer than we *can* suppose." His oft-cited remark applies admirably to modern astronomical research. We now catalog a host of cosmic objects — pulsars, quasars, and supernovae, for example — that were unimaginable a scant century ago. Those who expect the aliens, and their effect on humankind, to be less than extraordinary should take warning.

ensure it. Furthermore, by the time a detection is confirmed, a large number of people will be aware of the discovery (and some of these are likely to be at foreign observatories). Finally, the rapid and occasionally embarrassing spread of information about previous "close calls" shows that the tendency among researchers is to err on the side of bravado rather than caution. There is a far greater chance that an unconfirmed signal will be mistakenly reported as a detection than that a bona fide result will be covered up.

At the February 1996 meeting of the American Association for the Advancement of Science, a British reporter asked Dan Werthimer, a SETI researcher at the University of California, Berkeley, whether his team had investigated any of the star systems recently found to have planets. Werthimer responded in the affirmative that "yes, we have. The results were consistent with noise." This standard radio astronomer jargon was misinterpreted by the reporter, who assumed that "noise" meant "signal." A story soon appeared in a London paper that the aliens had been heard.

Indeed, if the leaks to the press that preceded the recent story about possible microfossils in a martian meteorite are a good example, the orderly sequence of detection, confirmation, and announcement outlined in the declaration is likely to be muddled. The Internet will be throbbing with rumors, and excited but incomplete reports will wash over radio and television.

In Contact

From this chaos, the facts of the detection will inevitably emerge. In view of the known technological limitations of SETI experiments, one can dare to predict what we would learn in those first, exciting days.

Although Project Phoenix's candidates are nearby (they are all star systems within 200 light-years of Earth), a simple detection by the 140-foot telescope would not prove that E.T.'s home was in orbit around the target star. After all, it's possible that, by lucky happenstance, a highly remote civilization was in the telescope's beam (which, at 1.4 gigahertz, covers a patch of sky about half the area of the full moon). But this uncertainty about the source of the signal could be quickly dispelled. Observations with the much larger Arecibo radio telescope, for example, would substantially narrow the field of sky from which the transmission came. Interferometers could resolve, both literally and figuratively, any remaining ambiguity in the source direction.

This would prompt optical astronomers to zero in on the aliens' star system and initiate a search for orbiting worlds that might prod their sun into a detectable wobble. Radio astronomers would be busy looking for slow variations in the frequency of the signal that would give a first indication of the spin and orbital parameters of the aliens' planetary home.

But what about the message? Wouldn't the SETI Institute's staff cryptographer leap into the breech to decode E.T.'s transmission? Alas, there is none. A popular misconception about SETI experiments is that they are intended to pick up actual messages, when in fact they are designed to find steady, or slowly pulsing, narrow-band signals that would pack a lot of energy into a small slice of the radio dial. The counterpart of this type of signal in earthly practice is the carrier wave that underpins most radio and television broadcasts. The carriers have the highest signal-to-noise of any part of a transmission, and would be thousands or millions of times easier to detect than the modulation, or message.

Does this mean that we would get the dial tone, but not the call? Not necessarily. Philip Morrison, the MIT physicist who was among the first to point out the possibility of interstellar communication by radio, thinks that a great deal of information might be packed into the transmission. According to Morrison, "We will be able to read the intention of the senders in the characteristics of the signal itself."

The fact that only slow variations could be easily detected is something that the extraterrestrials will foresee, says Morrison. Nonetheless, "They won't send an empty carrier. It will have a lot of structure, it will have clues, for after all it is the purpose of the transmission to inform us." One possibility is that the detected signal will simply tell us where on the dial we should go to find the real message. This presupposes that the signal was deliberately sent, but Morrison believes that we are more likely to pick up a powerful, intentional transmission than to stumble upon the aliens' internal radio traffic.

Talking Back

Deciphering any possible message might take a long time. Indeed, it might take forever. This discouraging prospect derives from the argument that our chances of hearing a technological civilization are proportional to the average lifetime of such societies. If we do overhear someone, then we can expect that they are enormously more advanced than we are and probably far beyond our understanding.

However, while a SETI detection would mark *our* first contact with another galactic civilization, the same won't be true for those on the transmitting end. They will likely be an old society, one that has seen, and presumably dealt with, many other inhabited worlds. So once we make our existence known, perhaps the aliens will spoon-feed us by sending signals we can understand.

Such thoughts naturally raise the question of whether we should encourage such interplanetary pedagogy by letting the aliens know that we got their transmission. Broadcasting a reply may seem superfluous in light of the plethora of television and radar signals that have been streaming willy-nilly off Earth for half a century. But a directed response would be far stronger than this leakage, and presumably at a frequency the aliens will monitor. It might make sense to do this, if our intention is to join an interstellar chat group.

However, a deliberate reply raises some difficult political and social questions. Who will decide what to "say," or whether anything should be said at all? John Billingham, a SETI pioneer and scientist at the SETI Institute, is attempting to address this thorny, if hypothetical problem. He heads a committee of the International Academy of Astronautics that wishes to extend the declaration to treat the matter of deliberately transmitting to extraterrestrials. "Some people think that at this stage it's not worth indulging in such thinking," says Billingham. "But the significance of transmitting from Earth is potentially so profound in its implications for our civilization that we should consult extensively at the international level beforehand." The principles espoused by Billingham's committee are that any message sent to our cosmic neighbors should reflect the will and wisdom of all Earth's people.

Such diplomatic niceties may seem to be more science fiction than science. The protocols for establishing interstellar discourse are, after all, of no use today. We are, to our knowledge, still alone in a vast cosmos. But a single night's observing in the corrugated hills of West Virginia could change all of that.

Seth Shostak is the public program scientist at the SETI Institute in Mountain View, California. His previous feature article for Astronomy *was "Listening for Life" in the October 1992 issue.*

The best cosmology there is

For three decades, the Big Bang Universe has been the generally accepted model of how the world we know came into being: between 10 and 20 billion years ago, everything sprang into being out of nothing. The impulse of that event has kept the Universe expanding ever since.

As theories go, the Big Bang has been a great success. It provides a framework within which most observations of, say, external galaxies can be accommodated. It has even accommodated entirely unexpected phenomena, notably the ubiquitous microwave background radiation discovered by Penzias and Wilson (*Astrophysical Journal* **142**, 419–421; 1965).

Little is said about the initial event, except that its immediate result was a great concentration of energy, in the form of high-temperature radiation and of particles of matter (consisting of the unstable particles of high-energy physics). Nothing is said, or assumed, about the cause of the event in the sense in which that word was used by David Hume.

Thereafter, the very early evolution of the Universe (occupying just a few seconds) is simply a matter of the succession of intrinsically energetic by intrinsically less energetic particles. Eventually, ordinary atoms (mostly of hydrogen) are formed, perhaps a few seconds after the initial event. At a later stage, perhaps 100,000 years afterwards, the Universe becomes transparent to radiation, which then continues cooling at a rate determined by the expansion of the Universe as a whole.

As a framework for the accommodation of observations, which are mostly of objects such as galaxies formed long after the high-density and high-temperature phase of the Big Bang, this model is not distinguishable from solutions of Einstein's relativity equations that describe an expanding Universe and whose parameters are chosen to match the observed large-scale properties of the Universe, its density for example.

There are two respects in which the observable Big Bang Universe is linked directly with events in its earliest history. The strongest link between the present and the Big Bang is the abundance of deuterium (and other light nuclei) in the Universe as now observed, which is correctly predicted (from known nuclear cross-sections) by the events in the first few seconds of the Big Bang. The present deuterium content of the Universe cannot be the product of nucleosynthesis in the stars, where it is (with hydrogen) a fuel rather than a product, but the Big Bang requires some of the extant ^{3}He and ^{4}He should also be primordial. These arguments imply that the material in the present Universe must at some early stage have been through a high-temperature, high-density phase, and are thus the most direct support for the Big Bang (see figure).

The microwave background radiation, which fills even the corners of the Universe, would psychologically have been more compelling evidence for the Big Bang if it had been predicted before its discovery in 1965. That it was not is something of a surprise, which is nevertheless now irrelevant. The microwave background is a true relic of the time at which the Universe became transparent to radiation.

The uniformity over the sky of the microwave background radiation bears more directly on conditions in the early Universe than its precise temperature, measured as 2.73 K. Given the inevitable uncertainties of the calculation of the course of events after the Big Bang, the temperature of the radiation must be less critically diagnostic of conditions in the early Universe.

This does not mean that the Big Bang Universe is trouble-free. There is, for example, a problem about the present density of matter in the Universe, estimated from the luminosity and dynamics of visible galaxies to be about 5 per cent of what is called the critical density, that required if gravitational self-attrac-

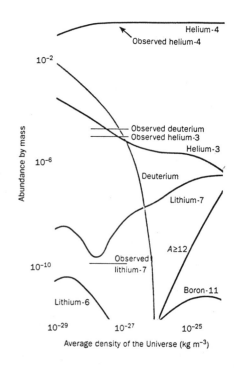

Predicted abundances of light elements as a function of the average density of matter in the Universe, showing agreement with observations. Figure adapted from *The New Physics*, ed. P. Davies, Cambridge University Press, 1989.

tion is to decelerate the expansion of the Universe asymptotically to zero. (For a Hubble constant at the lower end of the usual estimated range, the critical density is estimated as 5×10^{-27} kg m^{-3}, or roughly one atom of hydrogen per litre).

The difficulty is not that the figure is too far from unity, but that it is uncomfortably close to it. Small departures from equality at the Big Bang would by now have been magnified hugely, so that the Universe would either be conspicuously empty of matter, or conspicuously over-full. What this argument implies is that the Big Bang would have had to have begun with a matter density almost but not quite equal to the critical density to reproduce the present state of affairs. But nobody likes initial conditions so carefully prescribed.

One of the first difficulties to be recognized, in the early 1970s, was the capacity of a Universe expanding as a consequence of an initial impulse to magnify the scale of density fluctuations soon after the initial event in proportion to the expanded scale of the Universe as a whole. The consequence of that would be that the matter in the initial fireball would quickly have clumped, by virtue of its self-gravitation, into galaxies much larger than those now observed. Conversely, the present large-scale uniformity implies that the density fluctuations in the initial fireball must have been much smaller (perhaps by a factor of 10^{15}) than would be expected for a particle gas in equilibrium.

Both problems have now been resolved, at least formally, by a device called 'inflation', originally due to Alan Guth, for which, since the development of the theories of the weak and strong nuclear interactions, there is now physical justification of a kind. Guth argued, in 1980, that there would be a point in the hierarchical conversion of one kind of particle into another when something like a phase transition would take place between one regime and the next in the hierarchy, with the release of the equivalent of latent heat sufficient to make the still tiny Universe expand geometrically by a factor of 10^{50} or more in an interval of time no greater than 10^{-32} seconds.

One result of inflation is that it would drive the density of the Universe towards the critical density, whatever the initial conditions. In other words, if there were an inflationary phase in the history of the Universe, the true density of the Universe should now be 20 times greater than it has been found to be. But 5% is uncomfortably far from 100% and becomes the more so as the search for 'missing mass' remains unsuccessful.

Inflation also irons out the expected fluctuations of the initial high-temperature Big Bang, so that the large-scale uniformity of the Universe becomes explicable, but at the cost of putting an impenetrable observational barrier between the pre-inflation epoch and the present. In other words, it ceases to be possible to tell, from where we are now, the initial conditions for the Big Bang before the onset of inflation. There is a sense in which the universe began only when inflation ended.

Holes in the Big Bang

THE cosmology that has held sway for the past 30 years commands general support, but that is not to say that it is free from defects. How could it be, a mere three decades after being formulated, and in a field in which the inadequacy of the data is as evident as in cosmology?

What follows is a list of some outstanding problems with the Big Bang picture of the Universe, derived from the published literature. To have compiled it is not to suggest that the Big Bang is an incorrect picture of how the Universe has evolved, but may equally be taken as a recitation of what needs to be done to complete the picture.

Missing mass. This is a problem only if it is supposed that the density of the Universe is equal to the critical density, p_c say. (Coles and Ellis have recently assembled evidence to suggest that the supposition is untenable[1].) It is conventional (and convenient) to use dimensionless quantities. If the density is p and the critical density is p_c, the test of whether the density is more or less than the critical value is determined by whether the density parameter $\Omega = p/p_c$ is greater or less than unity.

The extent of the missing mass is very large. The luminosity of the observable galaxies betokens a mass equivalent to less than 1 per cent of the critical density. On the other hand, accurate measurements of the rotational dynamics of galaxies, and in particular of the rotational velocity of visible material as a function of radius, have shown that the effective masses of galaxies are typically 10 times greater, perhaps enough to account for $0.05 p_c$.

The unseen mass may plausibly consist of extinct stars, or stars too small to be luminous, in the haloes of galaxies. A recent study of the galaxy NGC5907 has revealed that this may indeed by reality[2].

On the assumption that $\Omega = 1$, where is the remaining material, and of what does it consist? One possibility is that it is simply primordial gas, mostly hydrogen, left over from the Big Bang and not yet formed into galaxies. Such material may lie at the heart of galaxy clusters, providing the mutual gravitational attraction that binds them together. But that seems to be unlikely: the dynamical effects of such large amounts of matter would be more pronounced. In any case, calculations of the formation of ^{4}He in the Big Bang[3] place an upper limit on the proportion of primordial matter that can consist of nucleons, which is at least an order of magnitude less than p_c. Another possibility is that the missing mass may be embodied in particles such as neutrinos, supposedly entirely devoid of mass. But if all neutrinos had a very small mass, or if some had a slightly larger one, the difficulty of the missing mass would be resolved, given the ubiquity of neutrinos in the Universe. Suggestions[2] that there may be a neutrino with a mass equivalent to 17 keV were originally welcomed as a means of reaching the critical density, but the case for such a mass (needlessly large for cosmological purposes) is now discounted. In the past 15 years, other material particles have been considered as candidate particles for the missing mass, only to be rejected, usually on the grounds of implausibility. The question of the missing mass remains obdurately open.

Hubble constant. From our present vantage point, the expansion of the Universe is simply described by the Hubble relation between the recessionary velocity of a galaxy (v), perhaps measured by the redshift of a spectral line, and its distance (r). The velocity is a multiple of the distance, or $v = Hr$, where the Hubble constant, H, is measured in units of km s^{-1} per megaparsec. (One parsec, defined as the distance at which the radius of the Earth's orbit about the Sun would subtend one second of arc, is numerically about 3.26 light-years: a megaparsec (Mpc) is 10^6 times as much.) To avoid the repetition of the units, it is again convenient (and conventional) to put $h = H H_0$ where H_0 is taken as 100 km s^{-1} Mpc^{-1}, so that h becomes a dimensionless number.

From the outset, in the 1930s, there has been great uncertainty about the value of h. That persists. The values measured over the years seem to have clustered around 0.5 and about 1.0. The origins of the uncertainty are well understood: the difficulty of constructing a distant-scale that will span the whole Universe. That requires some means of recognizing in distance galaxies objects whose astrophysical properties are sufficiently well understood within the Galaxy for them to serve as 'standard candles'.

The issue is important because the actual value of h is linked directly to quantities such as the age of the Universe, but in ways that are influenced by the model used to describe the general expansion. At its simplest, h will have decreased since the origin of the Universe because of the way the expansion is decelerated by mutual gravitational attraction (so that the value of Ω will also be relevant).

What has now emerged from measurements of Cepheid variable stars in distant galaxies of the Virgo cluster of galaxies is that the value of h is more like 1.0 than 0.5. Pierce et al.[5] have determined a value of 9.87 ± 0.07, a result confirmed by measurements with the Hubble Space Telescope[6]. That is inconveniently (but not impossibly) high for the conventional cosmology.

Age of the Universe. Like other properties, the age of the Universe is directly dependent on h (but also on Ω, which determines the deceleration of the expansion). Coles and Ellis use[1] the emerging consensus on such a value of h to conclude that Ω must be considerably less than 1 if the age of the Universe as a whole is not to be considerably less than the ages of the stars of the globular clusters in the Galaxy, which lie in the range 14–16 billion years. The alternative is that if $h = 0.95$ and Ω 1, there must be a contradiction between the theory of the inflationary

54. Holes in the Big Bang

Big Bang and such data as the cosmologists have been able to assemble.

Light Isotopes. Even the strongest source of empirical support for the Big Bang raises difficulties still to be resolved. Nucleosynthesis in the Big Bang would have begun as the temperature fell into the equivalent of 10 MeV. In present circumstances, free neutrons decay to protons and electrons (plus an antineutrino) with a half-life of just over 10 minutes in a weak interaction transformation involving an energy of 1.293 MeV. As a consequence, at a temperature of 10 MeV, both nucleons are in equilibrium with each other and with electrons and positrons. That allows for the formation of deuterium (^{2}H) and tritium (^{3}H) nuclei and, by combinations between the products, of the nuclei of ^{4}He, ^{3}He and ^{7}Li, proportions of which are then frozen out as the temperature of the equilibrium gas falls to the equivalent of 1 MeV or less.

The most recent recalculation of these reactions[3] is based on more accurate measurements of the nuclear reaction rates now available, as well as on recent estimates of primordial abundances (deuterium in the solar wind, helium isotopes in carbonaceous chondrites and ^{7}Li in population II stars of the galactic halo, for example). The measured abundances can be reproduced, but only if the density of baryons (nucleons) in the Big Bang at 10 MeV is an order of magnitude less than the critical density.

Interestingly, the same calculation yields a limit to the number of types of neutrino allowed; if there were four or more, there would be uncomfortably large amounts of primordial ^{4}He.

Inflation. The inflationary phase of the Big Bang remains a puzzle, chiefly because its driving mechanism is not unambiguously specified or identifiable in what is known of particle physics. The argument starts from the assumption that the early Universe supported one or more quantum fields of the find known as 'Higgs fields', which have the property that the state in which the field is identically zero is not that in which the energy is least.

The phase transition responsible for the inflationary phase is therefore a transition from the 'false vacuum' of identically zero fields to the 'true vacuum' in which the numerical values of the field or fields are different from zero. From the latter, energetic particles emerge as the consequences of quantum fluctuations, and set in train the hierarchical sequence of particle transformations that lead, eventually, to the Universe as it is.

The concept is not really as bizzare as it may seem; just such a Higgs field sits astride current theories of particle physics (and the 'Higgs boson' is one of the particles that may yet be found at particle accelerators). Making the nature of the inflationary process more specific than it is at present must be an important goal for the future, but one that must depend on a better understanding of outstanding problems in particle physics.

Quantum gravity. A similar (but unrelated) difficulty concerns the treatment of the pre-inflationary phase of the Universe, where it is agreed that the energy density would be so high as to require a quantum treatment of the gravitational field. Much has been learned in the past 20 years of the difficulties of this task, but practical ways of making progress are still beyond the ken of those concerned.

Quasars. The discovery by 1960s radio astronomers of powerful, but apparently point-like, sources of radio energy at great redshift (and thus at an early stage in the evolution of the Universe) is an astrophysical problem with cosmological implications. Several questions arise. Is a quasar a stage in the early evolution of only some galaxies, in which case how are the relics of that past condition to be recognized in the present Universe? Or is a quasar a stage in the evolution of most galaxies, in which case what is its significance?

Both questions bear directly on the cosmological question, given the importance of galaxy formation as a stumbling block in the proper understanding of the evolution of the Universe without the invocation of inflation.

Alternatives? The formidable problems attending Big Bang cosmology do not imply that the picture must forthwith be abandoned. There are several ways in which the simple version of the Big Bang may be varied, many of which have been aired in the literature. But there seems no easy escape from the contradiction between, on the one hand, the arguments in favour of $\Omega = 1$ and, on the other, the low value of the baryon density required by nucleosynthesis and the disappointment, at least so far, of the search for non-baryonic dark matter.

It is true that Holye, Burbidge and Narlikar have recently[7] argued for what they call the "quasi-steady-state cosmology", which entails episodes of matter creation that tend to be concentrated at 40-billion-year intervals in the history of a Universe that is infinite in space and time.

That has the advantage over the earlier steady-state theory of Bondi, Gold and Hoyle in providing a means by which nucleosynthesis could proceed. It can also accommodate difficulties arising from objects in the present Universe that are older than the age of the Big Bang itself, which may be objects left over from an earlier phase of matter creation. The difficulty is that is hardly more specific about the quantum field supposed to be responsible for the appearance of matter than are the theories of the inflationary phase about the Higgs fields they require. Does the future of cosmology rest with the particle physicists? Other alternatives hang on the notion that string-like structures in the early Universe may have evolved into large-scale structure; "cosmic string" is the name.

Meanwhile, it is relevant that continuing uncertainty about the value of W means that we do not know whether the Universe will continue to expand indefinitely (if $\Omega < 1$), or will decelerate in its expansion to the point at which it collapses again under its own weight (if $\Omega > 1$).

Notes

1. Coles, P. & Ellis, G. *Nature* **370**, 609–615 (1994).
2. Sackett, P. D., Morrison, H. L., Harding, P. & Boroson, T. A. *Nature* **370**, 441–443 (1994).
3. Walker, T. P., Steigman, G., Schramm, D. N., Olive, K. A., & Kang, H. S. *Astrophys J.* **376**, 51–69 (1991).
4. Simpson, J. J. *Phys. Rev. Lett.* **54**, 1891–1893 (1985).
5. Pierce, M. J., Welch, D. L., McClure, R. D., van den Bergh, S. & Racine, R., *Nature* **371**, 385–389 (1994).
6. Freedman, W. L. *et al. Nature*, **371**, 757–762 (1994).
7. Arp, H. C. *et al. Nature* **357**, 287–288 (1992).

In Defense of the Big Bang

Yes, questions remain. But the big bang is the most successful theory ever put forth for the origin and evolution of the universe.

By Neil de Grasse Tyson

What, you might ask, could possibly induce a rational astrophysicist to believe that fifteen billion years ago, the universe—with all of its matter and energy—was packed into a primeval fireball smaller than a marble, and that it has been expanding ever since? The answer is simple: Regardless of what you may have read or heard to the contrary, the big bang is supported by a preponderance of evidence and has become the most successful theory ever put forth to explain the origin and evolution of the universe.

Scientific evidence in support of a theory sometimes takes you places where your senses have never been. Twentieth-century science has largely been built upon data collected with all manner of tools that enable us to see the universe in decidedly uncommon ways. As a consequence, while we have always required that a theory make mathematical sense, we no longer require that a theory make common sense. We simply demand that it be consistent with the results of observations and experiments. This posture facilitated the rise of counterintuitive, yet remarkably successful, branches of physics, such as relativity and quantum mechanics.

Of all the theories that describe the physical world, the big bang, first described in 1948 by physicist George Gamow, seems to intrigue the general public most consistently. Some people vehemently oppose the big bang, even when they are unaware of its fundamental tenets. Others like to claim that the big bang is "just a theory" and should therefore be discounted.

Don't be fooled. The beginning of the twentieth century saw the end of describing successful theories as "laws." This change of vocabulary came when new experimental domains revealed that the predictions of previous physical laws were incomplete. The adoption of the term *theory* came with the humble recognition by physicists that data from newer and better equipment might provide a deeper understanding of the physical world. This is why before 1900 we had Kepler's *laws* of planetary motion, Newton's *laws* of gravity, and the *laws* of thermodynamics, whereas after 1900 we have Einstein's *theory* of relativity, the *theory* of quantum mechanics, big-bang *theory,* and so forth.

A well-constructed theory must explain some of what is not understood, predict previously unknown phenomena, and, to be successful, have its predictions consistently confirmed. Furthermore, skeptics should not hesitate to question every possible assumption, no matter how basic.

Confidence in the big bang is derived from the strengths of many arguments. Let us start with Edwin Hubble's 1929 observation that we are part of an expanding universe in which distant galaxies recede from us faster than the near ones in direct proportion to how far away they are. Further support came from Einstein's theory of gravity, better known as the general theory of relativity. One of its solutions predicted a universe that expands according to precisely the pattern found by Hubble. Since Einstein's theory preceded Hubble's discovery (by thirteen years), it was to provide independent corroboration of the idea.

If you happen to have a gripe with the big-bang theory's claim that objects with high recession velocity are farther away than objects with low velocity of recession, then consider the existence of gravitational lenses. As first predicted by Einstein, the gravity of a high-mass object in an observer's foreground can act as a lens. It distorts space in its vicinity so that light from a background object along the same line of sight is split into three or more images. Such optical antics have been observed for dozens of quasars all around the sky. In each case,

55. In Defense of the Big Bang

the "lensed" object always has a higher recession velocity than the object whose gravity is serving as the lens itself.

Could it be an illusion that very distant galaxies are receding from us at very high velocities? We can test for this because objects moving at very high speeds should measurably exhibit the effects of "time dilation" predicted in Einstein's theory of relativity. Indeed, supernovae recently discovered in distant galaxies do take more time to explode and to decline in luminosity than comparable supernovae in nearby galaxies.

The most significant supporting argument for the big bang derives from the existence of "cosmic microwave background." Shortly after the notion of a hot, explosive origin for the universe was first proposed, physicists Ralph Alpher and Robert Herman, invoking simple principles of thermodynamics, inferred that the density of matter and energy of the universe must have been much higher in the past. They were forced to conclude that the universe should betray some sign of leftover emergy from its earlier, much hotter existence. As the universe expanded, such a signal should have cooled appreciably to become an omnidirectional bath of microwaves with a characteristic temperature of a few degrees on the Kelvin absolute temperature scale. In 1965, part of this background signal serendipitously revealed itself in data from microwave antennae used by two Bell Laboratories physicists, Arno Penzias and Robert Wilson. For this finding, they were jointly awarded the 1978 Nobel Prize for physics.

If you are skeptical of the claim that some accidentally discovered microwaves are the cooled remnant of a youthful, hot universe, then consider that the big bang predicts a specific mixture of energy with a characteristic temperature. (Similarly, the Sun's mixture of energy, which includes specific proportions of infrared, visible, and ultraviolet light, also has a characteristic temperature (6,000 kelvins) at its surface. The Cosmic Background Explorer satellite measured the cosmic background in every direction and indicated a single temperature of 2.726 kelvins.

You might be skeptical about whether this single-temperature assortment of microwaves actually came from the early universe. Perhaps you prefer to think they were created by your neighbor's microwave oven or by a police radar gun or by some microwave-emitting wall of interstellar material nearby in space. One proof of a distant source uses the hot gas embedded in galaxy clusters, which we expect to slightly increase the temperature of cooler energy that passes through. And when we look toward these distant clusters, we do see an increase in the temperature of the microwave energy along our line of sight, implying that the microwave background indeed hails from beyond the clusters and not from a source in the foreground.

If you are unconvinced that the universe was hotter in the past than it is today, consider that because of the time it takes their light to reach through intergalactic space, we see galaxies not as they are but as they once were. If big-bang cosmology is correct, these distant galaxies were once bathed in a hotter cosmic background than they are at present. To test this notion, we can use selected molecules as "thermometers" that allow us to infer a temperature for the background microwaves that once bathed distant galaxies. Sure enough, the measured temperature is in precise accord with the predicted temperature of the universe at the time the light left those galaxies.

Just for fun, let's turn back the big-bang clock and use current laws of physics to extrapolate the behavior of the observable universe at a time when it was much smaller, denser, and hotter—when the background was upward of a trillion degrees. (Our current theories of physics actually allow us to describe the behavior of the universe from the first 0.000000000000000000000000000000000000000001 second of its existence all the way up to fifteen billion years and beyond.) At these early times and high temperatures, all atoms were broken apart into their component nuclear particles, which themselves were broken into their quark subcomponents.

Combining all that we know of quantum mechanics and all we have learned from busting atoms to smithereens in particle accelerators, we conclude that as the cosmic soup expanded and cooled, nuclear particles recombined to make a specific and predictable assortment of atoms: the universe forged most of its elemental mass into hydrogen and about 25 percent into helium. These are bold extrapolations, but surveys of the most helium-deficient galaxies (those that

> While a theory must make mathematical sense, we can no longer demand that it make common sense.

have undergone very little star formation) routinely find between 22 and 27 percent helium, in good agreement with big-bang predictions.

A few other light elements are predicted to have formed in trace amounts during the first moments of the universe. Among these are "heavy" hydrogen (which is simply a proton and a neutron), "light" helium (helium that is missing a neutron from its nucleus), and lithium (the third-lightest element on the periodic table of elements). The measured quantities of these light elements in the universe are also consistent with big-bang predictions.

We didn't just make this stuff up. The acceptance of the big bang represents an unprecedented marriage of astrophysics and particle physics, one in which a coherent cosmic picture emerges from a minimum of assumptions and measure-

5. IDEAS, HYPOTHESES, AND THEORIES: The Big Bang Theory

ments. Yes, the galaxy velocities are real, galaxy distances are real, the expanding universe is real, and the big bang is real. Whenever very different experiments support the same theory, then the confidence you have in the theory is greatly enhanced.

But alas, all is not perfect in paradise. A few holes remain in the big bang.

Most worrisome is that the mass density of today's universe would have to be remarkably close to the critical value—the point at which the universe is delicately balanced between recollapse and infinite expansion. This scenario requires a little too much fine-tuning among several cosmological parameters in the early universe.

When we go beyond the simple extrapolations from the big bang, we also find that the microwave background is far too uniform from one patch of the sky to the next to have emerged from the conditions we think were present in the early universe. And the subsequent rapid expansion of the universe does not leave enough time for the galaxies to form as we think they should.

Moreover, the big bang offers us no insight into what the universe was doing before time began or, for that matter, why the laws of physics are what they are.

Do we throw out the big bang along with the bath water because of these complications? Or do we retain the theory's successful predictions and see if there is room to modify the details? These sorts of questions have arisen before. In the mid-sixteenth century, the Polish astronomer Nicolaus Copernicus proposed a model of the known universe that placed the Sun, rather than Earth, at the center of all motion. This heliocentric model was much, much simpler than the competing geocentric model because it removed the need for complex epicycles to account for the motions of the planets. But there was a problem. The predicted paths of the planets did not always conform to their actual paths in the sky. Should Copernicus have therefore discarded the entire idea of a Sun-centered system, or should he have modified some of the model's details? His heliocentric view was, of course, basically correct. The problems arose because he naïvely assumed that planets orbited the Sun in perfect circles rather than in ellipses. It would be two centuries before Newton conceived of his universal law of gravitation, which supplied a bigger picture that modified and subsumed Copernicus's view of the world.

Progress has been made toward resolving some of the outstanding problems with the big-bang model. The most significant contribution, introduced in the early 1980s by the American physicist Alan Guth, is known as inflation, which posits that the energetics of the very early universe passed through a phase that spontaneously triggered a period of extremely rapid expansion. When the details are worked out, inflation naturally explains the embarrassingly fine-tuned "critical" density and also allows the cosmic microwave background to be as uniform as it is measured to be. A consequence of the principles of quantum mechanics applied to the fabric of space and time in the early universe, inflation has no household analog. It predicts a universe that was born at critical density and remains there. Unfortunately, astrophysicists have not been able to track down the requisite amount of mass, but current observations have come up with anywhere from 20 to 40 percent of it. Of course, inflation enthusiasts are fervently looking for the rest.

One class of inflationary theories describes a megauniverse with multiple areas of expansion, each of which looks like a big-bang universe from within and sustains laws of physics that may differ from the ones we know. If this model can be tested and supported, then inflation will have subsumed the entire big bang into a larger cosmological picture.

If you choose to discard the big bang entirely, then step lightly. You will be forfeiting an impressive set of successful predictions—far more than most theories in progress enjoy. Nearly everyone in the community of astrophysicists has chosen to work with the big bang, while recognizing that it may one day become the core idea of something even bigger.

Neil de Grasse Tyson, an astrophysicist, is the Frederick P. Rose Director of the Hayden Planetarium at the American Museum of Natural History. He is also a research scientist at Princeton University.

Imagine yourself floating in space, watching the universe come to life in the Big...
BANG.

A blinding flash suddenly erupts out of an immense black void. Seconds later clouds of outrushing debris engulf you.

It's a powerful vision, but not an accurate one. Because before the universe was born, there was no space to float in, no perspective from which to view the Big Bang.

You don't need to perform strenuous mental gymnastics to understand the Big Bang. All you need to do is ask the right questions. That's what our readers have done, and we've collected ten of the best. We hope the answers provide you with fresh insights into the nature of this remarkable universe.

Everything You Wanted to Know About the Big Bang

Richard Talcott

If the universe was created in a Big Bang, why doesn't the cosmic background radiation come from one point in the sky?

For the same reason that you couldn't have seen the Big Bang from the outside. The Big Bang did not occur at some point in pre-existing space we can now turn our telescopes toward. Rather the entire cosmos, including space and time themselves, originated with the Big Bang. There was no "outside" in which to stand and watch.

The cosmic background radiation is light that dates from a few hundred thousand years after the Big Bang. Before that time, the universe was a seething cauldron of energetic subatomic particles and photons of light. The photons could travel only short distances before interacting with one variety of these subatomic particle, free electrons. But when the temperature of the expanding universe fell to about 5,000 kelvins, electrons combined with atomic nuclei to form atoms. With few free electrons left to interact with, photons could travel unimpeded and the universe became transparent to light.

These photons literally filled the universe then and still do now, which is why the background radiation comes from every direction and not from one point. But in the roughly 15 billion years since the universe became transparent to light, the universe's expansion has cooled this radiation from 5,000 kelvins to a frigid 2.73 kelvins (–454.76° F).

Is space expanding or are galaxies moving apart?

This is a tricky question with no definite answer. At first the question seems little more that a matter of semantics. After all, does it matter whether a distant gal-

5. IDEAS, HYPOTHESES, AND THEORIES: The Big Bang Theory

Do Galaxies Move or Does Space Expand?

When astronomers look at a distant galaxy, it appears to be moving away from us. Is this because space remains constant and the galaxy moves through it or because space itself expands, taking the galaxy along for the ride?

axy cluster is speeding away from us through space or just riding along as space itself expands? In some instances, it does. Few people would equate walking through a large airport with standing on a moving walkway. But in the first instance you are powering the move through the airport while in the second case the airport (walkway) takes you along for the ride. And even though each method gets you to the same destination, the amount of energy you expend shows that the two are not equivalent.

A large majority of astronomers would agree that space itself is expanding. Einstein's theory of general relativity—the theoretical backbone for most cosmological theories—defines the geometry of spacetime and how it expands over time.

Some people argue that if space expands, then everything in it should expand too, including the rulers you use to measure distance. Then you would never be able to detect the expansion in the first place because the ruler would increase in size at the same rate as the expansion you wanted to measure!

To see why this isn't so, imagine gluing several coins to a balloon and then blowing it up. The expanding balloon represents the expansion of space and the coins represent galaxy clusters. As the balloon inflates, the coins go along for the ride, taking part in the expansion of the balloon but not themselves expanding. Molecular forces hold the coins together, keeping them from expanding. In the same way, gravity holds the galaxy clusters together so they don't expand.

How do astronomers know how old and how big the universe is?

They don't, precisely. But astronomers have developed several ingenious methods for getting at the age of the universe, which seems to be between 10 billion and 20 billion years old.

One estimate comes from what's called the "Hubble age," an age derived from the expansion of the universe. In the 1920s, American astronomer Edwin Hubble observed scores of distant galaxies and discovered two remarkable features: All the galaxies appear to be moving away from Earth, and the farther away the galaxy is, the faster it recedes. Hubble's discovery paved the way for measuring the universe's age—figure out how fast the universe is expanding and run the video in reverse back to the time when the expansion began.

The technique is the same you would use to calculate how long it takes to drive from point A to point B: Take the distance between the points and divide by the speed of your car. In the universe's equivalent, astronomers measure the distances to many galaxies, divide by how fast they move away from Earth, and then average the results. In theory, the technique works great. In practice, measuring the distances to galaxies is fraught with uncertainty, making the age derived from the results debatable.

Some astronomers measure the expansion rate as 15 kilometers per second per million light-years; others find 30. (An expansion rate of 15 means that a galaxy 10 million light-years away recedes at 150 km/sec while a galaxy 100 million light-years away speeds along at 1,500 km/sec.) The inverse of these expansion rates yields a Hubble age between 10 billion and 20 billion years.

The Hubble age measures how old the universe would be if it contained no matter. But you and I offer living proof that the universe is not empty. The gravity of all the objects in the universe slows down the expansion, so the expansion must have moved faster in the past. This in turn means that the universe didn't have to expand for as long to reach its present state, so the Hubble age sets an upper limit to the actual age of the universe.

If the universe is at the critical density—having just enough mass to halt the expansion in the distant future but not enough to cause the universe to contract again—then the real age of the universe must be two-thirds the Hubble age. Many cosmologists think that's

56. Everything You Wanted to Know about the Big Bang

Is the Milky Way the Center of the Universe?

In an expanding universe, all observers see themselves as the center of expansion.... Although each view starts with the same distribution of clusters and experiences the same rate of expansion, both observers see more distant clusters receding faster.

precisely how old the universe is. Depending on which expansion rate you want to believe, that makes the universe 7 billion or 14 billion years old. Other cosmologists argue that the density of the universe is significantly less than the critical density, which would make the universe closer to the Hubble age. The age debate continues among cosmologists and between cosmologists and stellar astronomers, who argue that the oldest stars in our Galaxy are more ancient than the cosmologists' universe (see "The Age Paradox," June 1993).

Sizing up the universe is even more difficult than finding its age. The most distant objects we see are quasars, which appear to lie 9 billion to 18 billion light-years from Earth, again depending on how fast the universe is expanding. To find the size of the universe, astronomers must estimate how much the universe expanded between the Big Bang and the birth of quasars 1 billion to 2 billion years later.

If during this time the expansion rate changed only because gravity slowed it down, then the universe's size is just a bit more than the distance to the farthest quasars. That would make the universe approximately 10 billion to 20 billion light-years in radius.

But many, if not most, astronomers think something much different happened right after the Big Bang. In 1981 Massachusetts Institute of Technology physicist Alan Guth proposed a model that attempted to explain two conundrums of big-bang cosmology: why the universe started off exactly uniform everywhere and why it is so near its critical density.

His "inflationary" model and subsequent refinements to it suggest that the universe expanded by some one thousand billion billion billion times (10^{30}, or a 1 followed by 30 zeros) during an exceedingly brief interval just after the Big Bang. If this scenario is correct, the radius of the universe must be vastly larger than 10 or 20 billion light-years. In effect most of the universe is out of sight, beyond the horizon denoted by the cosmic background. As time goes on, our horizon will expand and we'll start to see more of the rest of the universe.

If all galaxies move away from us, doesn't that make the Milky Way the center of the universe?

No. This is another misconception that comes from thinking of the Big Bang as an explosion occurring before your eyes. Instead of thinking of yourself as an outside observer, imagine yourself swept up in the Big Bang.

A good way to envision this is with an analogy. Picture a magical balloon made of so much rubber that it can be blown up as much as you want. Now imagine a two-dimensional creature living on the surface of the balloon. All it can see is what lies on the balloon's two-dimensional surface, which is the creature's entire universe. As the balloon inflates, its neighbors move farther away as the rubber between them stretches. And the more distant the neighbors, the more rubber there is to stretch, so they appear to move away from each other faster.

The two-dimensional creatures of our analogy are the galaxies in our universe. As the universe expands (the balloon inflates), all the galaxies (creatures) move away from each other, and the more distant they are, the faster they move. Finally, the universe (the surface of the balloon) has no center.

How could galaxies form if during the Big Bang everything moved away from everything else?

Gravity is both strong and pervasive. As Isaac Newton showed over 300 years ago, every particle with mass in the universe attracts every other particle, whether they move away from each other or not. In the chaotic aftermath of the Big Bang, incredible numbers of elementary particles interacted, creating pock-

5. IDEAS, HYPOTHESES, AND THEORIES: The Big Bang Theory

Where Is the Edge of the Universe?

The edge of the universe is an edge in time, not space. At any time, an observer can see only those objects close enough that their light has had time to travel to the observer. As the universe ages, the observer's horizon expands, and ever-more-distant objects come into view.

ets of slightly higher density. These formed the seeds for the later development of galaxy clusters.

Despite gravity's key role in organizing the universe, the exact mechanism by which it forged the galaxies and galaxy clusters we see today remains elusive. In fact it's one of the major mysteries of modern cosmology. Measurements from the Cosmic Background Explorer satellite show that the seeds for galaxy formation existed by the time the cosmic background radiation began its journey through space (see "COBE's Big Bang," August 1992). But we still don't know whether the seeds grew around cold dark matter, warm dark matter, cosmic strings, or something else.

If the universe is expanding, what is it expanding into?

To answer this question, we need to get into the geometry of four-dimensional spacetime. And the easiest way to do that is to return to the analogy of two-dimensional creatures living on an inflating balloon.

In this analogy, we can see that the balloon is expanding—but into a third dimension undetectable to the two-dimensional creatures on its surface. In our own case, we are three-dimensional creatures living in a four-dimensional spacetime. And like the balloon inhabitants, our universe is expanding into a higher dimension, the fourth. Unfortunately visualizing four dimensions is exceedingly difficult—I've never met anyone who could—so it's hard to get a good mental image of what we're expanding into.

Is there an edge to the universe, and if so, what lies beyond?

There is an edge to the universe, but it lies in time, not space. As we look farther out into space, we look further back in time. Look at the Moon on some clear night and you see it not as it is now, but as it was a bit over a second ago. Look at the Sun and you see it as it was eight minutes ago. Look at Sirius, the brightest star in the sky, and you see it as it was more than eight years ago.

You can carry this analogy as far as you want. You see nearby galaxies as they were a few million years ago; distant galaxies as they were billions of years ago. Eventually as you look out ever farther, you should be able to see to a time before the universe had formed any stars or galaxies.

When astronomers probing the universe with the Hubble Space Telescope or the 10-meter Keck telescope say they hope to see to the edge of the universe, they mean they want to observe back to a time when the first stars and galaxies were forming, probably some time between 1 billion and 2 billion years after the Big Bang. (The cosmic background radiation is, of course, even more distant, but astronomers usually don't mean that when they speak of the edge.)

What lies beyond the edge of the most distant galaxies? Not much. Before stars turned on, the universe should have been profoundly dark. There's not much that astronomers could even hope to observe out there.

Did the universe expand faster than the speed of light during the inflationary period?

If you accept Alan Guth's inflationary hypothesis, then the answer is yes. The universe expanded by some one thousand billion billion times in about one-thousand billion billion billionths of a second, growing from roughly one centimeter in diameter to several hundred billion light-years across in a small fraction of a second. That's a speed far greater than the speed of light.

How can this be so? Many people know that Einstein's theory of special relativity implies that nothing can move faster than the speed of light. But fewer people realize that this is a restriction only on how fast matter and energy can move through space. Although special relativity restricts any object from traveling faster than the speed of light, it places no restrictions on how space itself can behave. Inflation demands that space expands, so the elementary particles that made up the early universe did not themselves move, but rather space carried them along.

What's the fate of the universe?

Does any question stir the imagination more than this? The surprising thing is that to a great extent we can answer it. Astronomers have a good idea of what will happen to the universe depending on how much matter it contains. The problem is that measurements of how much matter exists vary so much that none of the three possibilities can be ruled out entirely.

When astronomers look at the amount of matter visible in the universe—typically stars and galaxies—they tally only about one percent of the total needed to reach the critical density that will just barely stop the universe's expansion in the distant future. But when they look at the motions of stars in individual galaxies and of galaxies in clusters, they calculate that ten times more material must be exerting gravity than we can see. The amount of this dark matter will determine the universe's fate. If there is not enough to boost the universe's density to the critical value, then the universe is considered "open" and will expand forever. There simply wouldn't be enough matter to exert the gravitational pull needed to stop the expansion. In this case, the universe will continue to expand and cool. Stars will go on forming for a long while, but the amount of material available for star formation will progressively dwindle. Eventually all the mass of the universe will be contained in dead planets and stars. The end of the universe will be the exact opposite of its beginning: a huge, utterly black corpse of a universe containing no light or heat.

If the universe has slightly more than the critical density of material, the universe is said to be "closed." The material will exert enough gravity to eventually halt the expansion and start a contraction.

> **The Big Bang seems like such a simple idea: Everything in the universe, from the building blocks of matter to space and time themselves, burst forth at the same instant. Despite its simplicity, the theory describes the birth and evolution of the universe amazingly well. It only goes to prove that simple ideas are often the most powerful.**

Ultimately all the matter in the universe would collapse back to a state of extraordinary temperature and density—the Big Crunch!—mimicking the conditions in the Big Bang that started the universe. What would happen after the Big Crunch? One possibility is that we live in an oscillating universe where a Big Bang always follows a Big Crunch. Or perhaps the entire universe would disappear into a mammoth black hole, destroying space and time just as the Big Bang created it.

If the universe has just enough dark matter to reach the critical density, as Guth's inflationary theory demands, then the universe will also expand forever but at a rate that decreases essentially to zero. Although this universe falls in between the open and closed universes, the fate of the stars and galaxies will be much the same as in the open universe.

What ignited the Big Bang?

If we live in an oscillating universe, then a Big Crunch gave birth to the Big Bang. But in any case, learning what triggered the Big Bang is one area where cosmologists are still searching for answers.

Conflict over the age of the Universe

M. Bolte & C. J. Hogan

The ages of the oldest stars in our Galaxy can be estimated by comparing stellar populations in globular clusters to calibrated stellar models. The best data give cluster ages of about 15.8 ± 2.1 Gyr, which conflicts with the age of the Universe estimated from measurements of the Hubble constant and the 'standard' cosmological model of a flat, matter-dominated universe (8–13 Gyr).

THE cosmic abundances of the light elements and the precise black-body character of the microwave background radiation provide strong evidence that the expansion of the Universe began a finite time (t_0) ago, with the Big Bang[1]. Determining just how long ago the Big Bang took place is a fundamental question of cosmology, but a difficult one to answer. The current rate of expansion of the Universe is quantified by the Hubble constant (H_0), the ratio of the speed with which distant galaxies appear to be receding from us to their distance. This rate is slowed with time by the mutual gravitational attraction of all the matter in the Universe[2], and possibly accelerated by the energy density of the physical vacuum, the effect of which is usually quantified as the 'cosmological constant'[3,4]. By determining precisely H_0 and t_0, we can investigate the mass density in the Universe (how much dark matter there is), attempt to establish whether the cosmological constant has a finite value, and test the idea that the Universe went through a period of rapid expansion ('inflation') in its early stages.

The standard cosmological model (on grounds of elegance as well as inflation) is the Einstein–de Sitter model, in which the cosmological constant is set to zero and the mass density is such that the Universe is 'flat', expanding at precisely the velocity needed to sustain that expansion indefinitely. In this model, the age of the Universe is simply related to H_0 by $t_0 = \frac{2}{3}H_0^{-1}$. The global value of H_0, and hence the Einstein–de Sitter age, has long been uncertain; by current estimates[5-7], the Universe may be as old as 13 Gyr (for $H_0 = 50$ km s^{-1} Mpc^{-1}) or as young as 8 Gyr (for $H_0 = 80$ km s^{-1} Mpc^{-1}).

An independent constraint on t_0 comes from studies of stellar evolution, since the Universe must be older than its oldest stars. But herein lies the problem. The ages of the oldest stars in our Galaxy are estimated to be about 16 Gyr, significantly older than the Einstein–de Sitter age for current estimates of H_0. If the Einstein–de Sitter model is to remain the cosmological model of choice, we must then re-examine the possible errors in our model-derived estimates of the ages of the oldest stars.

Here we look at the procedure by which stellar ages are determined to see how much scope there is for reducing these estimated ages. We find that a significant revision to the ages of the oldest stars is improbable and conclude that this age test rules out a flat, matter-dominated Universe.

Estimating stellar ages

The oldest known stars lie in globular clusters (Fig. 1). Reflecting this antiquity, their abundance of elements heavier than helium (the 'metallicity') is extremely low. The Big Bang produced hydrogen, helium and trace amounts of lithium, but all heavier atoms have been created as a result of thermonuclear burning in the centres of stars. These heavy elements are distributed throughout space by the winds from evolved stars and, particularly in the early Universe, as a result of supernovae; therefore on average, the earlier a star formed, the less heavy elements it contains, because the gas from which it formed had not been significantly enriched with these heavy elements.

Stars provide a relatively accurate chronometer because their structure, a delicate quasi-equilibrium where the inward force of gravity is balanced by the pressure of the gas heated by compression and kept heated by nuclear fusion in the core, changes as the core uses up the hydrogen available for burning. After spending most of its life at an essentially constant surface temperature and luminosity (on the 'main sequence'), a star evolves rapidly to a new equilibrium phase with a lower surface temperature and higher luminosity (the giant phase, of which there are many types). Plotting the 'colours' of cluster stars against their brightness (a colour-magnitude diagram; Fig. 2), a distinctive bend in the distribution clearly marks the transition from the main-sequence to giant-branch phases of evolution. This bend in the colour-magnitude plane is called the main-sequence turnoff (MSTO). After correction for interstellar 'reddening' (the property of grains to scatter blue light more effectively than red light), the colour of a star is used to derive its effective surface temperature (T_{eff}) and, once the distance to a cluster is determined, the brightness of a star gives its luminosity (M_{bol}). The key to using stars as clocks is that for fixed composition, the main-sequence surface temperature, luminosity and lifetime of a star are all functions of its mass alone. By deriving the surface temperature and luminosity of the cluster stars at the MSTO we can use stellar models to derive their mass and main-sequence lifetime. Since these stars are just completing their main-sequence phase of evolution, we have derived the age of the cluster.

Although the details of this procedure are complicated, the individual steps are now well calibrated and realistic estimates of the errors have been made. This does not mean that the method is foolproof, but any modification needs to pass several calibration tests.

Stellar models and calibrations

We live at a convenient time in the evolution of the Universe: stars now near the MSTO of globular clusters (approximately one solar mass, 1 $M_\odot$ and 'metal-poor') are those for which our models are the most reliable[8-12]. Stars of around 1 $M_\odot$ are intrinsically less complicated than more or less massive stars. They have radiative cores, relatively simple atmospheres and, throughout most of their interiors, the equation of state does not deviate far from that of a perfect gas. Stars more massive than about 1.5 $M_\odot$ have core convection and 'overshooting' (the tendency of convective momentum to propel gas beyond its equilibrium depth)—phenomena that have yet to be completely understood. For stars less massive than about 0.5 $M_\odot$ the equation of state becomes increasingly complex, and the transformations from the T_{eff}–M_{bol} plane of stellar interior theory to the equivalent observational quantities of colour index and absolute magnitude in a particular bandpass are uncertain owing to the difficulties of accurately calculating model atmospheres for relatively low temperatures where molecules are plentiful.

Model predictions are insensitive to current uncertainties in stellar 'physics' (interaction cross-sections for fusion reactions, and the equation of state at different radii), and to the details of the numerical algorithms used to solve the stellar structure equations[8,11]. Recent improvements include the use of non-solar abundance ratios, most importantly the over-abundance of oxy-

57. Conflict Over the Age of the Universe

FIG. 1 Globular clusters are discrete clusters of a million or more stars, which are held together by the mutual gravitational attraction of all the stars. Approximately 200 globular clusters orbit the centre of the Milky Way. This cluster, M92, contains stars which are among those having the lowest abundance of elements heavier than helium of any stars observed; the abundance of iron in these stars is a factor of 150 less than in the Sun. Precise photometry of M92 provides our best estimate of the oldest stellar ages.

gen compared to scaled solar values for metal-poor systems (which has reduced the inferred ages of clusters by about 15%)[13,14], hydrostatic effects and mixing due to internal rotation, and the effects of helium diffusing downwards through low-mass stars—another ~5% reduction[15–17]. Remaining shortcomings of the models (the parametrized treatment of convection by the so-called 'mixing-length' theory, and an inadequacy in calculating opacities in intermediate temperature regimes of about 10^5 K) primarily affect the calculated radii and therefore effective temperatures and colours of model stars. There is recent progress in both areas[18,19] but model luminosities and the mass–lifetime relations, particularly for the most metal-deficient models, are insensitive to the treatment of convection and opacity.

The main reason to believe that systematics are well controlled is that models pass several calibration tests. With very accurately known values for its age, luminosity, effective temperature and abundance parameters (although helium is not measured directly), the Sun provides a precise test of stellar models for stars near $1 M_\odot$. At relatively high metallicity, solar models suffer much more from the inadequacy of current opacity tables than do models appropriate to globular-cluster stars. Although the solar-neutrino problem has been the source of significant consternation to stellar modellers, it seems likely that the resolution of that mystery will not be found in changes to the solar standard model, but rather in our understanding of neutrino physics[20]. (Note that neutrino losses in any case only contribute to a few per cent of the solar luminosity). Certain eclipsing binary stars (whose orbits are thereby known to be viewed nearly edge-on), for which velocity curves can be derived for both stars in the binary system, allow very accurate determinations of both the individual stellar masses and the distance to the system. Studies of the known handful of such objects have provided strong confirmations of the mass–luminosity and colour–luminosity relations predicted by the models, even off the main

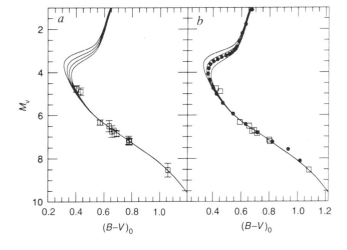

FIG. 2 a, The [Fe/H] = −2.26 zero-age main sequence (squares) formed by the solar-neighbourhood subdwarfs. These are the subdwarfs with trigonometric parallax measures for which the errors are 30% or less and [Fe/H] ≤ −1.3. Small corrections determined by interpolation in the models of Bergbusch and VandenBerg[14] have been made to the colours of the subdwarfs with [Fe/H] values other than −2.26, to bring them to a 'mono-metallicity' sequence. The solid lines are models for [Fe/H] = −2.26, Y = 0.235, [O/Fe] = +0.7 and ages (in descending order) of 12, 14, 16 and 18 Gyr. b, As a with the addition of the binned data from M92 plotted as filled circles. The M92 sequences have been corrected for reddening (0.02 mag in $E(B-V)$) and adjusted vertically to give the best fit to the subdwarf sequence between the colours 0.55 and 1.1.

5. IDEAS, HYPOTHESES, AND THEORIES: The Age Paradox

TABLE 1 Typical error budget for age determinations at [Fe/H] = −1.6

Input parameter	Typical errors	(δ age/age) × 100	
		Colour (MSTO)	M_V (MSTO)
Y	0.02	5%	2%
[Fe/H]	0.2	16%	6%
[O/Fe]	0.2	13%	6%
α	0.3	8%	<1%
D	25%	—	22%
$E(B-V)$	0.02 mag	16%	2(8)%*
$\delta V, \delta(B-V)$	0.02 mag	16%	2(8)%*

Column 1 shows the input parameter, column 2 the typical errors in each parameter for nearby clusters, column 3 the fractional error in the age estimate as a result of the parameter uncertainty if ages are determined by comparing models to main-sequence turnoff (MSTO) colours and column 4 the fractional error in derived ages if MSTO luminosities are used to derive ages[9]. Symbols used: Y, helium mass fraction; [Fe/H], decimal logarithm of iron abundance relative to solar; [O/Fe], oxygen abundance relative to solar; α, 'mixing length' parameter for convection; D, distance in linear units; $E(B-V)$, colour excess due to intervening dust; $\delta V, \delta(B-V)$, typical uncertainties in photometric zero-points. The asterisk indicates quantities that depend on the method used to derive cluster distance.

sequence[21,22]. Most convincing is the case of AI Phoenicis, a double-lined eclipsing system where one star is a subgiant and the other just evolved above the zero-age main sequence. The models successfully predict the same age for the two components with a precision[23] of ~1%—a remarkable coincidence if ages are systematically wrong, as they would have to be wrong by exactly the same amount for two stars of very different mass.

Finally, the models predict accurately the colour-luminosity relation (including the metal dependence[10,14]) for the unevolved metal-poor low-mass stars (which have just joined the main sequence) in globular clusters and, even better, in the local versions of globular-cluster main-sequence stars, the so-called subdwarfs. These are members of the Galactic halo which happen to be near enough to the Sun to have measurable trigonometric parallaxes. The nearby subdwarfs are also thereby the linchpin of the globular-cluster distance scale, and yield a distance calibration confirmed by RR Lyrae stars in the Large Magellanic Cloud—the foundation of the extragalactic distance scale[24,25].

Age estimates and errors

To estimate an age we need to determine the mass of those stars at the MSTO given the observed apparent brightness and colour of the MSTO. The observed magnitude and colour of the MSTO is converted to absolute magnitude M_V and colour $(B-V)_0$ with knowledge of the distance to, and reddening towards, the cluster. These quantities are converted to the magnitude of the total energy produced in fusion reactions in the core M_{bol} and the effective temperature of the stellar photosphere T_{eff} through the use of a model stellar atmosphere for the appropriate abundances ([Fe/H], [O/Fe] and helium mass fraction Y) and convolution of the calculated spectrum with the response function of the filter system, telescope and detector. The values of M_{bol} and T_{eff} for stars at the MSTO then give the mass and lifespan when used in a stellar-interiors model with the correct abundances. Although this is a fairly complex procedure with a formidable list of input parameters (as discussed above), for the case of metal-poor, low-mass stars, the stellar-interiors and stellar-atmosphere models are well understood. The important questions are: how do uncertainties in determining the input quantities propagate through the determination of cluster age, what is the best age estimate and how large are the random and systematic errors in this age?

We can use the models to derive a table of the influence of the various input parameters on age estimates. Table 1 summarizes the results for the cases of estimating ages by MSTO colour and MSTO luminosity.

It is clear from Table 1 that, in addition to stellar colours being the most uncertainly predicted quantity in the stellar models, uncertainties in virtually every input parameter contribute significantly to the uncertainty in derived age when using the colour of the MSTO as an age indicator, which is why turnoff colours are to be avoided as an age diagnostic. On the other hand, if the MSTO luminosity is used to derive cluster ages, the only significant source of error is in cluster distance determinations. This is the crux of the age issue.

The local subdwarfs can directly tie cluster distances (and ages) to the most reliable distance scale in astronomy, that provided by trigonometric parallax, or triangulation using the Earth's orbit as a baseline[26-28]. For each subdwarf we therefore have a 'standard candle' at a particular metal abundance and colour which, when compared to main-sequence stars of the same colour in any cluster, gives the distance to the cluster from the inverse square law of radiation. Specifically, after correcting for interstellar reddening the colours and magnitudes of the stars on the cluster unevolved main sequence, the vertical shift required to match the cluster main sequence to the sequence defined by the local subdwarfs gives the distance to the cluster. (For comparison with any one cluster, each subdwarf's observed colour must also be 'corrected' to give the colour that star would have at the metallicity of the cluster.)

The sample used for the past 10 years comprises only eight metal-poor stars with distances and M_V values known to 30% or better. Moreover, until recently the nearby subdwarfs, even corrected to a mono-metallicity sequence, did not form a particularly tight fiducial for deriving distances[29-31], contributing most of the uncertainty (~15%) in the distance modulus given in Table 1. But with the availability of the newest compilation of trigonometric parallax measures in 1991[32] and with new measurements of [Fe/H], V and $B-V$ for some of the local subdwarfs, it appears that much of the scatter seen in the previous subdwarf sequence was due to observational inaccuracies. Figure 2a shows the sequence of subdwarfs with distance errors <35% and colours corrected to [Fe/H] = −2.26 (squares); superposed are two model isochrones from Bergbusch and VandenBerg[14] for this metallicity and ages of 12, 14, 16 and 18 Gyr. The model colours were uniformly adjusted by 0.015 mag (to the red) to match the colour of the subdwarf with the largest parallax, HD 103095 (Groombridge 1830)[32], but other than that there has been no further adjustment of model parameters to fit the subdwarf data. The agreement not only encourages confidence in the models, but provides a precise fix on cluster distances.

The age of M92

The most accurate photometry for a globular cluster with high galactic latitude (to avoid reddening uncertainties) and low metallicity (where the models are best and the clusters possibly oldest) remains the study of M92 by Stetson and Harris[31]. Figure 2b shows as filled circles the mean position of the locus of points for the M92 stars after correcting for reddening (0.02 mag in $E(B-V)$) and making a vertical registration based on a least-squares fit of the unevolved M92 main sequence to the subdwarfs of Fig. 2a. This gives a distance of 8.5 kpc to M92 and puts the bluest point of the MSTO region at $M_V = 4.0$. Superposed on the M92 data are isochrones from ref. 14 for [Fe/H] = −2.26, [O/Fe] = +0.7, $Y = 0.235$ and for 12, 14, 16 and 18 Gyr. The agreement between the isochrone shapes and the M92 sequences is impressive, but the most reliable age estimate comes from comparing the observed turnoff, M_V of the MSTO for M92, with the model-derived M_V–main-sequence lifetime relation for the metallicity of M92. Consistent with the 'fit' in Fig. 2b, the age of M92 based on M_V (MTSO) = 3.95 from the [Fe/H] = −2.26 models is 15.8 Gyr. The random errors as in Table 1 can be propagated to give a standard deviation σ = 2.1 Gyr with the uncertainty shared equally between the distance error and the uncertainty in [Fe/H]. This value for σ does not include any systematic error in the models.

It is a much easier task to determine relative cluster ages, and it has been shown convincingly that the age difference between M92 and essentially all of the other nearby clusters with [Fe/H] ≲ −2 is <0.8 Gyr (ref. 33), and that M92 is probably a good representative of the metal-poor halo of the Galaxy as far as age is concerned.

Although there are no other comparably precise methods of estimating ages for objects as old as globular clusters, the cooling ages of white dwarfs in the disk of the Milky Way provide an independent check on the globular-cluster ages. The disk of the Milky Way probably formed much later than the globular clusters, so the white-dwarf ages determined by models of their cooling (between 7.5 and 11 Gyr; ref. 34) are consistent with the globular-cluster ages. The white-dwarf ages also coincide with the ages for the oldest 'open clusters' in the disk of the Galaxy[35].

The age of the Universe

The expansion age of the Universe, defined as the inverse Hubble constant $H_0^{-1} = 9.78 \times 10^9 (H_0/100 \text{ km s}^{-1} \text{ Mpc}^{-1})^{-1}$ yr, sets the timescale of all cosmological models. It is simply the relation between the expansion velocities of distant galaxies and their distances, $v = H_0 r$, and therefore even an approximate agreement with stellar ages (as seen) confirms the basic premise of a Big Bang at the beginning of an evolving Universe, as the expansion age is completely independent of the ages of the stars.

The increasingly reliable measurements of the Hubble constant seem to be converging on a value around 70 to 80 km s^{-1} Mpc^{-1} (refs 5, 6, 36–41). Although some dispute remains about the calibration of type Ia supernovae (which have been used to argue for a smaller value of H_0) as standard candles[42–46,50], and about the structure and depth of the Virgo cluster[51], the formal uncertainty of many determinations is $\chi \pm 10$ km sec^{-1} Mpc^{-1}. We probably know H_0 to about the same precision—something like ±25% at the 2σ level—that we know the ages of the oldest stars.

The current data already significantly constrain models that had higher expansions rates in the past, particularly the 'standard' Einstein–de Sitter cosmology (with Ω = 1, and zero cosmological constant) which predicts an age of $t_0 = (2/3)H_0^{-1}$; this differs from the stellar ages by about 3σ for H_0 = 70 km s^{-1} Mpc^{-1}. Even for H_0 = 50 km s^{-1} Mpc^{-1}, the value favoured by some groups using type Ia supernovae as distance indicators[42-44], the Einstein–de Sitter cosmology predicts an age of only 13 Gyr, which is still hard to reconcile with the ages of the oldest stars. If we allow 1 Gyr for the first stars to form, they should be no older than 12 Gyr now, which is about 2σ younger than M92. Within the context of general relativistic cosmological models, those with low mass density (Ω < 1) or models dominated by a cosmological constant would now seem to be favoured by the data. Unless there is some important missing ingredient in the models for stars around 1 $M_\odot$—and such an ingredient would have to leave the mass–luminosity and luminosity–colour relations essentially unchanged—we conclude that the age test already rules out the standard cosmological model.

It is improbable that in the near future a strong case could be made on the basis of the age test for a nonzero cosmological constant, or that we will be able to differentiate between flat and open cosmological models, as realistic errors on t_0 and H_0 will still permit some uncertainty. Further progress will require 'nonlocal' tests of the change with time of the expansion rate of the Universe, which may be possible using the distances to very distant galaxies (perhaps using supernovae as standard candles[47]) or other methods that are sensitive to global geometry[48,49]. Perhaps these studies will offer clues to resolving the current conflict. □

M. Bolte is at Lick Observatory, University of California, Santa Cruz, California 95064, USA. C. J. Hogan is at the Astronomy and Physics Department, University of Washington, Seattle, Washington 98195, USA.

1. Peebles, P. J. E., Schramm, D. N., Turner, E. L. & Kron R. K. Nature **352**, 769–776 (1991).
2. Coles, P. & Ellis, G. Nature **370**, 609–615 (1994).
3. Carroll, S. M., Press, W. H. & Turner, E. L. A. Rev. Astr. Astrophys. **30**, 499–532 (1992).
4. Weinberg, S. Rev. mod. Phys. **61**, 1–23 (1989).
5. Freedman, W. L. et al. Nature **371**, 757–762 (1994).
6. Pierce, M. J. et al. Nature **371**, 385–389 (1994).
7. Hogan, C. J. Nature **371**, 374–375 (1994).
8. Rood, R. T. in Astrophysical Ages and Dating Methods (eds Vangioni-Flan, E. et al.) 313 (Frontieres, Gif sur Yvette, 1990).
9. Renzini, A. in Observational Tests of Inflation (eds Banday, T. & Shanks, T.) 131 (Kluwer, Dordrecht, 1991).
10. Demarque, P., Deliyannis, C. P. & Sarajedini, A. in Observational Tests of Inflation (eds Banday, T. & Shanks, T.) 355 (Kluwer, Dordrecht, 1991).
11. VandenBerg, D. A. in The Formation and Evolution of Star Clusters 183 (ed. Janes, K.) (ASP Conf. Ser. Vol. 13, 1991).
12. VandenBerg, D. A. Astrophys. J. **391**, 685–709 (1992).
13. Salaris, S. C., Chieffi, A. & Straniero, O. Astrophys. J. **414**, 580–600 (1993).
14. Bergbusch, P. A. & VandenBerg, D. A. Astrophys. J. Suppl. Ser. **81**, 163–220 (1992).
15. Deliyannis, C. P., Demarque, P. & Kawaler, S. D. Astrophys. J. Suppl. Ser. **73**, 21–65 (1990).
16. Proffitt, C. R. & VandenBerg, D. A. Astrophys. J. Suppl. Ser. **77**, 473–514 (1991).
17. Chaboyer, B. & Demarque, P. Astrophys. J. **443**, 510 (1994).
18. Chieffi, A., Straniero, O. & Salaris, M. Astrophys. J. **445**, L39–L42 (1995).
19. Canuto, V. M., Minotti, F. O. & Schilling, O. Astrophys. J. **425**, 303–325 (1994).
20. Bahcall, J. N. & Bethe, H. A. Phys. Rev. D**47**, 1298–1301 (1993).
21. Andersen, J. Astr. Astrophys. Rev. **3**, 91–126 (1991).
22. Swenson, F. J., Faulkner, J., Rogers, F. J. & Iglesias, C. A. Astrophys. J. **425**, 286 (1994).
23. Andersen, J. et al. Astr. Astrophys. **196**, 128 (1988).
24. Walker, A. R. Astrophys. J. **390**, L81–L84 (1992).
25. Carney, B. W., Storm, J. & Jones, R. V. Astrophys. J. **386**, 663–684 (1992).
26. Sandage, A. R. Astrophys. J. **162**, 841 (1970).
27. Laird, J. B., Carney, B. W. & Latham, D. W. Astr. J. **95**, 1843 (1988).
28. Ryan, S. G. Astr. J. **104**, 1144–1155 (1992).
29. Fahlman, G. G., Richer, H. B. & VandenBerg, D. A. Astrophys. J. Suppl. Ser. **58**, 225 (1985).
30. Bolte, M. Astrophys. J. **319**, 760–771 (1987).
31. Stetson, P. B. & Harris, W. E. Astr. J. **96**, 909 (1988).
32. van Altena, W. F., Lee, J. T. & Hoffleit, E. D. The General Catalogue of Trigonometric Parallaxes: A Preliminary Version (electronic preprint, Yale University, 1991).
33. VandenBerg, D. A., Bolte, M. & Stetson, P. B. Astr. J. **100**, 445–468 (1990).
34. Wood, M. A. Astrophys. J. **386**, 539–561 (1992).
35. Janes, K. A. & Phelps, R. L. Astr. J. **108**, 1773–1785 (1994).
36. Schmidt, B. P. et al. Astrophys. J. **432**, 42–48 (1994).
37. Pelt, J., Kayser, R., Refsdal, S. & Schramm, T. Astr. Astrophys. (submitted).
38. Schild, R. E. & Thomson, D. J. Astr. J. (in the press).
39. Fukugita, M., Hogan, C. J. & Peebles, P. J. E. Nature **366**, 309–312 (1993).
40. Jacoby, G. et al. Publs astr. Soc. Pacif. **104**, 599–662 (1992).
41. van den Bergh, S. Publs astr. Soc. Pacif. **106**, 1113–1119 (1994).
42. Saha, A. et al. Astrophys. J. **425**, 14–34 (1994).
43. Sandage, A. et al. Astrophys. J. **423**, L19–L22 (1994).
44. Saha, A. et al. Astrophys. J. **438**, 8–26 (1995).
45. Phillips, M. Astrophys. J. **413**, L105–L108 (1993).
46. Riess, A. G., Press, W. H. & Kirshner, R. P. Astrophys. J. **438**, L17–L20 (1995).
47. Perlmutter, S. et al. Astrophys. J. **440**, L41–L44 (1995).
48. Rix, H.-W., Maoz, D., Turner, E. L. & Fukugita, M. Astrophys. J. **435**, 49–54 (1994).
49. Mao, S. & Kochanek, C. S. Mon. Not. R. astr. Soc. **268**, 569–580 (1994).
50. Hamuy, M. et al. Astr. J. **109**, 1–13 (1995).
51. Mould, J. R. et al. HST key project preprint (1995).

ACKNOWLEDGEMENTS. This work was supported by the US NSF and NASA.

Breakthroughs: Crisis Redux

NGC 6752: ITS STARS are getting younger, but they're still too old.

WHEN DATA COLLECTED LAST YEAR by the Hubble Space Telescope indicated the universe was 8 to 12 billion years old, the news troubled astronomers. The problem was that our galaxy contains stars that seem to be much older—14, 18, maybe 20 billion years old, depending on whom you ask. Contradictions don't get more serious than that. But photos of a globular star cluster taken recently by a telescope on the space shuttle may help narrow the gap: they put the ages of the oldest stars at the young end of the spectrum of estimates.

Globular clusters are known to contain some of the oldest stars in the universe, but measuring those ages is far from straightforward. What astronomers actually measure is a star's brightness—or rather its apparent brightness on Earth. If they know the distance to the star, they can calculate how much the light has been dimmed on its way to us, and thus how bright the star really is. From a theoretical model of how a star's brightness evolves over time, they can determine the age of the star.

"The biggest uncertainty in calculating ages has always been the distance," says Wayne Landsman at NASA's Goddard Space Flight Center. Landsman and his colleagues tried to reduce this uncertainty by looking at NGC 6752 with the Ultraviolet Imaging Telescope, which was carried into orbit by the space shuttle *Endeavour* in March. The hottest stars in the globular cluster produce ultraviolet light, which cannot penetrate Earth's atmosphere but which can be seen from orbit. Astronomers think that such hot, ultraviolet-emitting stars in all globular clusters have the same intrinsic brightness. If so, then the brightness measured by a telescope is a good indicator of the stars' distance.

Landsman's team found that the UV-emitting stars in NGC 6752, and thus the whole cluster, are around 14,000 light-years away—1,200 light-years farther than previously thought. The UV stars themselves are young. But if the oldest stars in the cluster, which are too cool to radiate in the UV, are also 1,200 light-years farther away, then they must be bigger, brighter, and younger than previously estimated. Landsman says that the oldest stars in NGC 6752—and the universe—are only 14 billion years old.

That still isn't as young as the estimated age of the universe, but it is closer. Landsman says that it's hard to see how the new distance estimate to the globular cluster could be very far off. Now it's up to cosmologists to reevaluate the Hubble's estimates for the age of the universe. Says Landsman: "I don't think this end is going to move much more."

Younger than they look

Cheryl Jones, Sydney

THE paradox that the oldest stars appear to predate the Universe has troubled astronomers for years. But now researchers in Australia claim that estimates for the age of the oldest stars are out by as much as 7 billion years.

A team led by Alex Rodgers of the Australian National University (ANU) in Canberra used a new technique to measure the age of a group of pulsating stars, existing as globular clusters in the Milky Way, which are among the oldest in the Universe. From measurements of the stars' luminosity—the amount of light they give off—the researchers calculated that the stars are between 9 and 12 billion years old, a figure in line with the age of the Universe.

Previously, the oldest stars were estimated to be up to 16 billion years old. But in 1994, Wendy Freedman and her colleagues at Carnegie Observatory in Pasadena, California, used the Hubble Space Telescope to estimate the value of the Hubble constant, a measure of the speed at which galaxies are moving away from each other. This suggested the Universe was born no more than 12 billion years ago. The discrepancy is one of astronomy's most disputed puzzles.

The luminosity of a hydrogen-burning star like the Sun changes with age. The luminosity can be deduced from the star's apparent brightness—how bright it appears from Earth—but only if its distance from Earth is known: a far-off star that appears faint may nevertheless be extremely luminous.

'In the past, scientists based their estimates on convoluted measurements of brightness and distance'

In the past, says Rodgers, researchers based their estimates of luminosity on convoluted measurements of brightness and distance, which carry a large potential for error. Rodgers believes his team has come up with a better way of measuring luminosity. He concentrated on a group of RR Lyrae stars in the Large Magellanic Cloud. These pulsating bodies, which swell and contract continuously, have contemporaries in the Milky Way. They exist in clusters with, and are thus the same age as, the oldest stars. The team found about 70 RR Lyrae stars with a special property—they pulsated simultaneously in two modes, at the fundamental frequency and the first overtone. In RR Lyrae stars, pulsation in these modes together occurs only within a well-defined temperature range.

The mass of a pulsating star can be calculated from the period of its pulsations. Stars that have a high mass pulsate quickly because gravity forces them to contract before they expand too far. By using a computer model for the density of RR Lyrae stars, Rodgers calculated their radii and surface areas, and from these—knowing their temperatures—the stars' luminosities. The work suggests that the RR Lyrae stars are more luminous, further away and younger than previously thought.

Jeremy Mould, director of the ANU's Mount Stromlo Observatory, says Rodgers's method is "much cleaner" than previous attempts to calculate the age of stars. "But the debate is sure to intensify," he adds. The team's results have been submitted for publication in the *Astrophysical Journal*.

Ages of the Oldest Clusters and the Age of the Universe

Sidney van den Bergh

One of the most heated debates in the history of astronomy focuses on the numerical value of the Hubble parameter H_o. This parameter is of fundamental importance because it gives the scale-size of the universe and provides constraints on world models and the age of the universe. The most direct path to the determination of the extragalactic distance scale is through the Cepheids in the Virgo cluster. The relative merits of other techniques for determining H_o have recently been reviewed in great detail by Jacoby et al. (1), van den Bergh (2, 3), and Kennicutt et al. (4). The table lists the true distance moduli μ_o, which is the apparent magnitude corrected for absorption that a star of absolute magnitude $M = 0.0$ would have, in four spiral galaxies in the Virgo region in which Cepheid variables have been observed so far. The distances of all four of these spirals are in excellent agreement. The data in this table yield a formal weighted mean distance modulus $\langle\mu_o\rangle = 31.02 \pm 0.08$ (mean error) for the Virgo cluster. To this quoted mean error should be added a 0.1-magnitude (mag) systematic uncertainty resulting from possible errors in the calibration of the zero-point of the Hubble Space Telescope (HST) photometry and an uncertainty of ~0.1 mag in the distance modulus of the Large Magellanic Cloud relative to which the Virgo distances were determined. In the subsequent discussion, it will be assumed that the true distance modulus of the Virgo cluster is μ_o(Virgo) = 31.02 ± 0.2 ($D = 16.0 \pm 1.5$ Mpc). This distance modulus for

| Distance moduli of spirals in Virgo region |||||
Galaxy	m_o	D (Mpc)	Telescope	Reference
NGC 4321 (M100)	31.00 ± 0.20	15.8	HST	Farrarese et al. (9)
NGC 4496	31.10 ± 0.15	16.6	HST	Saha et al. (10)
NGC 4536	31.05 ± 0.15	16.2	HST	Saha et al. (10)
NGC 4571	30.91 ± 0.15	15.2	CFHT	Pierce et al. (11)

HST, Hubble Space Telescope
CFHT, Canada-France-Hawaii Telescope

four Virgo spiral galaxies is consistent with the value μ_o(Virgo) = 31.12 ± 0.26 that Whitmore et al. (5) have recently determined with HST by comparing the luminosity function of globular clusters in the Virgo elliptical galaxy M87 with that for globular clusters in the Milky Way system and in the nearby Andromeda galaxy. Because M87 lies at the center of the Virgo cluster, this observation appears to rule out the possibility that the spirals listed in the table lie a significant distance in front of the core of the Virgo cluster.

Tanvir et al. (6) have used HST observations of Cepheids in NGC 3368 to derive a distance modulus $\mu_o = 30.32 \pm 0.16$ for the Leo I cluster. In conjunction with a difference $\Delta\mu_o = 0.99 \pm 0.15$ between the distance moduli of the Virgo and Leo I clusters this yields μ_o(Virgo) = 31.31 ± 0.22. This indirect distance determination is also consistent with, but slightly larger than, the value μ_o(Virgo) = 31.02 ± 0.20 derived above from Cepheids observed in four Virgo spirals. It is concluded that the distance of the Virgo cluster is now well determined.

Because both the peculiar motion of the Virgo cluster and the magnitude of the retardation of the Local Group by the Virgo supercluster remain controversial, it is safest to determine the Hubble parameter from the Coma/Virgo distance ratio and the Coma velocity relative to the microwave background. The difference in the distance moduli of the Virgo and Coma clusters is well determined. From 12 concordant determinations, van den Bergh (2) finds $\Delta\mu_o = 3.71 \pm 0.05$. Adopting $\Delta\mu_o = 3.71 \pm 0.05$, in conjunction with a distance modulus μ_o(Virgo) = 31.02 ± 0.20, yields μ_o(Coma) = 34.75 ± 0.21, corresponding to a distance D(Coma) = 89 ± 9 Mpc. Durret et al. (7) found a mean redshift $\langle V \rangle = 6901 \pm 72$ km s^{-1} for the Coma cluster. With a correction of $+258 \pm 10$ km s^{-1} to place Coma in the cosmic microwave background frame, this yields a true velocity V(Coma) = 7159 ± 73 km s^{-1}. From these values, one obtains $H_o = V$(Coma)$/D$(Coma) = 81 ± 8 km s^{-1} Mpc^{-1}. For a matter-dominated $\Omega = 1$ Einstein–de Sitter universe (Ω is the ratio of the actual density of the universe to the critical density), the corresponding age $t_o = (2/3)H_o^{-1} = (2/3) \times 9.78 \times [100/(81 \pm 8)] = 8.0 \pm 0.8$ billion years. Such a short age conflicts with the age of 15.8 ± 2.1 billion years that Bolte and Hogan (8) obtained from main sequence fitting of the metal-poor globular cluster M92 to the subdwarf main sequence derived from trigonometric parallaxes. If the oldest galactic globular clusters have ages of ~16 billion years, and if the time interval between the "Big Bang" and the formation of the first globular clusters was ~1 billion years, then the age of the universe is ~17 billion years. This value is twice as large as the value 8.0 ± 0.8 billion years previously found from H_o in an Einstein–de Sitter universe.

References and Notes

1. G. H. Jacoby et al., Publ. Astron. Soc. Pac. **104**, 599 (1992).
2. S. van den Bergh, ibid., p. 861.
3. ———, ibid. **106**, 1113 (1994).
4. R. C. Kennicutt, W. L. Freedman, J. R. Mould, Astron. J. **110**, 1476 (1995).
5. B. C. Whitmore, W. B. Sparks, R. A. Lucas, F. D. Macchetto, J. A. Biretta, Astrophys. J., in press.
6. N. R. Tanvir, T. Shanks, H. C. Ferguson, D. R. T. Robinson, Nature **377**, 27 (1995).
7. F. Durret et al., in Observational Cosmology, G. Palumbo, Ed. (in press).
8. M. Bolte and C. J. Hogan, Nature **376**, 399 (1995).
9. L. Farrarese et al., Astrophys. J., in press.
10. A. Saha, private communication.
11. M. J. Pierce et al., Nature **371**, 385 (1994).
12. I thank A. Saha for permission to cite his new HST distance determinations.

The author is at the Dominion Astrophysical Observatory, National Research Council of Canada, Victoria, British Columbia, V8X 4M6, Canada.
E-mail: vandenbergh@dao.nrc.ca

Index

absolute magnitude, 126, 133
active galactic nucleus (AGN), 23, 24, 25, 38, 39
adaptive optics, 193
age of universe, conflict over, 208–209, 214–215, 218, 221, 222, 223, 224
Agrippina, 104
Aldebaran, 134, 136, 138
ALH 84001, 112–113
Alpher, Ralph, 211
Alvarez, Walter, 115
amino acids, 57, 109
Angel, J. Roger P., 12–13
Antares, 136, 138
Antony, Mark, 103
apparent magnitude, 126, 133
Arcturus, 136, 138
Arecibo, 41–42, 45–46
ASCA, 24, 25, 146
ASEPS (Astronomical Studies of Extrasolar Planetary Systems), 193
asteroids 58, 100; colliding with Earth, 114–118, 119–121; dinosaur extinction and, 116–118
astrometric method, of star tracking, 10
astronomical units (AU), 98
atmosphere: ice and, of Earth, 53–56; of outer planets, 78–79
Australia Telescope (AT), 43–44
Aztec Empire, 104

BACODINE (BAtse COordinates DIstribution NEtwork), 32
bars, 160
Beppo-SAX satellite, 28, 29, 32
Beta Pictoris, 10
Betelgeuse, 134, 136, 138
Big Bang theory, 157, 213–217; controversy over, 206–207, 208–209, 210–212
Black, David C., 12
black dwarfs, 135
black holes, 23, 30, 134, 139, 142, 144–147, 167, 178–180; galactic, 146; quantum theory and, 188, 189, 190, 192; stellar, 145–146
Blum, Jürgen, 82, 83, 84
Bohr, Niels, 192
bolometric correction, 126
bolometric magnitude, 126
Borucki, William J., 12, 194–195
Boss, Alan P., 9, 13
Boudicca, 104
brown dwarfs, 12, 128–129, 134
Bruno, Giordano, 10
buckyballs, 58
bulk convection pattern, 81
Burrows, Adam, 11
Burst and Transient Source Experiment (BATSE), 29, 30, 31, 32
Busse, Fritz, 80
Butler, R. Paul, 8, 9, 11, 12, 13, 127

Caesar, Augustus, 104
Caesar, Julius, 61, 103–104
Caesarian party, 103, 104
calendars, 60–62
Callisto, 84, 85, 90, 92
Caltech Sub-Millimeter Telescope, 42

Campbell, Bruce, 11
Cassini, Haygens, 89
Cepheus, 136–137, 138, 208, 224
Chandrasekhar limit, 135
charge-coupled device (CCD), 11, 99
Charon, 101
Chicxulub, 114, 115
Chinese calendar, 62
Chyba, Christopher, 58–59
Claudius, Emperor, 104
Close Encounters of the Third Kind, 198
Cochran, William D., 11
cold-dark-matter theory, 158
collapse picture, 165–166
color, of starlight, 132–133
Comet Hale-Bopp, 107–110
Comet Hyakutake, 109
Comet Shoemaker-Levy 9, 115, 118
comets, 55, 58; life on Earth and, 107–110
COMPTEL (COMPton TELescope), 26–27
computer simulations, 12, 99, 100, 101
conquistadors, 104
Copernican principle, 197, 198
Copernicus, 10
Cortés, Hernan, 104
Cosmic Background Explorer satellite, 211, 216
Cosmic Dust Aggregation Experiment (CODAG), 83
cosmic microwave background, 211
cosmic X-ray background (CXB), 25
CPT (charge-parity-time) invariant, 191, 192
Crab Nebula, 24–25
critical density, missing mass problem of Big Bang, 206–207, 208, 212, 214–215
Cruikshank, Dale P., 108, 110
Cryobot, 92–93
cyanogen gas, 104–105
Cygnus X-1, 24
Cygnus X-2, 22

dark halo, 149
dark matter, 152–156, 157
DARWIN, 195–196
Decaspec, 184
"Declaration of Principles Concerning Activities Following the Detection of Extraterrestrial Intelligence," 202
degeneracy pressure, 188–189
dinosaurs, extinction of, 114–118
DNA (deoxyribonucleic acid), 57
Doppler shift, 10
Drake equation, 57, 59, 199–200
Duncan, Martin J., 99
DUO (Disk Unseen Objects), 153–154
dwarf galaxies, 150
dynamical friction, 150

Earth: atmosphere of, 53–56; comets and life on, 107–110; measuring size of, 52; meteorites from Mars and Moon on, 111
Edgeworth, Kenneth Essex, 98
Eggen, Olin, 151
Einstein, Albert, 188, 191, 192
Einstein Observatory, 23
Einstei–de Sitter model, 218, 221
electrons, 188–189

Energetic Gamma Ray Experiment Telescope (EGRET), 32
Epicurean philosophers, 10
Eratosthenes of Alexandria, 52
EROS (Expérience de Recherche d'Objets Sombres), 153–154
Europa, 84, 85, 86, 87–89, 90–93
exobiology, 198
exosphere, 53
extrasolar planets, 9, 10, 11, 12, 13; planet formation and, 127–131

Fabricant, Dan, 182, 184–185
Fernández, Julio A., 98–99
51 Pegasi planet, 11, 12, 127, 128
5145 Pholus, 100, 101
53W002, 168
"flaming ear of corn," 104
Follow-Up Detection Device (FUDD), 201, 202
Ford, Eric B., 12
4U1728-34, 22
Fraunhofer, Josef von, 18
FRESIP (Frequency of Earth-Sized Planets), 194–195

galactic black holes, 146
Galactic plane, 149
galaxies: dwarf, 150; elliptical, 172–173, 175; evolution of, 170–177; formation of, 163–164, 165–169; lenticular, 171–172; quasars and, 178–180; radio, 167; satellite, 150; spacing of, 181–185; spiral, 149, 172, 174, 175–176. *See also* Milky Way
Galileo spacecraft, 81, 88, 92
Gamma Ray Large Area Space Telescope (GLAST), 32
gamma-ray astronomy, 26–27, 28–33
Ganymede, 84, 85, 86, 90, 92
Gatewood, George, 11
G-dwarf problem, 159–160
Geller, Margaret, 181–185
general relativity, 189, 190
Giant Metrewave Radio Telescope (GMRT), 44
giant stars, 134, 136
Gliese 229, 12
Global Oscillation Network Group (GONG), 70, 75–76, 77
Global Surveyor, 95–96
globular star clusters, 160–161; age of universe and, 222, 223
"Goldilocks orbit," 12
Goldin, Daniel S., 13, 94–95
GOLF, 75, 76
gravity, 10, 12, 80, 81, 98, 100–102, 188, 192
GRB 970228, 28, 30
Great Disappointment, day of, 105
Great Red Spot, 81
Greeley, Ronald, 82
Green Bank Telescope (GBT), 41, 45–46
Gregorian calendar, 61
Guth, Alan, 215, 216–217

Halley, Edmund, 106
Halley's comet, 98, 103, 105, 106
Hartle, James B., 191
Hawking, Stephen W., 188–192

HD 114762, 128–129, 131
Hectospec, 184–185
Herman, Robert, 211
Hertzsprung-Russell (H-R) diagrams, 20, 132–133
High Energy Transient Explorer (HETE), 28–29
Hipparchus, 124
Holeman, Matthew J., 100
Hubble constant, 208–209, 218, 221, 223, 224
Hubble Deep Field (HDF), 14–15
Hubble, Edwin, 170, 210
Hubble Space Telescope (HST), 9, 14, 16, 17, 28, 164, 173, 176, 216, 222, 223, 224
Huchra, Hohn, 181, 183
Hulse-Taylor binary pulsar, 189
hurricanes, 81
hydrobot, 93

ice, atmospheres and, 53–56
icy planetesimals. See comets
Independence Day, 198–199
inflationary hypothesis, on expanding universe, 215, 216–217
infrared astronomy, 34–37, 38–39
Infrared Space Observatory (ISO), 38–39
INTEGRAL, 27
interferometers, 13, 195–195
Io, 84, 85, 86, 88–89, 90, 91–92
iridium, 115

Jansky, Karl, 40, 47
Jewish calendar, 61
Joss, Paul C., 98
Joule, James, 179
Jupiter, 9; atmosphere of, 78, 79, 82; possibility of life on moons of, 84–86, 87–89, 90–93; winds on, 80

Keck Telescope, 30
Keller, Gerta, 116–117
Kepler spacecraft, 12
kinematics, 150
krypton, 54
K-T boundary, asteroid impact and, extinction at, 114–118
Kuiper belt, 97–102
Kuiper, Gerard P., 97

lab modelling, 81
Lalande 21185 planets, 11, 129–131
Large Magellanic Cloud, 24, 140, 141, 150, 157, 161
Large Millimeter Telescope (LMT), 42
Latham, W., 11
Leonid meteor storm, 150
Lick telescope, 9
life: comets and, on Earth, 107–110; on Mars, 112–113; origins of, 57–59, 193–196; possibility of, on moons of Jupiter, 84–86, 87–89, 90–93; search for extraterrestrial, 193–196, 197–200, 201–205
light cones, 189
light years, 133
Lin, Douglas N. C., 12
Liouville's Theorem, 190
luminosity, of star, 19–20, 132–133, 136
Lynden-Bell, Donald, 151

MACHOs (Massive Compact Halo Objects), 152–156, 157
magnetic fields, life and, 84–86
magnetometer, 85–86

Magnum Mirror Telescope (MMT), 184
main-sequence stars, 133–134
main-sequence turnoff (MSTO), 218, 220
Malhotra, Renu, 102
Manneville, Jean-Baptiste, 80, 81
Marchant, Edgar, 105
Marcy, Geoffrey, 8, 9, 10, 11, 13
Mars: atmosphere of, 53, 55; life on, 9, 112–113; meteorites from, on Earth, 111, 112–113; NASA exploration of, 94–96; storms on, 82
Mars Volatiles and Climate Surveyor, 96
Marsden, Brian G., 100
Martian Surface Wind Tunnel (MARS-WIT), 82
Mayor, Michel, 11, 127
meteorites: from Mars and Moon on Earth, 111, 112; life on Mars and, 112
Medium Deep Survey, 177
M81, 24
M87, 146, 224
MERLIN array, 44
metallicity, 150
Metrodorus of Chios, 10
Meyers, Steven, 81
M51, 36, 37, 149
Milky Way, 9, 10, 148–151; evolution of, 158–162
Miller, Stanley L., 58
Miller, William, 105
Millerite comet, 104–105
Millerites, 105, 106
Millimeter Array (MMA), 44–45
millimeter-wave arrays, 44–45
missing mass problem, of Big Bang, 206–207, 208, 212, 214–215
M92, 219, 220–221
Montezuma II, 104
Moon: charting of, with naked eye, 65–67; meteorites from, on Earth, 111; watching, 63–64
Mu Cephei, 136–137, 138
Multichannel Astrometric Photometer (MAP), 194
Multiple Mirror Telescope (MMT), 184
multiplexing, 184
multipole moments, 189
Muslim calendar, 61–62

Naderi, Ferouz, 13
NASA (National Aeronautics and Space Administration), 16–17, 94–96, 193, 195
Neptune, 78, 79, 98, 102
Nero, 104
neutrinos, 70–73, 139, 143, 157
neutron stars, 10, 11, 21–22, 134, 137, 142, 145, 189, 195
neutrons, 188–189
Newton, Isaac, 106, 192
NGC 2346, 36–37
NGC 3377, 146
NGC 3379, 146
NGC 4486b, 146
NGC 6752, 222
NICMOS, 169
1995 DA_2, 102
no-boundary proposal principle, 191, 192
novae, 135
nucleic acids, 57
null surface, 189

objective reduction (OR), 191
Octavian, 103–104
Oemler, Augustus, 170–171, 173, 174, 177

OGLE (Optical Gravitational Lensing Experiment), 153–154, 156
Olson, Peter, 80, 81
Omega Nebula (M17), 35
omens, comets as, 103–106
Oort cloud, 98
Oort, Jan H., 98
optical interferometry, 13
Origins Program, 13
Orion Nebula, 35–36, 37
OSSE (Oriented Scintillation Spectrometer Experiment), 26

parallax, 133
parsec, 133
Pathfinder, 81, 94, 95
Penrose, Roger, 188–192
Penzias, Arno, 211
phase space, 190
photon, wave function of, 190
Planck scale, 191
planet formation, 127–131
Planet X, 97
planetary nebula, 135, 138–139
plankton, asteroid impacts and, 117
Plutinos, 102
Pluto, 98, 102
Pogson ratio, 124
Poppe, Torsten, 82, 83, 84
Project Phoenix, 201, 202, 204
proteins, 57
Proxima Centuri, 132
PSR B1257+12, 195
PSR 1913+16, 189
Ptolemy, Claudius, 124
pulsars. See neutron stars

QB1, 99, 100
quantum field theory, 189, 190
quantum theory, 192; black holes and, 188–189
quasars, 146, 167, 178–180, 209, 215
quasi-periodic oscillations (QPO), 21–22
Queloz, Didier, 11, 127
Quetzalcoatl, 104
Quinn, Thomas, 99

radio astronomy, 10, 40–46, 47–49, 167–168
Rasio, Frederic A., 12
Rayleigh numbers, 81
Reber, Grote, 40, 47
Red Spot, Great, 81
redshifts, 14–15, 134, 137, 163
relativity, 188–192
religion, effect of discovery of extraterrestrial life on, 203
Rho Cancri, 11
Ring Nebula (M57), 36, 37
RNA (ribonucleic acid), 57
Roman society, comets and, 103–104
Rosentgen Satellite (ROSAT), 23, 24
Rossi X-ray Timing Explorer (RXTE) satellite, 21, 147
RR Lyrae stars, 223

sacrifices, human, Aztec Empire and, 104
Sandage, Allan, 151
satellite galaxies, 150
Saturn, 78, 79
Schrödinger's cat, thought experiment, 190–191, 192
Scorpio X-1, 23, 24
search for extraterrestrial intelligence (SETI), 193–196, 201–205

Seneca, 104, 106
70 Virginis, 9, 11, 12
Shoemaker-Levy 9, 101
sidus Julium, 103–104
singularities, of space-time, 190
Sirius, 137
Sirius B, 137
16 Cygni B, 11
Small Magellanic Cloud, 150, 161
Smith, Bradford A., 10
Society of Amateur Radio Astronomers (SARA), 49
SOI/MPI, 75, 76, 77
Sojourner, 94, 96
Solar and Heliospheric Observatory (SOHO), 70, 73, 75, 76, 77
solar system, 12, 13, 80, 82, 83, 97, 98, 99, 102
Sommeria, Joel, 81
space, and time, nature of, 188–192
Space Interferometry Mission, 13
space-time, 188–192
special relativity, 189
spectroscopic method, measurement of starlight and, 10, 11
star formation, 26–27, 38–39
stars, 18, 132–135, 136–139, 140–143; luminosity of, 19–20, 132–133, 136, 136, 137–138; origins of, 138–139; spectral classes of, 18–19; stellar magnitude, system of classifying, 124–126; temperature of, 136, 137–138. *See also* age of universe; Sun; specific star; specific type

stellar black holes, 145–146
stellar halo, 149, 151
stellar luminosity. *See* luminosity
stellar populations, 150
Stern, S. Alan, 101
stickman, 181, 182
Sub-Millimeter Array (SMA), 45
Sudbury Neutrino Observatory (SNO), 71, 72
Suetonius, 104
Sun, 74–77, 80, 135; neutrino problem and, 70–73
supergiant stars, 134, 138
Superkamiokande neutrino detector, 71, 72
Supernova 1987A, 141
supernova remnants (SNR), 24–25
supernovae, 24–25, 26–27, 134, 135, 139, 140–143, 211; type Ia, 26, 135, 221; type Ib, 134; type Ic, 134; type II, 134
Swinney, Harry, 81, 82

Tacitus, 104
Tenochtitlán, 104
Terrestrial Planet Finder, 13
Terrile, Richard J., 10
time, and space, nature of, 188–192
time dilation, 211
Titan, 88
Toltecs, 104
Toon, Owen, 114, 115, 116
Tremaine, Scott D., 99
Triton, 88–89, 101
2060 Chiron, 100, 101

UHURU, 23
Ultraviolet Imaging Telescope, 222
uncertainty principle, Heisenberg's, 189
Uranus, 78
Urey, Harold C., 58

Very Large Array (VLA), 43–44, 45
very long baseline interferometer (VLBI), 44
V404 Cyg, 145–146, 147
virgin-sacrifice hoax, and Halley's comet, 105, 106
VIRGO, 75, 76

Walker, Gordon A. H., 11
wave function, of photon, 190
Weyl, Hermann, 191
Weyl tensor, 191, 192
white dwarfs, 135, 137, 138–139, 141, 155, 157
Wilson, Robert, 211
WIMPs (Weakly Interacting Massive Particles), 152–155
winds, 80, 81, 82
Wisdom, Jack L., 100
Wollaston, William, 18
Wolszczan, Alexander, 10
Woolf, Neville J., 12–13

xenon, 54–55
X-ray astronomy, 21–22, 23–25, 144–147

years, different calendars and, 60–62

Zahnle, Kevin, 114, 115

Credits/Acknowledgments

Cover design by Charles Vitelli

1. Data-Gathering Techniques
Facing overview—© 1997 PhotoDisc, Inc.

2. The Earth and Moon
Facing overview—© 1997 PhotoDisc, Inc. 56—NASA/USGS photo.

3. The Solar System
Facing overview—© 1997 PhotoDisc, Inc. 75—Illustrations courtesy SOHO SOI/MDI Consortium and Philip Scherrer. All SOHO images also courtesy of Bernhard Fleck. 82—Illustration by Chris Draper. 87–88—NASA photos.

4. The Universe
Facing overview—© 1997 PhotoDisc, Inc. 137—*Astronomy* magazine graphics by Lee Vande Visse. 139—*Astonomy* magazine graphics by Elisabeth Rowan. 183—© 1994 by the Smithsonian Astrophysical Observatory.

5. Ideas, Hypotheses, Theories
Facing overview—© 1997 PhotoDisc, Inc. 202—Photo by Seth Shostak. 222—NASA photo.

PHOTOCOPY THIS PAGE!!!

ANNUAL EDITIONS ARTICLE REVIEW FORM

■ NAME: _____ DATE: _____

■ TITLE AND NUMBER OF ARTICLE: _____

■ BRIEFLY STATE THE MAIN IDEA OF THIS ARTICLE: _____

_____

■ LIST THREE IMPORTANT FACTS THAT THE AUTHOR USES TO SUPPORT THE MAIN IDEA:

_____
_____
_____
_____

■ WHAT INFORMATION OR IDEAS DISCUSSED IN THIS ARTICLE ARE ALSO DISCUSSED IN YOUR TEXTBOOK OR OTHER READINGS THAT YOU HAVE DONE? LIST THE TEXTBOOK CHAPTERS AND PAGE NUMBERS:

_____
_____
_____
_____
_____

■ LIST ANY EXAMPLES OF BIAS OR FAULTY REASONING THAT YOU FOUND IN THE ARTICLE:

_____
_____
_____
_____

■ LIST ANY NEW TERMS/CONCEPTS THAT WERE DISCUSSED IN THE ARTICLE, AND WRITE A SHORT DEFINITION:

_____
_____
_____
_____

*Your instructor may require you to use this ANNUAL EDITIONS Article Review Form in any number of ways: for articles that are assigned, for extra credit, as a tool to assist in developing assigned papers, or simply for your own reference. Even if it is not required, we encourage you to photocopy and use this page; you will find that reflecting on the articles will greatly enhance the information from your text.

We Want Your Advice

ANNUAL EDITIONS revisions depend on two major opinion sources: one is our Advisory Board, listed in the front of this volume, which works with us in scanning the thousands of articles published in the public press each year; the other is you—the person actually using the book. Please help us and the users of the next edition by completing the prepaid article rating form on this page and returning it to us. Thank you for your help!

ANNUAL EDITIONS: ASTRONOMY 98/99
Article Rating Form

Here is an opportunity for you to have direct input into the next revision of this volume. We would like you to rate each of the 60 articles listed below, using the following scale:

1. Excellent: should definitely be retained
2. Above average: should probably be retained
3. Below average: should probably be deleted
4. Poor: should definitely be deleted

Your ratings will play a vital part in the next revision. So please mail this prepaid form to us just as soon as you complete it.
Thanks for your help!

Rating	Article	Rating	Article
	1. In Golden Age of Discovery, Faraway Worlds Beckon		32. The Day the Dinosaurs Died
	2. Learning from Hubble's Deep Field		33. A Cosmic Collision
	3. Beyond the Hubble Space Telescope		34. The Stellar Magnitude System
	4. The Spectral Types of Stars		35. The Strange New Planetary Zoo
	5. Shrill Notes from the Stars		36. Life and Times of a Star
	6. Recent Advances of X-Ray Astronomy		37. Extreme Stars: At the Edge of Stellar Behavior
	7. In Line for a New Mission		38. Ka-Boom! How Stars Explode
	8. Gamma-Ray Bursts		39. New Findings Suggest Massive Black Holes Lurk in the Hearts of Many Galaxies
	9. An Infrared View of Our Universe		
	10. Cool Gaze at Heartless Galaxies		40. The Milky Way
	11. Radio Astronomy in the 21st Century		41. The Dark Side of the Galaxy
	12. A Radio Map of the Milky Way		42. Is the Dark Matter Mystery Solved?
	13. Measuring the Size of the Earth		43. The Evolution of Our Galaxy
	14. How the Earth Got Its Atmosphere		44. Seeing How Galaxies Form
	15. What Makes a Planet a Friend for Life?		45. Before Galaxies Were Galaxies
	16. A Day in the Life of a Year		46. What Makes Galaxies Change?
	17. Moon Watching: An Experiment in Scientific Observation		47. Galactic Engines
			48. Beyond the Soapsuds Universe
	18. Charting the Moon by Eye		49. The Nature of Space and Time
	19. Unsolved Mysteries of the Sun—Part 1		50. Searching for Alien Earth
	20. Unsolved Mysteries of the Sun—Part 2		51. Is Anybody (Like Us) Out There?
	21. Atmospheric Dynamics on the Outer Planets		52. When E.T. Calls Us
	22. Planet in a Bottle		53. The Best Cosmology There Is
	23. Magnetic Fields on Distant Moons Hint at Hidden Life		54. Holes in the Big Bang
	24. Life in a Deep Freeze?		55. In Defense of the Big Bang
	25. Water World		56. Everything You Wanted to Know about the Big Bang
	26. The Stars of Mars		57. Conflict over the Age of the Universe
	27. The Kuiper Belt		58. Breakthroughs: Crisis Redux
	28. Comets That Changed the World		59. Younger than They Look
	29. The Comet's Gift: Hints of How Earth Came to Life		60. Ages of the Oldest Clusters and the Age of the Universe
	30. Bits of Mars and Pieces of the Moon		
	31. Life from Ancient Mars?		

(Continued on next page)

ABOUT YOU

Name _____ Date _____
Are you a teacher? ❑ Or a student? ❑
Your school name _____
Department _____
Address _____
City _____ State _____ Zip _____
School telephone # _____

YOUR COMMENTS ARE IMPORTANT TO US !

Please fill in the following information:
For which course did you use this book? _____
Did you use a text with this ANNUAL EDITION? ❑ yes ❑ no
What was the title of the text? _____
What are your general reactions to the Annual Editions concept?

Have you read any particular articles recently that you think should be included in the next edition?

Are there any articles you feel should be replaced in the next edition? Why?

Are there any World Wide Web sites you feel should be included in the next edition? Please annotate.

May we contact you for editorial input?

May we quote your comments?

ANNUAL EDITIONS: ASTRONOMY 98/99

BUSINESS REPLY MAIL
First Class   Permit No. 84   Guilford, CT

Postage will be paid by addressee

Dushkin/McGraw·Hill
Sluice Dock
Guilford, CT 06437

No Postage
Necessary
if Mailed
in the
United States